Reactions of the Citric Acid Cycle

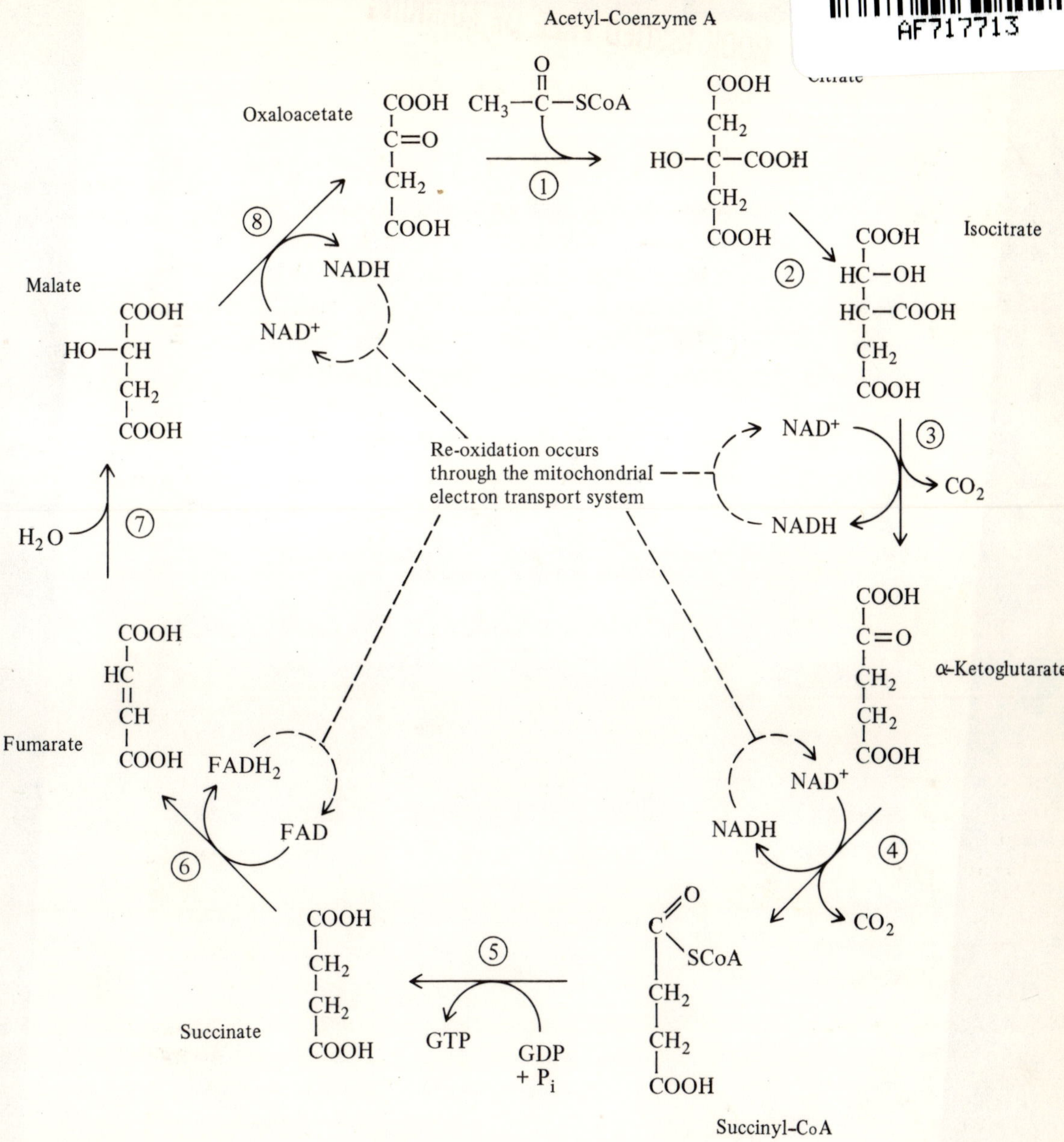

Enzymes

1. Citrate synthetase
2. Aconitase
3. Isocitrate dehydrogenase
4. α-Ketoglutarate dehydrogenase complex
5. Succinyl-CoA synthetase
6. Succinate dehydrogenase
7. Fumarate hydratase (fumarase)
8. Malate dehydrogenase

Southeastern Louisiana University
Textbook Rental Policy

Chem 281 10/81 New

Course and Number

Price of book is 19.40 and **WILL BE CHARGED TO STUDENT IF UNRETURNED BY THE FRIDAY PRECEDING REGULAR REGISTRATION OF NEXT SEMESTER, DAMAGED, OR WRITTEN IN.**

1. This book is the property of the University and must be returned undamaged on or before **THE LAST DAY OF FINALS. A FINE OF $5.00 PER BOOK WILL BE CHARGED FOR THE DELINQUENT RETURN OF TEXTBOOKS AFTER THE DEADLINE FOR RETURNS.**
2. Books are issued by bar code number. You must return the book number that was issued to you. If the book number checked in does not correspond to the number checked out, Textbook Rentals will confiscate the book for return to its rightful lessor and you will be required to pay for a lost book.
3. You must have your class schedule/invoice and SLU I.D. to change, drop, or add a book.
4. This textbook may not be held past the last day of finals without prior permission of Textbook Rentals.

PRINT SEMESTER, NAME, AND SOCIAL SECURITY NUMBER ON LINES PROVIDED BELOW

SEMESTER	NAME	SOC.SEC.NO.

...of the Bookstore.

Fundamentals of Biochemistry

Fundamentals of Biochemistry

Gordon L. Kilgour
Portland State University

D. Van Nostrand Company
New York Cincinnati Toronto London Melbourne

Cover: Courtesy of Science Related Materials, Inc.

D. Van Nostrand Company Regional Offices:
New York Cincinnati

D. Van Nostrand Company International Offices:
London Toronto Melbourne

Library of Congress Catalog Card Number: 80-52348

ISBN: 0-442-25756-2

Published by D. Van Nostrand Company
135 West 50th Street, New York, N.Y. 10020

10 9 8 7 6 5 4 3 2 1

PREFACE

The short introductory course in biochemistry, taken early in the curriculum, has become increasingly desirable for students enrolled in biological programs. At the same time, the field of biochemistry has grown so rapidly and so extensively that a book for a one-term course can reflect only a very limited range of topics if the student is to have any hope of mastering the material. Therefore, the presentation of the fundamental topics in this text are dealt with in sufficient depth for understanding rather than trying to deal with a wider range of material. *Fundamentals of Biochemistry* provides students with a solid knowledge of basic biochemistry within the limitations of a one-semester or one-quarter course.

The organization of *Fundamentals of Biochemistry* is traditional in the sense that it deals first with the structure of proteins and uses this information to lead into the study of enzymes, the one truly unique facet of biochemistry. These chapters are followed by an examination of certain aspects of biochemical energetics, a brief review of carbohydrate structure and nomenclature, and a long, careful look at the metabolic reactions of carbohydrates and how these processes lead to the production of energy in biologically useable forms. Next there is a similar examination of the structures, nomenclature, and metabolism of common lipids. The chapters on nitrogen metabolism represent a brief introduction to a very large body of material and serve to introduce some important concepts and to indicate how the metabolism of nitrogen-containing compounds is related to other areas of metabolism. The chapters that follow explore the basis of nucleic acid structure and function and the expression of genetic information. The final chapter emphasizes the interdependent nature of the many processes of a cell by looking at the various means by which the cellular chemistry may be controlled.

The chapters detailing the structures and nomenclature of carbohydrates, lipids, and nucleic acids are in the appropriate places through the book rather than grouped together at the beginning. This arrangement permits a closer relationship between structure and function and an early introduction to enzymes, bioenergetics, and similar topics central to the study of biochemistry.

Fundamentals of Biochemistry deals with the reversal of metabolic pathways in the chapters in which they are first introduced. It also shows many of the organic acids involved in metabolism in the un-ionized forms (even though they may be totally ionized at pH 7) in order to smooth the transition from organic chemistry and make the reaction sequences more readily understandable. Similarly, the names of these acids and of their ions are used interchangeably since the student must be able to recognize the use of either form in order to read the existing literature.

There are end-of-chapter review questions with answers at the back of the book and a comprehensive glossary of technical terms at the end of the book.

The Additional Readings at the end of each chapter are arranged in order of increasing complexity or level of sophistication. Many of the entries are annotated to show the nature of the book or article and its relationship to the chapter at hand. An Instructor's Manual, available from the publisher, includes comprehensive chapter overviews and over 100 questions for testing or assignment to students.

I would like to express my appreciation to the many colleagues who gave so willingly of their advice, suggestions, and corrections on various portions of the manuscript. A special note of thanks goes to Robert Bell, University of Rhode Island, Ronald Watanabe, San Jose State University, and Delano Young, Boston University, who reviewed the manuscript in its preliminary stages and offered helpful suggestions and comments. I also thank Joann Loehr for many valuable suggestions and productive discussions. I have been fortunate in having the assistance of Harriet Serenkin, who as Editor has patiently converted a manuscript into a book. Finally, and most importantly, a word about the key role of my wife, Jewel, who not only put up with me through the whole project and kept me somewhere close to schedule but also did all the typing and helped prepare the final manuscript.

CONTENTS

Chapter 1

A Few Words of Introduction 1

Chapter 2

Proteins, Peptides, and Amino Acids 9

Chapter 3

The Structures of Proteins 29

Chapter 4

Enzymes I: The Nature of the Beast 45

Chapter 5

Enzymes II: Kinetics and Inhibition 67

Chapter 6

Biochemical Energetics: An Introduction 81

Chapter 7

Chapter 8

Chapter 9

Chapter 10

Electron Transport and Oxidative Phosphorylation 157

Chapter 11

The Pentose Phosphate Pathways—And Their Function in Photosynthesis 171

Chapter 12

Carbohydrate Storage and Some Related Reactions 187

Chapter 13

Lipids: Structures, Nomenclature, and Hydrolysis 201

Chapter 14

Fatty Acid Oxidation and Synthesis 217

Chapter 15

Lipid Biosynthesis 231

Chapter 16

Amino Acid Metabolism I. General Reactions 249

Chapter 17

Amino Acid Metabolism II. Specific Biosynthetic Pathways 267

Chapter 18

The Nucleic Acids: Structure and Biosynthesis 281

Chapter 19

DNA: The Genetic Inheritance of Cells 303

Chapter 20

Genetic Expression and Protein Synthesis 319

Chapter 21

Mechanisms of Metabolic Control 335

Fundamentals of Biochemistry

CHAPTER 1

A Few Words of Introduction

Biochemistry is fundamentally a study of the processes by which cells are able to survive and carry out their myriad functions. In a secondary sense, we are also concerned with the intercell relationships necessary to sustain multicellular organisms. In either case, we must recognize the special conditions that exist in the cell and the limitations and requirements that those conditions place on the chemical reactions involved.

1.1 The Need for Catalysis

First, most cells spend their lives at a temperature between zero and about 45°C. For the majority of cells, the range is much narrower than this. At the same time, a large number of reactions must be carried out at a speed that is seldom matched in the organic chemistry laboratory. A firefly that wants to flash simply cannot wait for a four-hour reflux time, even if he could stand the temperature. Thus there is the necessity to speed up reactions immensely without increasing the temperature. This, of course, brings us into the realm of catalysis, and the story of biochemistry is in large measure the story of the protein molecules, called **enzymes**, that catalyze cellular reactions.

A related consideration is the need to encourage only those reactions that are useful to the cell and discourage all side-reactions. The production of the "tar" so familiar to students of organic chemistry, or of toxic compounds, would be hazardous to any cell. Thus we find that essentially every reaction occurring in the cell is catalyzed, even such simple reactions as that of CO_2 with water to give bicarbonate ions.

1.2 The Compounds Involved

Along the line, we can expect to find many compounds undergoing reactions in metabolic systems that would be considered too unstable for the organic chemist to try to deal with. Because biochemical reactions take place in a very confined area and proceed so rapidly, even relatively unstable compounds may serve as important intermediates in the chemical processes of a cell. As one small example, free dihydroxyacetone phosphate will not be found in any catalog of organic chemicals, and yet it serves as an important intermediate in the metabolism of almost every living cell. The student who is accustomed to the considerations of synthetic organic chemistry will find the same types of reactions occurring, but must be prepared to adjust to new concepts of a "stable" intermediate and to some irritating but very real differences in terminology.

1.3 Organization and Control

Then there is the staggering task of organizing and controlling the thousands of separate chemical reactions to accomplish the changing needs of an organism efficiently and effectively. In general, this means in one way or another the control of the catalysts involved, the many enzymes of the cell. We will see that there are several mechanisms by which this type of control can be maintained for the metabolic processes of any one cell.

In the long view, of course, we also have to deal with the question of cellular differentiation in multicellular organisms as another type of control mechanism. All of the cells of a human body, for example, start with the same genetic inheritance, and yet develop the distinct characteristics of brain cells, muscle cells, or liver cells. This whole area of genetic expression and its control is one of the most active fields of research in biochemistry today.

1.4 The Structure and Nomenclature of Living Cells

Because we are dealing with the chemistry of the living cell, we must have some understanding of the functional organization and the terminology associated with different types of cells. Figure 1-1 shows a drawing of a typical cell characteristic of an animal tissue (a **eucaryotic** cell). This kind of drawing is obtained as a result of viewing a large number of actual pictures such as the electron micrograph of Figure 1-2. Note the characteristic

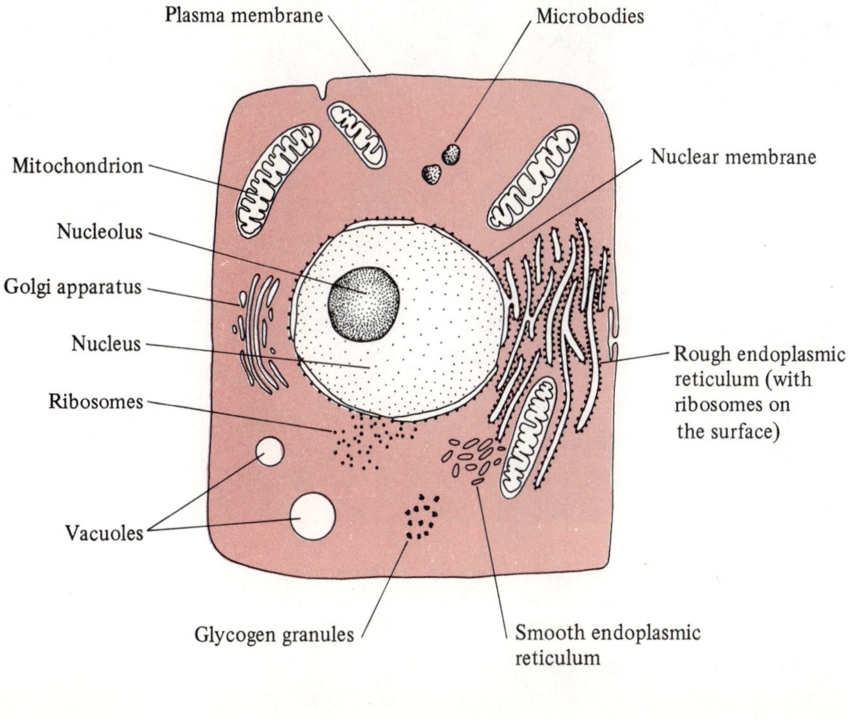

Figure 1-1
A drawing of the major features of a eucaryotic cell of animal tissue.

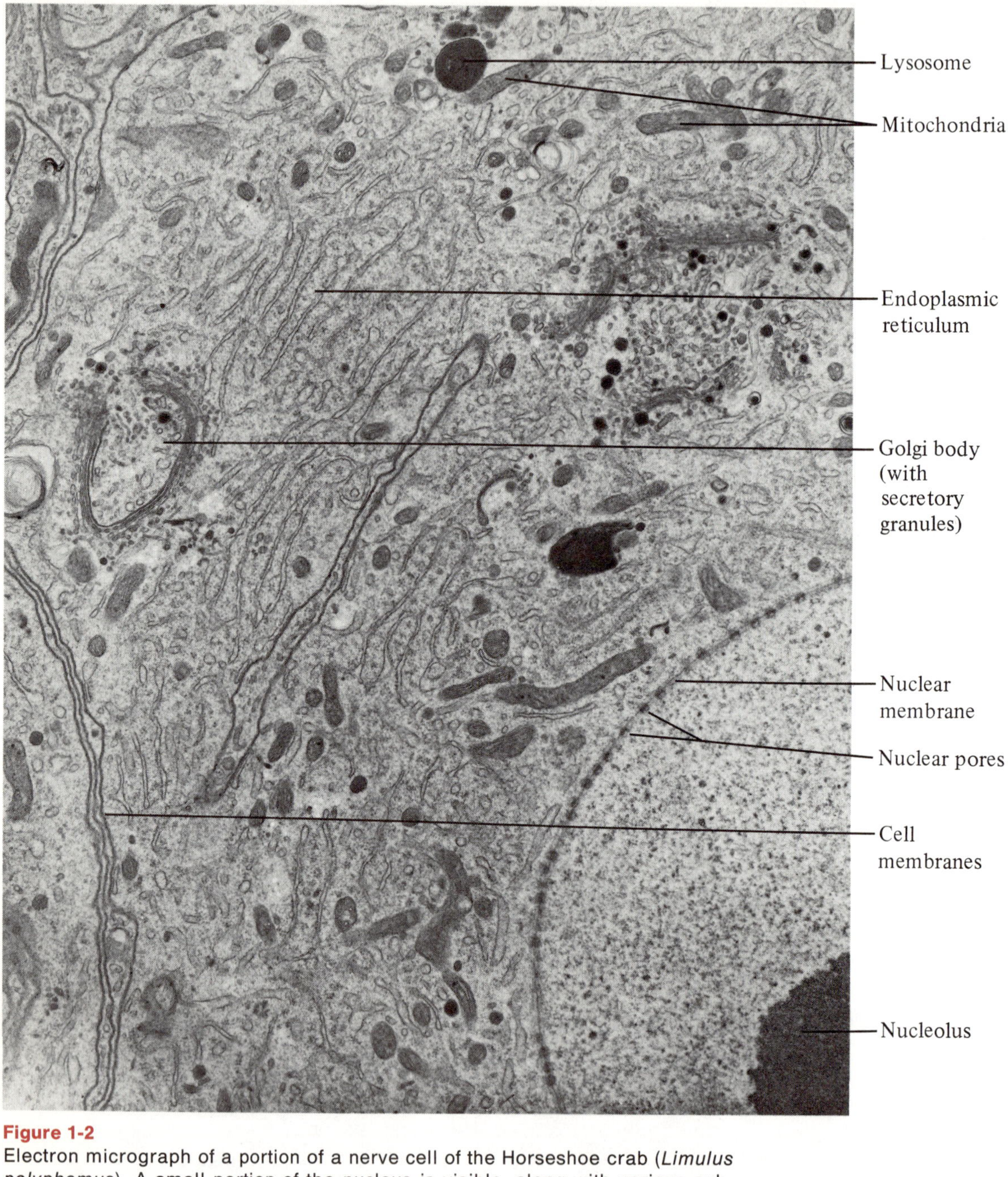

Figure 1-2
Electron micrograph of a portion of a nerve cell of the Horseshoe crab (*Limulus polyphemus*). A small portion of the nucleus is visible, along with various subcellular structures as indicated by the labels. ×23,000 (Courtesy of W. H. Farnbach, Oregon Regional Primate Research Center.)

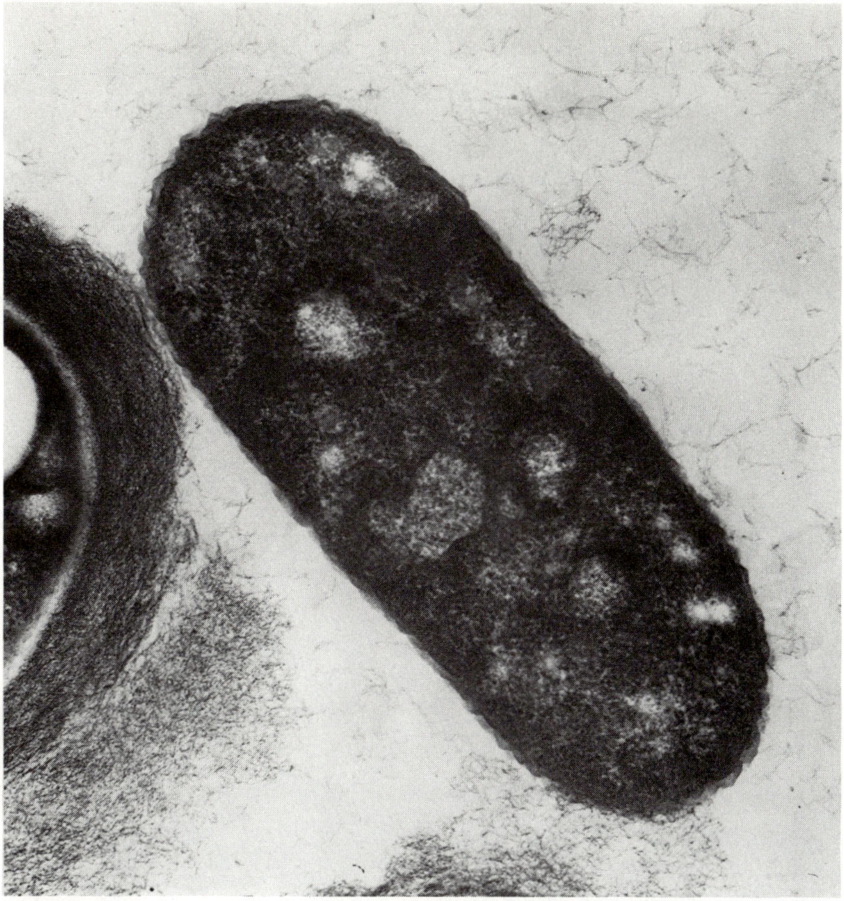

Figure 1-3
An electron micrograph of a section of a procaryote, the bacterium *Azotobacter vinelandii*. × 80,000 (Courtesy of J. W. Myers, Portland State University.)

presence of a central nucleus with nucleolus, as well as a number of organelles such as mitochondria, ribosomes, vacuoles, and various microbodies. Many bacteria show a much more primitive level of organization lacking a highly organized nucleus, mitochondria, and many of the other organelles of the eucaryotes. A typical example of an electron micrograph of one of these **procaryotic** organisms is shown in Figure 1-3.

The cells of the higher plants are basically eucaryotic in nature, but have some special characteristics, including a cellulosic "box" to enclose the cell and chloroplasts where appropriate to provide the capacity for photosynthesis. The basic structure of a photosynthetic plant cell is shown in schematic drawing as Figure 1-4.

It is important to bear in mind the constant relationship between structure and function that characterizes living systems, whether we are talking about single chemical compounds or the structures and functions of subcellular units, organs, or whole plants and animals. As a small example,

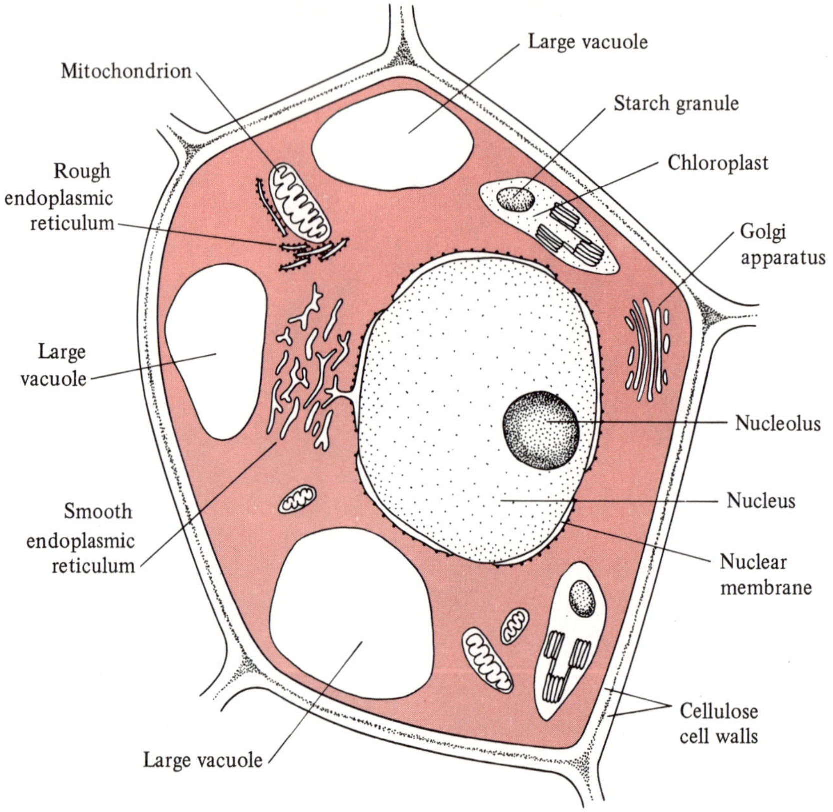

Figure 1-4
A drawing of the major structural features of a cell from a photosynthetic higher plant.

we shall see that within the eucaryotic cell, storage and replication of genetic information is largely the function of the **nucleus**, protein synthesis is related directly to the **ribosomes**, oxidative metabolism and conversion of the resulting energy into usable forms is largely the province of the **mitochondria**, while synthetic reactions occur either loose in the **cytosol** or within one or another of the specialized microbodies. In procaryotes, many of the functions that are carried out in a highly organized fashion by, for example, the mitochondrion of a eucaryote are achieved by organizing the necessary functional components on or in a specific region of the inner membrane of the cell.

1.5 A General Note on Nomenclature

Many of the compounds present in cells carry functional groups that are either partially or fully ionized at the normal pH levels of a living cell. For this reason, some people feel strongly that such compounds must always be shown in the ionized form, and whenever possible should be named as the appropriate ion. Thus acetic acid becomes acetate, L-glutamic acid becomes L-glutamate, and so on. However, complete devotion to this principle can lead to unnecessary confusion and, in some cases, a misleading view of the actual situation. Names and structures based upon both the free acids (or bases) and the related ions will be used more or less interchangeably throughout this book. The student who wants to know the precise state of ionization for a specific compound must always work from the pK values for the ionizing groups involved.

Additional Readings

For This Chapter

C. J. Avers, *Cell Biology* (D. Van Nostrand Co., New York, 1976).

General Biochemistry Texts and Reference Books

A. L. Lehninger, *Biochemistry*, 2nd edition (Worth Publishers, New York, 1975).

H. R. Mahler and E. H. Cordes, *Biological Chemistry*, 2nd edition (Harper and Row, New York, 1971).

R. W. McGilvery, *Biochemistry: A Functional Approach*, 2nd edition (W. B. Saunders Company, Philadephia, 1979).

D. E. Metzler, *Biochemistry: The Chemical Reactions of Living Cells* (Academic Press, New York, 1977).

A. White, P. Handler, E. L. Smith, R. L. Hill, and I. R. Lehman, *Principles of Biochemistry*, 6th edition (McGraw-Hill, New York, 1978).

References Dealing with Quantitative Problems

J. S. Finlayson, *Basic Biochemical Calculations* (Addison-Wesley Publishing Co., Reading, Mass., 1969).

R. Montgomery and C. A. Swenson, *Quantitative Problems in the Biochemical Sciences*, 2nd edition (W. H. Freeman and Company, San Francisco, 1976).

I. H. Segel, *Biochemical Calculations*, 2nd edition (John Wiley and Sons, New York, 1975).

CHAPTER 2

Proteins, Peptides, and Amino Acids

The group of compounds called "**proteins**" account, on the average, for about one-half of the total dry weight of living materials. While some proteins do serve as structural materials or as important components of extracellular fluids, the most striking characteristic of these compounds is their ability to function within living cells as biological catalysts. It has been estimated that as much as 90 percent of the total protein within a cell is enzyme protein.

The proteins are often referred to as macromolecules because of the high molecular weight of the individual molecules. The proteins are further characterized by the fact that they are all polymers; that is, they are chain-like molecules produced by the linking together of a number of similar small units. The small units of the proteins are chiefly *α-amino acids* of the following general structure:

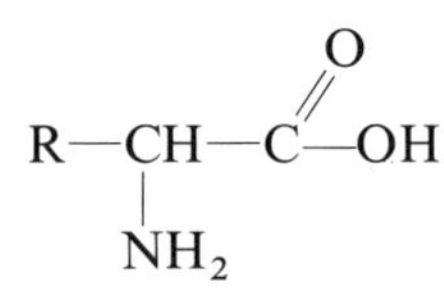

These units are joined through the carboxyl and amino groups to give a chain as shown below:

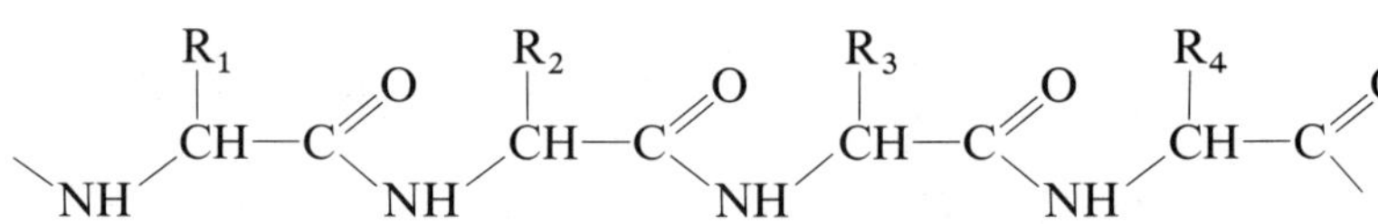

When two or more amino acids are joined in this manner, the resulting compound is called a **peptide**. The special kind of amide bond formed is called a **peptide bond**.

Peptides are often characterized by the number of amino acid units involved as "dipeptides," "tripeptides," and so on. The term "oligopeptide" is sometimes used as a general term to indicate "a few" amino acids. A **polypeptide** is a long chain of amino acids. All proteins contain one or more of these polypeptide chains.

The chain lengths of known proteins vary from about 50 to several thousand amino acid units. In addition, proteins differ in relative content of the 20 or so amino acids which are common constituents, and in the order in which the amino acids are linked together in the chain. It should be clear that with 20 different units to fill as many as several thousand places, the total number of possible proteins approaches infinity.

2.1 Amino Acids

Most of the amino acids which occur as components of proteins are α-amino acids and fit the general structural formula given above.

When we consider any of the α-amino acids other than glycine ($H_2N—CH_2—COOH$), we see that there are at least two possible optical isomers. The two forms of a typical amino acid can be represented as:

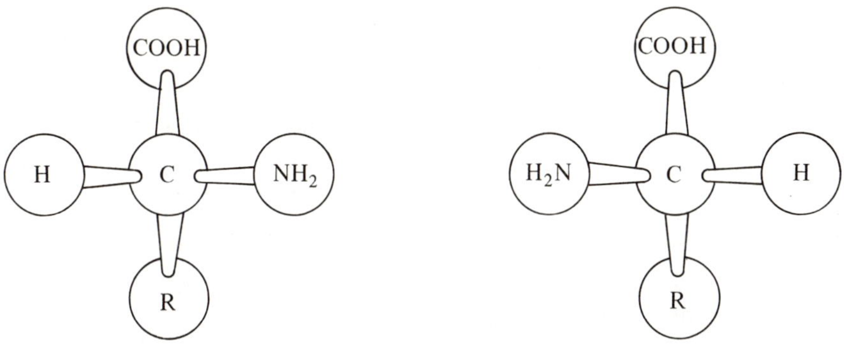

Because of the difficulty of representing three-dimensional molecules in the two dimensions available on a sheet of paper, the structures are usually presented in a conventional "projection" form proposed by Emil Fischer. To make such a projection, the first step is to arrange the structures as shown above, with the carboxyl group at the top and the hydrogen and amino groups above the plane of the paper. A projection of this structure onto the plane of the paper then appears as shown below, where the amino group in one case falls to the *right* of the carbon chain and in the other case to the *left*:

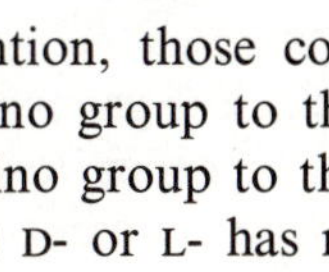

A D-amino acid

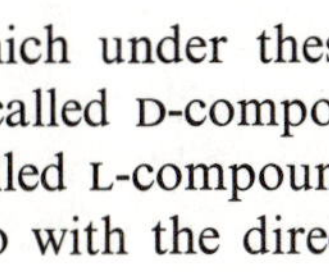

An L-amino acid

By convention, those compounds which under these circumstances have their amino group to the *right* are called D-compounds, and those having the amino group to the *left* are called L-compounds. It should be noted that this D- or L- has nothing to do with the direction in which a solution of the compound will rotate the plane of polarized light. This direction can be specified by a (+) or (−) designation:

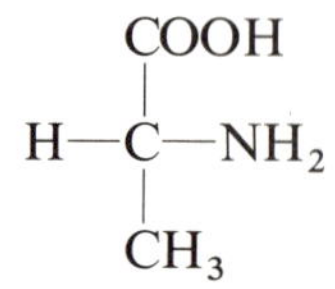

D-(+)-alanine

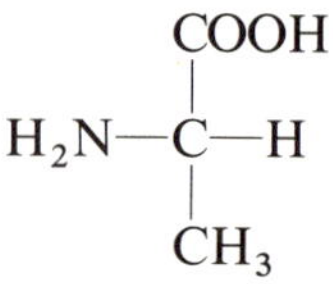

L-(−)-alanine

While some D-amino acids do occur in living cells, only the L-amino acids are used for protein synthesis. Similar specificities for one optical isomer are common in a wide variety of biochemical reactions.

Structures and Classification of Amino Acids Common to Most Proteins

The amino acids may be separated into groups on a number of bases. For some purposes, a classification based on the type of functional group in the side-chain (the R-group in the structures above) is quite valuable. However, when looking at the amino acids from a viewpoint of how they may affect the structure of a protein, the most useful single feature is the degree of polarity of the side-chains at neutral pH values. We will arrange our amino acids on that basis in Table 2-1.

At the same time, we must recognize that it is often the specific functional group of an amino acid (such as the alcohol of serine, thiol of cysteine, or imidazole group of histidine) that allows a protein to carry out its biological function.

The various amino acids are by no means equally common, nor are they evenly distributed in proteins. Table 2-2 shows the number of specific amino acid residues in each molecule for several specific protein molecules. It also shows the weight percentages of the various amino acids in the total proteins from several biological sources.

Acidic and Basic Nature of Amino Acids and Proteins

All free amino acids contain at least one acidic group (—COOH) and one basic group (—NH_2 or its equivalent). In order to understand the behavior of the amino acids, and eventually of proteins, we must first consider some of the properties of these groups.

TABLE 2-1
Amino acids which commonly occur in proteins.

A. *Amino Acids with Side-Chains That Are Nonpolar at pH 7.*

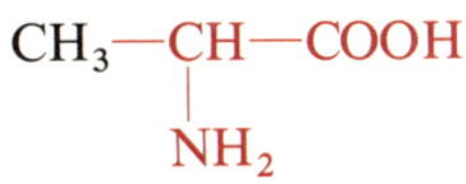
Alanine (ala)

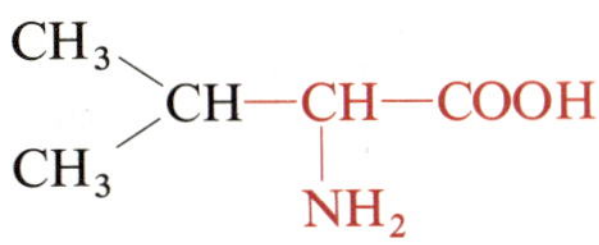
Valine (val)

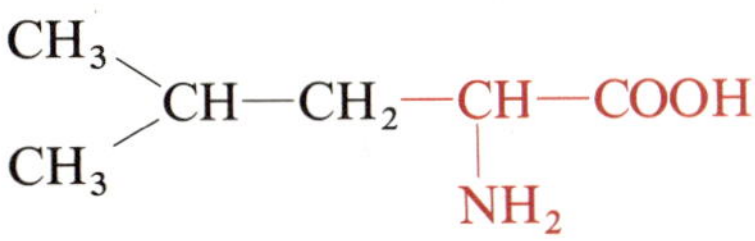
Leucine (leu)

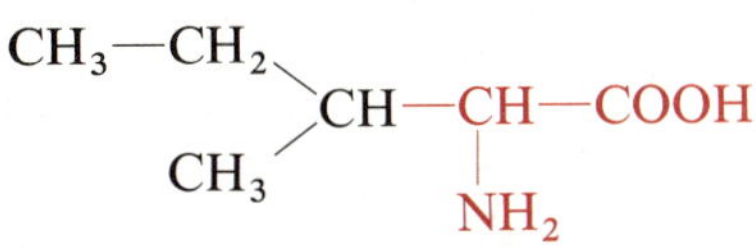
Isoleucine (ile)

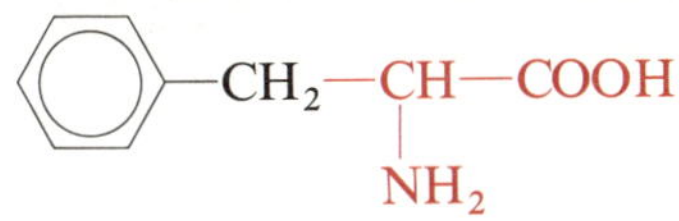
Phenylalanine (phe)

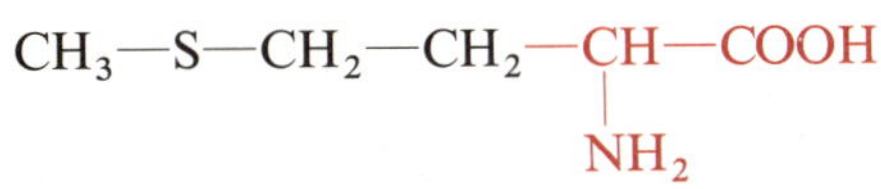
Methionine (met)

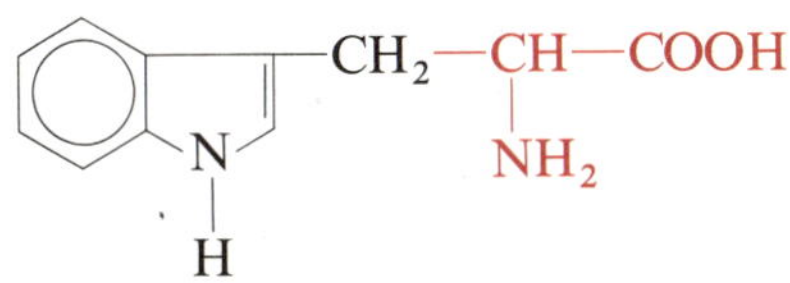
Tryptophan (trp)

B. *Amino Acids with Side Chains That Are Polar But Not Ionized at pH 7.*

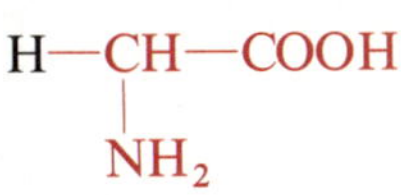
Glycine (gly)

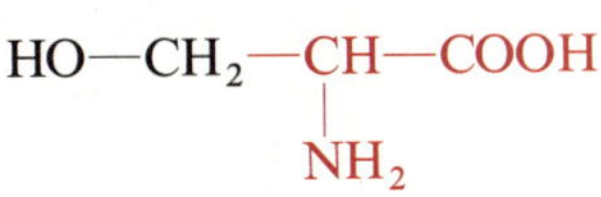
Serine (ser)

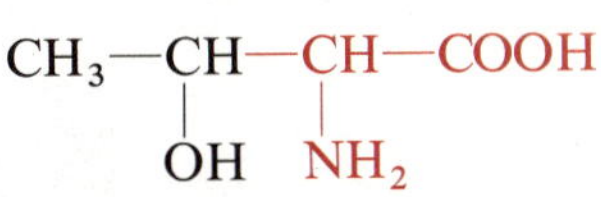
Threonine (thr)

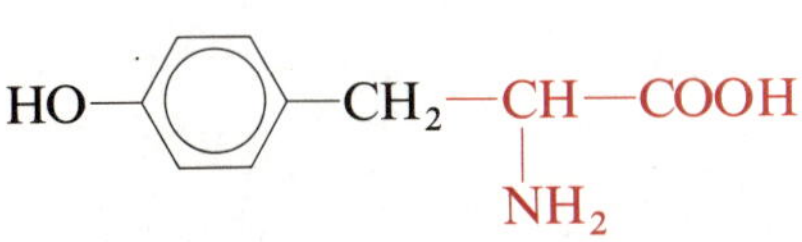
Tyrosine (tyr)

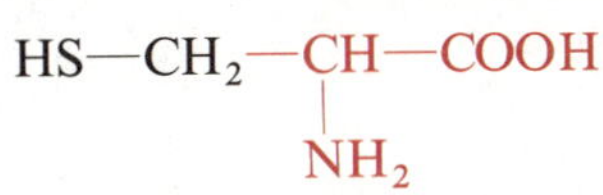
Cysteine (cys)

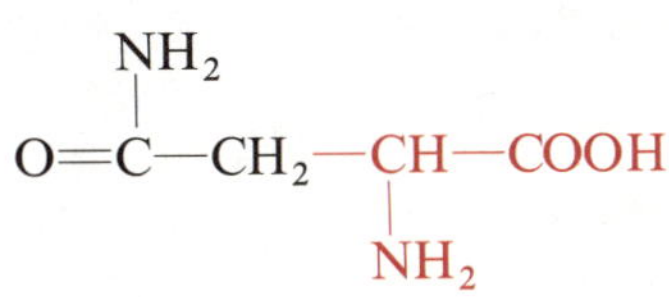
Asparagine (asn)

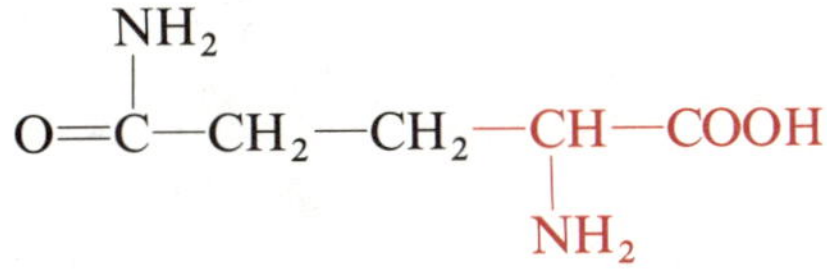
Glutamine (gln)

TABLE 2.1
continued

C. *Amino Acids with Side-Chains That Carry a Negative Charge at pH 7 ("Acidic" Amino Acids).*

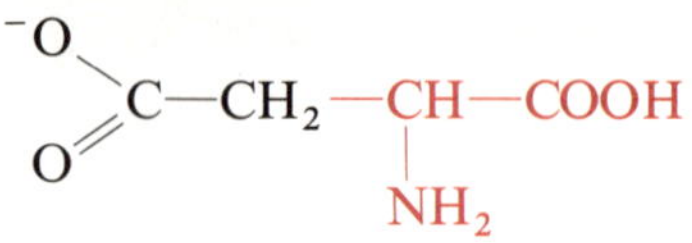

Aspartic acid (asp)

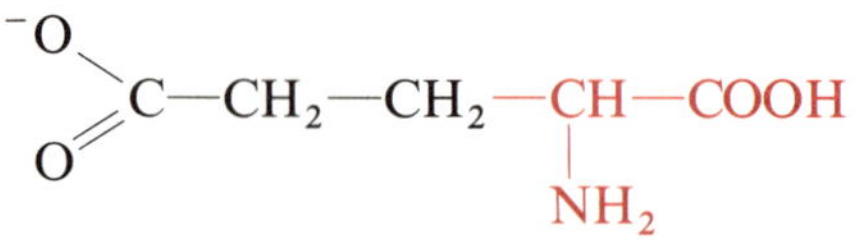

Glutamic acid (glu)

D. *Amino Acids with Side-Chains That Carry a Positive Charge at pH 7 ("Basic" Amino Acids).*

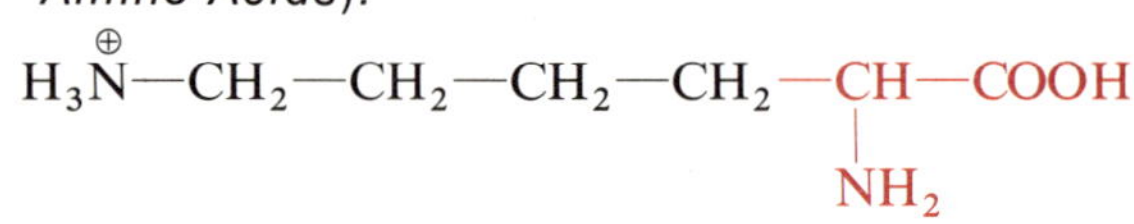

Lysine (lys)

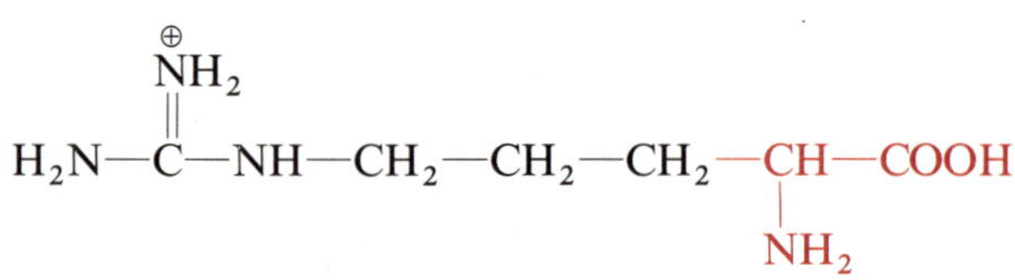

Arginine (arg)

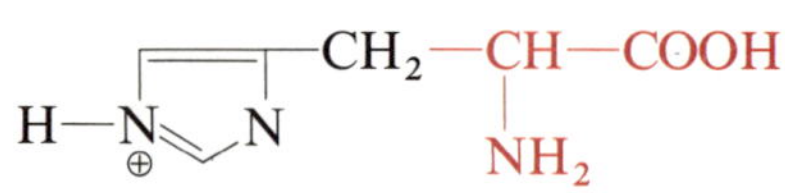

Histidine (his)

E. *A Special Case: A Cyclic Amino Acid.*

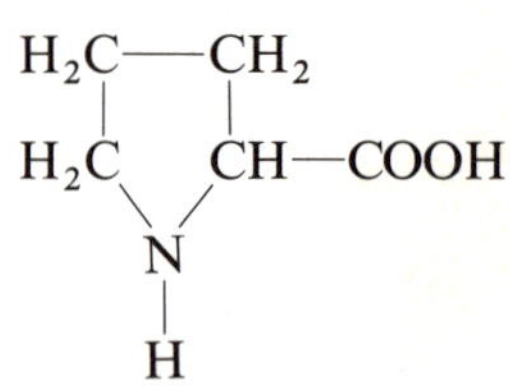

Proline (pro)

Note 1. Histidine has a pK_{a_R} of 6.4 and is thus only partially charged at pH 7.

Note 2. 4-hydroxyproline also occurs, but only in a very limited number of proteins.

Note 3. Parentheses after each amino acid enclose the standard three-letter abbreviation for that amino acid.

TABLE 2-2
Amino acid compositions of several specific purified proteins and of total protein from several biological sources. Content in the purified proteins is shown as amino acid residues per molecule. For total protein, the amino acid composition is given in percent by weight.

	AMINO ACID CONTENT			
	Weight Percent		Residues per Molecule	
	Cow's Milk	*Corn*	*α-Casein*	*Silk Fibroin*
Ala	3.6	10.0	9	45
Phe	5.1	4.5	8	4
Val	6.7	5.3	11	9
Leu	9.9	12.7	17	2
Ile	6.5	4.0	11	1
Met	2.4	1.8	5	0
Trp	1.3	0.7	2	1
Gly	2.1	3.5	9	74
Ser	5.2	4.4	16	17
Cys	0.9	1.6	0	2
Thr	4.7	4.1	5	2
Tyr	4.9	4.6	10	23
Asn	*	*	8	1
Gln	*	*	14	0
Asp	7.5	12.3	7	4
Glu	21.7	18.4	25	4
Lys	8.0	2.7	14	5
Arg	3.5	4.4	6	3
His	2.7	2.4	5	2
Pro	9.2	7.2	17	7
Total residues			199	206

* Included in aspartic and glutamic acid values.

Both carboxyl and amino groups take part in ionization reactions as follows:

$$\mathrm{R{-}C}\begin{matrix}\mathrm{{=}O} \\ \mathrm{{-}OH}\end{matrix} \rightleftharpoons \mathrm{H^+} + \mathrm{R{-}C}\begin{matrix}\mathrm{{=}O} \\ \mathrm{{-}O^-}\end{matrix}$$

$$\mathrm{R{-}NH_3^+} \rightleftharpoons \mathrm{H^+} + \mathrm{R{-}NH_2}$$

Since these are equilibrium expressions, the relative amounts of ionized and un-ionized compound present at any given time will be governed by the hydrogen ion concentration in the solution, in accord with the Mass Law for any chemical reaction. In quantitative terms, this can be expressed for the reactions just shown as

$$\frac{[\mathrm{H^+}] \times [\mathrm{R{-}COO^-}]}{[\mathrm{R{-}COOH}]} = K$$

and

$$\frac{[H^+] \times [R\text{—}NH_2]}{[R\text{—}NH_3^+]} = K$$

where the square brackets are used to indicate the molar concentrations of each of the species indicated, and the K's are the individual equilibrium constants (in this case ionization or dissociation constants). The exact value of K will depend on the particular compound involved. Typical values might be 1×10^{-5} for the dissociation of R—COOH and 1×10^{-8} for $R\text{—}NH_3^+$.

If we substitute these general values for the ionization constants into the equations above, along with any given value of the hydrogen ion concentration, we can obtain a value for the ratio of ionized to un-ionized molecules in the solution. For example, at pH 7 ($[H^+] = 1 \times 10^{-7}$ M) we obtain

$$1 \times 10^{-5} = \frac{(1 \times 10^{-7}) \times [R\text{—}COO^-]}{[R\text{—}COOH]}$$

that is,

$$10^2 = \frac{[R\text{—}COO^-]}{[R\text{—}COOH]}$$

That is, for every molecule of un-ionized R—COOH, there will be 100 molecules of $R\text{—}COO^-$. Another way to state this is to say that the acid is about 99 percent ionized at this pH.

Similarly, for the amine,

$$1 \times 10^{-8} = \frac{(1 \times 10^{-7}) \times [R\text{—}NH_3]}{[R\text{—}NH_3^+]}$$

or

$$10^{-1} = \frac{[R\text{—}NH_2]}{[R\text{—}NH_3^+]}$$

In this case, there are ten ionized molecules for each un-ionized one, or the amine is about 91 percent ionized. The manner in which percent ionization will vary with pH for each of these groups is shown in the form of a graph in Figure 2-1.

When the hydrogen ion concentration is equal to K, the equations reduce to the simple ratios

$$\frac{[R\text{—}COO^-]}{[R\text{—}COOH]} = 1 \text{ at } [H^+] = 10^{-5}\text{; that is, pH} = 5$$

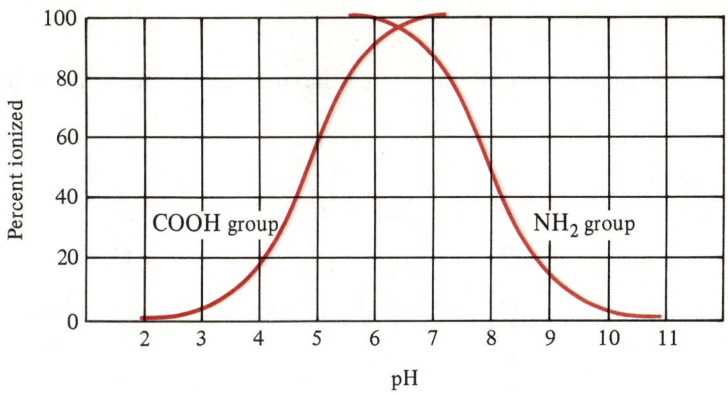

Figure 2-1
Graph of percent ionization versus pH for typical carboxyl and amino groups.

and

$$\frac{[\text{R—NH}_2]}{[\text{R—NH}_3^+]} = 1 \text{ at } [\text{H}^+] = 10^{-8}\text{; that is, pH} = 8$$

The pH at which this occurs is known as the p*K* for the particular ionizing group involved. Under these circumstances, the groups are half in the ionized and half in the un-ionized form. The p*K* is thus an expression of the hydrogen ion concentration present at half-ionization; in this way it serves as an indication of the relative ease with which the ionization takes place. That is, it indicates the strength of an acid or base.

The pK_a values of the carboxyl and amino groups of the amino acids are all in the range of 1.5 to 3 for the carboxyl and 9 to 11 for the amino group. In terms of the structure and function of proteins, we are more interested in the ionization of the side-chain groups of the acidic and basic amino acids. The pK_a values for these groups are shown in Table 2-3.

TABLE 2-3
Values of pK_a for ionizing side-chains of amino acids.

Amino Acid	pK_a
Acidic ionizations:	
Glutamic acid	
(γ-carboxyl group)	4.3
Aspartic acid	
(β-carboxyl group)	3.9
Cysteine (–SH group)	8.3
Tyrosine	
(phenolic hydroxyl group)	10.0
Basic ionizations:	
Histidine (imidazole group)	6.4
Lysine (ε-amino group)	10.5
Arginine (guanidino group)	12.5

Amino Acid Ionization

If we consider a typical amino acid having one of each type of group, we get

at low pH: 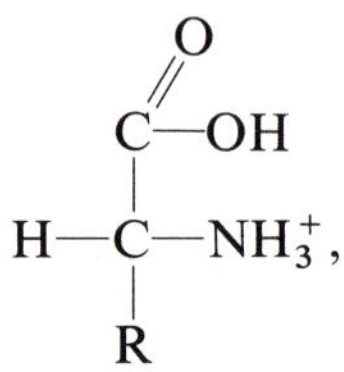, and at high pH: 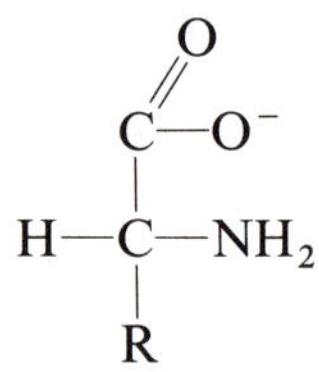

At some intermediate pH the number of molecules having the carboxyl group ionized and the number of molecules having the amino group ionized will be equal; at this point almost all of the molecules will be in the double ionized or *dipolar ion* (or "*zwitterion*") form:

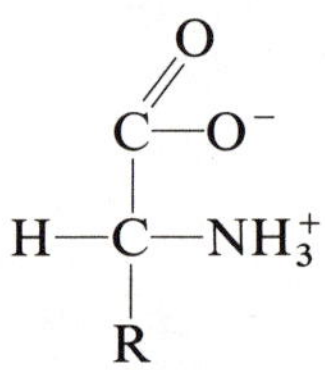

The pH at which this occurs is known as the *isoelectric point* of the particular compound. It is the pH at which there is no *net* positive or negative charge *on the molecule itself.* (Note: In the case of a simple amino acid, this is more properly called the "isoionic point," and you may see that term used in the literature.)

Protein Ionization

In proteins, of course, the alpha-amino and carboxyl groups of the amino acids are joined together in peptide linkages (except at the ends of chains) and thus are unable to ionize. However, if we look at the side-chains on a number of the amino acids, it is obvious that a variety of ionizing groups would be present as part of any protein. If we consider all of these ionizing groups as either carboxyl or amino groups, we can readily extend the above discussion to the protein molecule.

In this case we will wish to look not at large numbers of molecules, but at a single molecule with relatively large numbers of both kinds of ionizing groups. At extremes of pH, we would then get for a protein molecule:

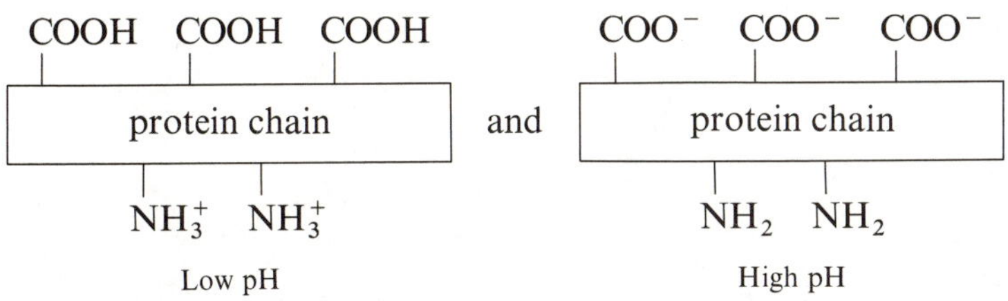

Somewhere in between these extremes, however, we would get a form such as is shown below; that is, with no net effective charge on the protein molecule:

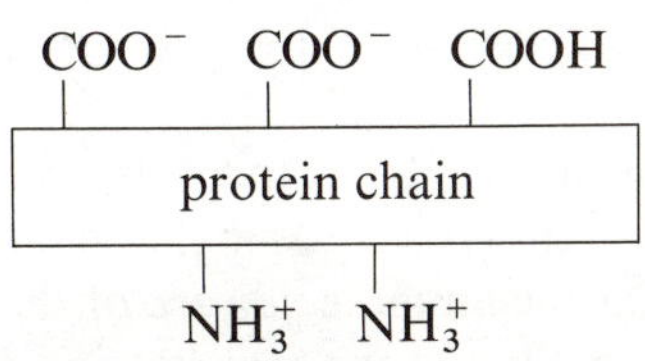

The pH at which this occurs is again referred to as the isoelectric point for that particular protein. It should also be clear that those proteins having an excess of carboxyl groups or their equivalent will tend to have a lower isoelectric point, while those having an excess of amino groups or their equivalent will tend to have a higher isoelectric point. For proteins especially, the isoelectric point is usually determined by finding the pH at which the molecules will not move toward either the anode or the cathode when subjected (in solution) to an electrical field.

Proteins are usually least soluble at their isoelectric points. This property is often used in protein separation and purification processes. As an example, the casein of milk is precipitated almost quantitatively at a pH of about 4.7. This is the first step in almost all types of cheese manufacture.

2.2 Peptides

Certain aspects of the structure and nomenclature of peptides may best be presented in conjunction with a specific example. Consider, then, a peptide composed of glycine, alanine, and serine with the structure shown below:

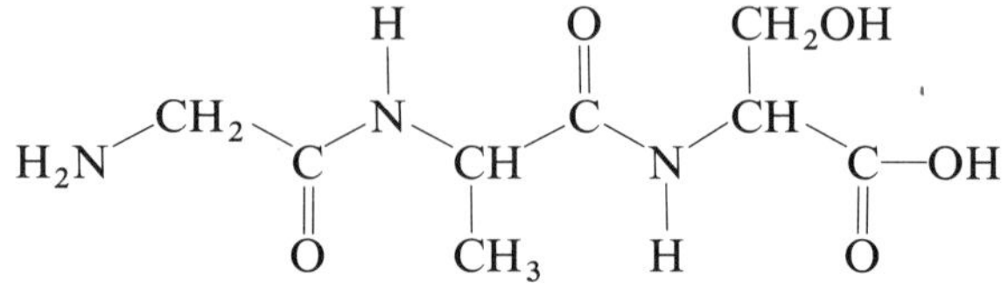

This is a tripeptide since three amino acids are present. Of course, this is only one of the six possible tripeptides that could be made from these three amino acids by varying their order in the chain.

The compound shown is **glycylalanylserine**. Note that the naming of the peptide begins with the amino acid having a free α-amino group and proceeds one amino acid at a time to the end of the chain having a free carboxyl group. Each amino acid unit except the last one is named as an acyl group. This involves the use of the root of the name for the amino acid and the suffix *-yl.*

In discussing naturally occurring peptides or those obtained by the cleavage of the natural compounds into smaller molecules, it is not normally necessary to indicate that the amino acids are of the L-series. This is simply understood to be the case. Should the specification of the optical isomerism be desirable, the compound shown above might be named as glycyl-L-alanyl-L-serine.

For longer peptide chains, the formal names become unwieldy, and the order of the amino acids is usually expressed in a shorthand form, using the standard three-letter abbreviation for each amino acid. Glycylalanylserine would become simply *gly-ala-ser.*

Note that the formation of two secondary amide or peptide bonds leaves only one free amino and one free carboxyl group per molecule, occurring at opposite ends of the peptide chain. The ends of the chain are referred to as the amino and carboxy terminals or commonly as the N— and C— terminals. These two terminal groups, one basic and one acidic, will be the only ionizable groups of any peptide chain unless amino acid units are present which have acidic or basic groups in their side-chains.

Small peptides are comparatively rare in biological material. However, a number of these are of sufficient importance to be mentioned here.

Glutathione is a widely distributed tripeptide of glutamic acid, cysteine, and glycine. The compound appears to function in certain oxidation–reduction reactions, although a variety of other specific roles have also been suggested:

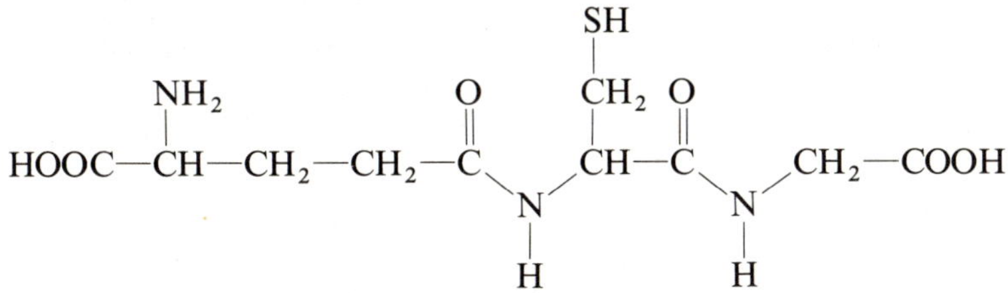

Glutathione
(γ-glutamylcysteinylglycine)

The important feature of this peptide appears to be its free thiol (—SH) group. Thiol groups undergo a reversible oxidation to form *disulfides*:

$$\text{R—SH} + \text{HS—R} \rightleftharpoons \text{R—S—S—R} + 2\,\text{H}\cdot$$

A substantial number of cellular proteins require for proper function the presence of a reduced thiol group. One major function that has been proposed for glutathione is serving to protect these thiol groups and to help regenerate them should they become oxidized.

Oxytocin and **vasopressin** are examples of a rather unusual group of cyclic peptides. These compounds are two of a number of peptide hormones produced by the pituitary gland. Oxytocin stimulates the contraction of uterine muscle during childbirth and the release of milk during lactation. Vasopressin increases the blood pressure by causing contraction of the walls of blood vessels; it also has a role in controlling excretion of water by the kidneys. The structures of these two peptides are shown in Figure 2-2.

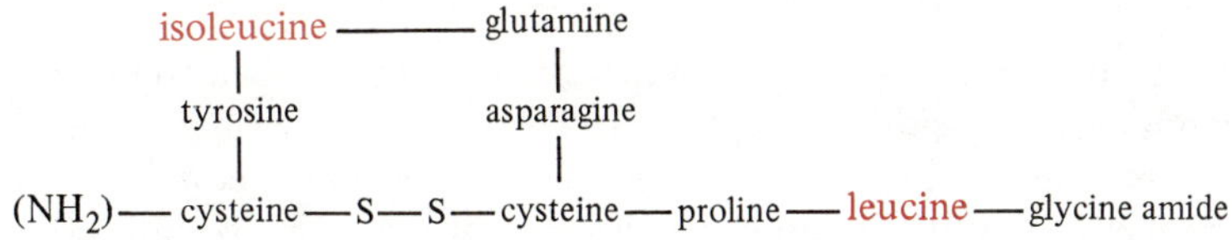

Oxytocin (human)

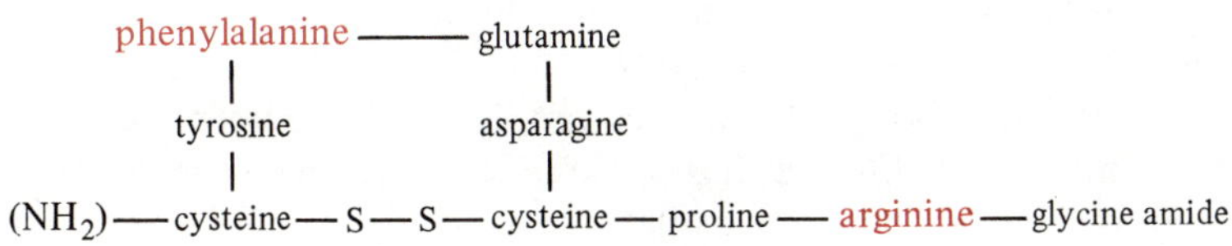

Vasopressin (human)

Figure 2-2
Structures of the peptide hormones *oxytocin* and *vasopressin*. The (NH_2—) has been added to show that this cysteine is the N-terminal amino acid of the peptide.

There are several points of interest in these structures. One of these is the cyclical nature of the compounds, resulting from the presence of a disulfide linkage between two cysteine units separated from each other in the peptide chain by four other amino acid units. Disulfide bonds of this type are also of considerable importance in the architecture of protein molecules.

Another point is derived from the comparison of the structures of the two compounds and their highly specific biological functions. You will have noted that vasopressin differs from oxytocin only in that leucine is replaced by a basic amino acid (lysine), and isoleucine is replaced by phenylalanine. Nevertheless, the two compounds have very different physiological activities, demonstrating that relatively minor differences in peptide structure can have far-reaching effects upon biological properties. This emphasizes the specificity of peptide structure with respect to amino acid composition and sequence. A given peptide, including the proteins as major examples, must have certain amino acids arranged in a certain order before its biological role can be fulfilled.

2.3 Amino Acid Sequence

When an amino acid is built into a peptide chain, we refer to that part of the molecule as an amino acid **residue**. The order in which the amino acid residues are joined together in a peptide chain is called the **amino acid sequence** of that chain. Since the polypeptide chains of proteins are often composed of hundreds of amino acid units, with the same ones occurring in many different places in the chain, determination of a complete sequence is an imposing task. However, several reactions exist that make it possible to determine the sequence of a peptide of limited length in a relatively short time.

There are three reagents that have commonly been used to identify the N-terminal amino acid of a peptide. In each case, the reagent reacts with the free amino group and forms a bond that is stable under conditions which disrupt one or all of the peptide bonds. Two of the reagents, fluorodinitrobenzene (FDNB) and dansyl chloride, require that the entire peptide be hydrolyzed (and thus destroyed) in order to release the derivative of the N-terminal amino acid for identification, as shown in Figure 2-3.

In spite of this limitation, Sanger and his associates used primarily the FDNB reaction in their classic elucidation of the amino acid sequence of insulin. Also, the dansyl derivatives are highly fluorescent and can be identified in very small quantities, making this a useful reagent when only small amounts of material are available.

By far the most commonly used procedure today, however, is that developed by Edman involving the use of phenylisothiocyanate. With this reagent, the derivative of the N-terminal amino acid can be removed from

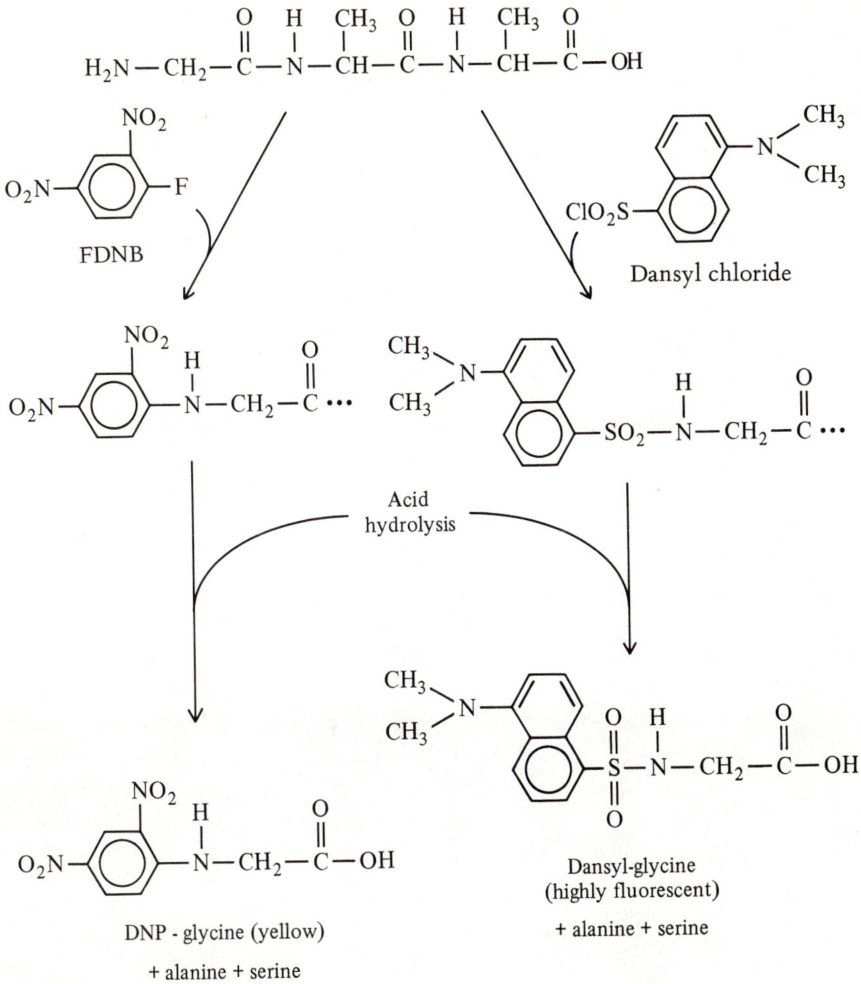

Figure 2-3
Use of the reagents fluorodinitrobenzene (FDNB) and dansyl chloride to determine the N-terminal amino acid of a typical peptide.

the peptide without disturbing the structure of the rest of the peptide. Thus the remaining peptide may be isolated, purified, and the process repeated to obtain the identity of the amino acid next to the N-terminus, and so on. The sequence of reactions involved in the Edman process is shown in Figure 2-4.

The Edman process has been automated for use in a machine (a "sequenator") capable of yielding the sequence of a peptide of 30 or more amino acids within a few days, under optimal conditions.

In every case, the amino acid derivatives produced must be identified by appropriate methods. These usually involve some form of thin-layer chromatography or gas–liquid chromatography.

Figure 2-4
The Edman reaction for determination of the amino acid sequence of a peptide. Removal of the N-terminal amino acid does not destroy the rest of the peptide, and the reaction can be repeated sequentially.

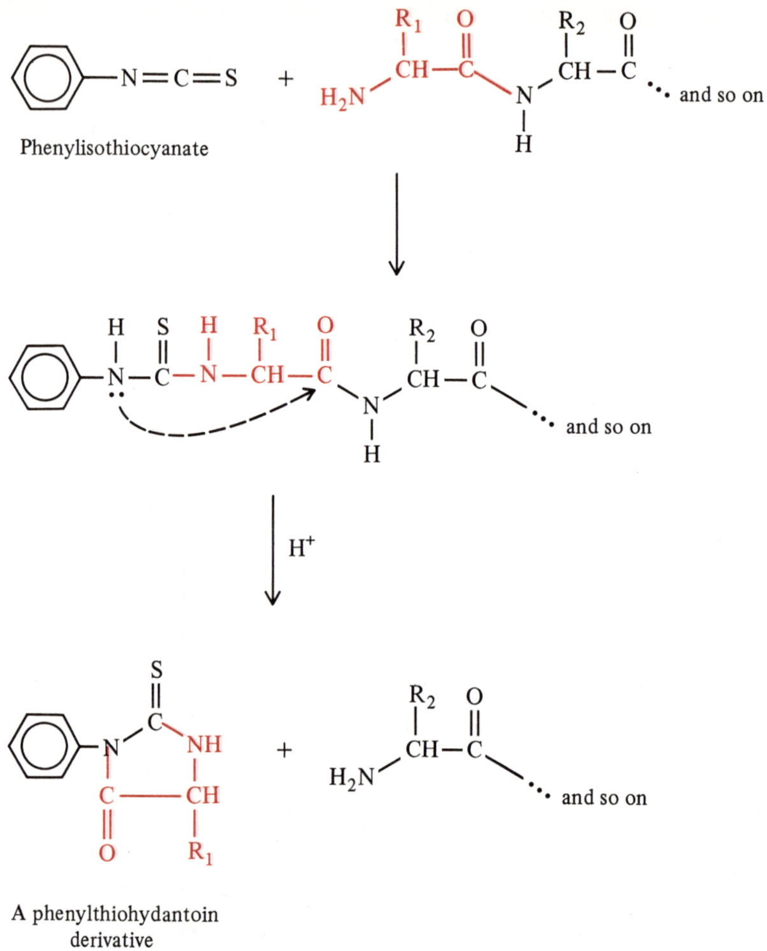

Determining the Sequence of a Polypeptide

Determination of the complete sequence of amino acids in a large polypeptide requires a systematic approach and a number of steps before using the reactions shown above. The process usually follows the following sequence:

1. Determination of the relative numbers of units of each of the amino acids in the peptide. This involves complete hydrolysis of the peptide, followed by quantitative separation of the constituent amino acids on an ion-exchange column.

2. Determination of the N-terminal and C-terminal amino acids of the peptide.
3. Splitting of any disulfide bonds by oxidation with performic acid. In many proteins, cysteine units are linked together by formation of disulfide bonds. These links create "loops" on the polypeptide which must be opened up to obtain a linear polypeptide. The reaction is as follows:

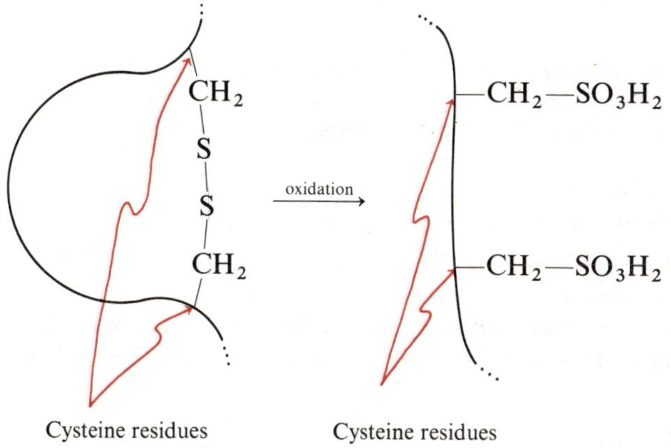

4. Cleavage of the polypeptide chain into a series of smaller peptides suitable for analysis by the Edman reaction. This is usually accomplished using enzymes that catalyze the hydrolysis of peptides at specific linkages only. Thus trypsin will catalyze hydrolysis only at the carboxyl group of arginine or lysine while chymotrypsin prefers the carboxyl side of large nonpolar side-chains such as those of the aromatic amino acids. A variety of other enzymes with different specificities are now available from plants and microorganisms. In addition, it is possible to cleave a peptide chain containing methionine through the action of cyanogen bromide, which splits the chain on the carboxyl side of the methionine moiety.

Note that it is necessary to apply several different cleavage procedures in order to get peptides that overlap one another. It is only in this way that it is possible to tell how the small peptides are joined to one another to make up the original polypeptide.

The effects of trypsin, chymotrypsin, and cyanogen bromide in splitting a typical portion of a polypeptide chain are illustrated in Figure 2-5.

Each of the peptides produced by whatever method of cleavage must be separated out and purified before analysis can proceed.

··· asp—arg—gly—ser—met—phe—ala—lys—met—val—tyr—ser ···

Trypsin ··· asp—arg | gly—ser—met—phe—ala—lys | met—val—tyr—ser ···

Cyanogen bromide ··· asp—arg—gly—ser—met | phe—ala—lys—met | val—tyr—ser ···

Chymotrypsin ··· asp—arg—gly—ser—met—phe | ala—lys—met—val—tyr | ser ···

Figure 2-5
Cleavage peptides formed by the action of trypsin, chymotrypsin, and cyanogen bromide on a typical peptide. Note that in this case any two of these agents would result in the necessary overlap of peptides.

5. Analysis of the cleavage peptides by Edman or similar procedure and the determination of the sequence for each. This is followed by the process of putting together the pieces of the jigsaw puzzle in the correct order to indicate the sequence of the original polypeptide. At this point it is also necessary to compare the final sequence with the amino acid analysis to be sure that no amino acid units have been "lost" along the way.

Application of the procedures has permitted determination of the amino acid sequences of an ever-growing number of proteins.

Chemical Synthesis of Polypeptides

Considerable progress has been made in recent years in the chemical synthesis of polypeptides. The general approach involves "blocking" the amino group of one amino acid and the carboxyl group of the other, then reacting these with a specific reagent to form the peptide bond. A typical reaction would be as follows:

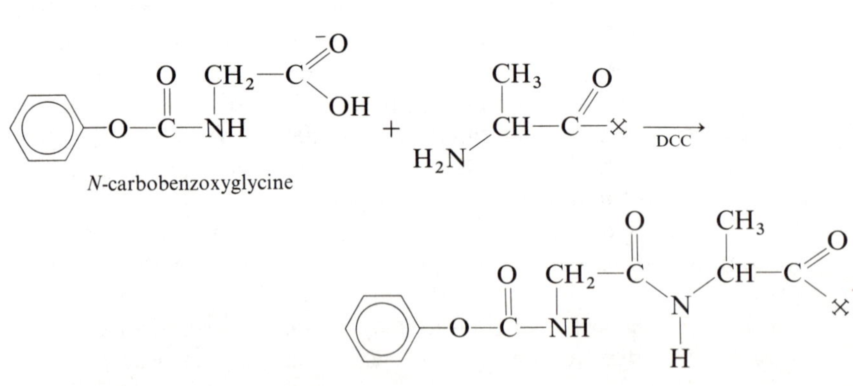

where DCC stands for dicyclohexylcarbodiimide, a reagent capable of bringing about the amide bond formation. The carbobenzoxy group can be removed without disturbing the peptide bond, and the process can be repeated with another amino acid to form a tripeptide, and so on. If the first amino acid is bound to an insoluble plastic bead (represented by the "×" above), the process can be automated and carried out sequentially by a machine. A number of peptides, including oxytocin, vasopressin, and growth hormone, have been synthesized in this way. In addition, it has proven possible to synthesize the insulin molecule (about 50 amino acids), although in very low yield.

Additional Readings

Scientific American articles

P. Doty, "Proteins," 197(3), 173 (September 1957).

W. H. Stein and S. Moore, "The Chemical Structure of Proteins," 204(2), 81 (February 1961).

R. B. Merrifield, "The Automatic Synthesis of Proteins," 218(3), 56 (March 1968).

W. A. Schroeder, *The Primary Structure of Proteins* (Harper and Row, New York, 1968). An intermediate level discussion of methods of protein sequence determination.

J. M. Stewart and J. D. Young, *Solid Phase Peptide Synthesis* (Freeman, San Francisco, 1969). A short (89 pp.) description of laboratory procedures used in peptide synthesis.

A. Meister, *Biochemistry of the Amino Acids*, 2nd edition (Academic Press, New York, 1965) (2 volumes). A thorough and advanced discussion of the chemistry and biochemistry of the amino acids.

Study Questions

1. Draw the possible stereoisomers for each of the following compounds:
(a) glycine (b) L-serine (c) L-threonine

2. Draw a structure for one example each of an amino acid having a side chain that contains:
(a) an aliphatic hydrocarbon group only
(b) a hydroxyl group
(c) a carboxyl group
(d) a sulfur atom
(e) an aromatic ring
(f) an amino group
(g) a nitrogen-containing heterocycle

3. Draw a tetrapeptide using some of your answers to question (2); do not use any one amino acid twice. Name the tetrapeptide.

4. A small protein contains 10 residues of glutamic acid and nine of lysine. From the data of Table 2-3, determine the isoelectric point for this protein.

5. In order to determine its sequence, a polypeptide is treated with trypsin and chymotrypsin. The resulting small peptides are then identified to be as follows:

trypsin: ser-tyr-met-gly-phe-gly-lys
thr-ala-ser-tyr-ala-arg
pro-glu-gly-arg
gly-ser-glu

chymotrypsin: gly-lys-thr-ala-ser-tyr
pro-gln-gly-arg-ser-tyr
ala-arg-gly-ser-glu
met-gly-phe

What was the sequence of the original peptide?

6. (a) In the organic chemistry laboratory, how would you proceed to make N-methylacetamide from acetic acid and methylamine (and whatever else you need)?
(b) Could you use the same reaction to synthesize glycylglycine? What difficulties would you expect? How might you get around them?

CHAPTER 3

The Structures of Proteins

The proteins may be viewed as very large peptides. A number of the properties of the proteins, however, are considerably different from those to be expected for a long chain of amino acids connected by peptide bonds. Such a chain should resemble a long, thin worm, continually twisting and turning in solution to assume different conformations, different shapes at different times. There is no doubt that peptide chains are a primary aspect of protein structure; yet each protein occurs in nature in a particular, relatively fixed, three-dimensional configuration. Some are essentially linear, others resemble coiled springs, and still others may have both linear areas and coiled areas and in addition may be complexly folded to give very compact, globular molecules.

The apparent discrepancy between the expected properties of a peptide chain and the actual properties of the protein is because of the presence in proteins of a variety of bonding interactions in addition to peptide bonds. These bonding forces hold the peptide chain in the naturally occurring or **native** conformation. A number of different types of bonding may be involved in creating and maintaining the overall structure of any particular protein. The major ones of general importance are disulfide bonds, hydrogen bonds, electrostatic attractions, and "hydrophobic bonding."

3.1 Major Bonds in Proteins

Disulfide Bonds

We have already seen that the thiol groups of two cysteine molecules may be reversibly oxidized to give the disulfide compound, cystine. When this reaction involves two cysteine units in a polypeptide chain or in different peptide chains, a strong covalent linkage is formed between the two portions of the molecule or molecules. Oxytocin has already been given as an example of the formation of a cyclic structure from a peptide chain by an internal disulfide. The polypeptide hormone, insulin, is an excellent example of the binding together of two peptide chains by disulfide bonds. The structure of this compound is given in Figure 3-1.

Hydrogen Bonds

You are no doubt familiar with the tendency of a proton involved in a polar bonding to some more electronegative atom to be attracted also to any other available unshared pair of electrons. The result of this attraction of the proton to two pairs of electrons on different atoms creates what we call a **hydrogen bond** between the attracting groups:

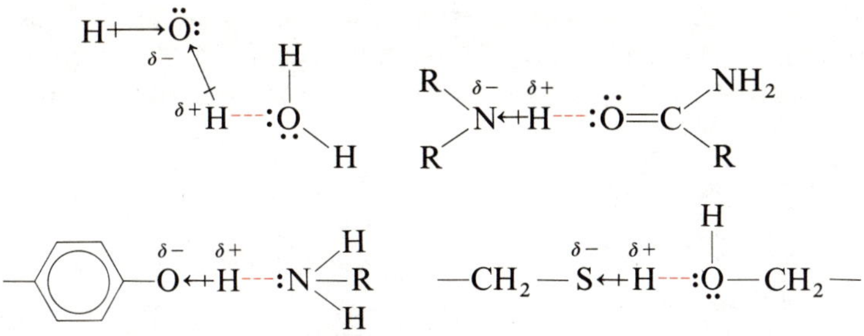

Examples of hydrogen-bonding interactions

The energy involved in any one of these bonds is small, on the order of 1 to 2 kcal (4 to 8 kJ) per mole. However, when there are many of them occurring between molecules (or within a molecule) the total energy can be quite large. Hydrogen bonding in proteins may typically involve dozens if not hundreds of interactions per polypeptide chain. We will look at some specific examples shortly.

Electrostatic Attractions

The side-chains of the acidic amino acids (aspartic and glutamic acids) will be charged negatively at neutral pH values. Those of the basic amino

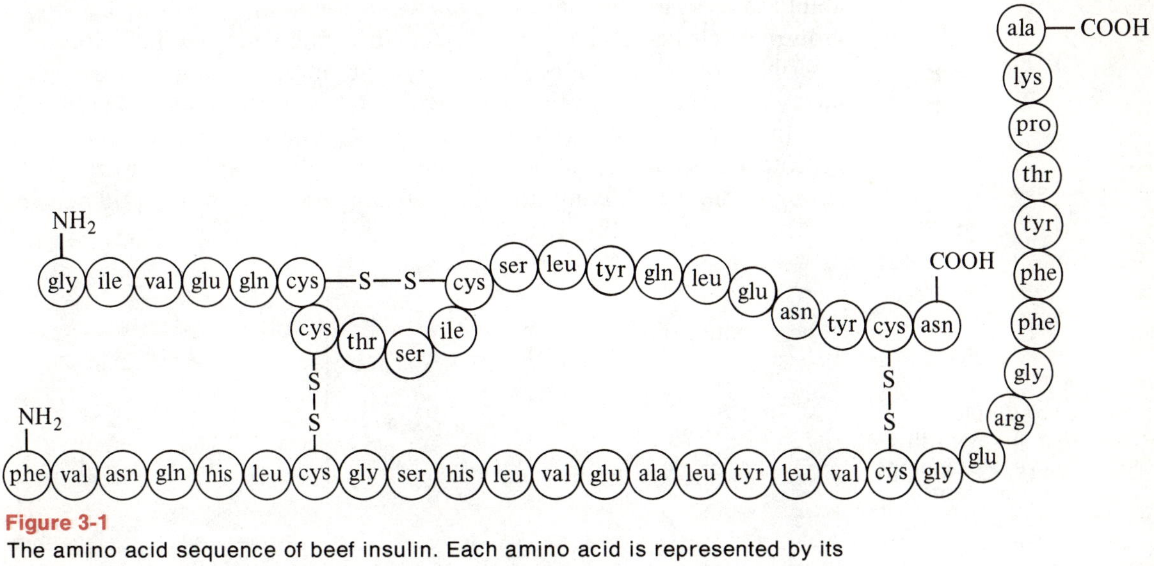

Figure 3-1
The amino acid sequence of beef insulin. Each amino acid is represented by its standard three-letter abbreviation. Glycine and phenylalanine are the N-terminal amino acids; asparagine and alanine are the C-terminals.

acids (lysine, arginine, and to some extent histidine) will be charged positively. If properly positioned, these groups may give rise to bonding between different portions of a given molecule or between two or more protein chains:

$$\sim\!-CH_2-CH_2-\overset{\overset{\displaystyle O}{\|}}{C}-O^{\ominus}\quad H_3\overset{\oplus}{N}-CH_2-(CH_2)_3-\!\sim$$

Electrostatic interaction between side-chains of glutamic acid and lysine

Hydrophobic Bonding

The amino acids that make up the nonpolar, uncharged group in the list of Table 2-1 have side-chains that are primarily hydrocarbon in nature. These side-chains are referred to as **hydrophobic** groups since they are unable to form hydrogen bonds with the solvent water molecules.

The presence of hydrophobic groups in a water solution has two effects. First, they prevent the water molecules from forming as many hydrogen bonds as they would be able to form in the absence of such groups. Second, they interfere with the organization of the water molecules and thus cause a change in the entropy of the solution. The net result is an increase in the total energy of the solution. This increase is minimized when the nonpolar groups are close together, thus presenting less surface to the water molecules. It is this ability to decrease the total energy of the solution that drives hydrophobic groups to cluster together. The term "hydrophobic bonding" does not describe a true attraction between nonpolar groups, but rather a response to interactions between solvent molecules. This effect is illustrated in Figure 3-2.

In any case, such aggregations of hydrocarbon side-chains are common in protein molecules in solution, and these associations often lead to relatively strong "bonding" holding together different portions of a peptide chain.

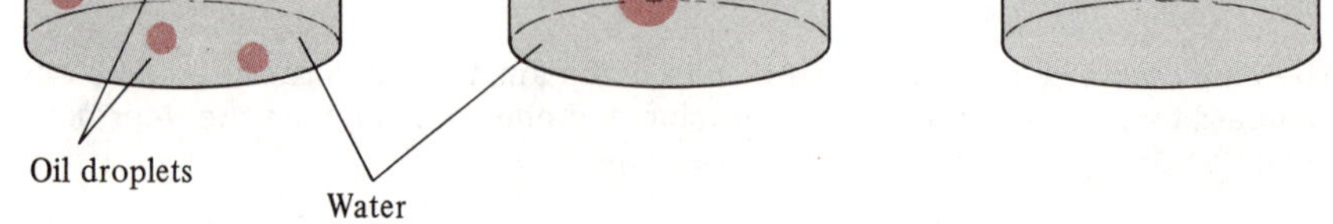

Figure 3-2
Nonpolar oil droplets suspended in water will tend to aggregate together in order to expose the minimum surface area possible to the polar solvent. The same basic forces are at work in "hydrophobic bonding."

3.2 Polypeptide Conformations

At first glance, a polypeptide chain would appear to have a great deal of freedom to rotate and assume just about any conformation imaginable. In reality, however, there are some powerful restraints on the range of possibilities.

First, the peptide bond has a structure intermediate between the following two resonance extremes:

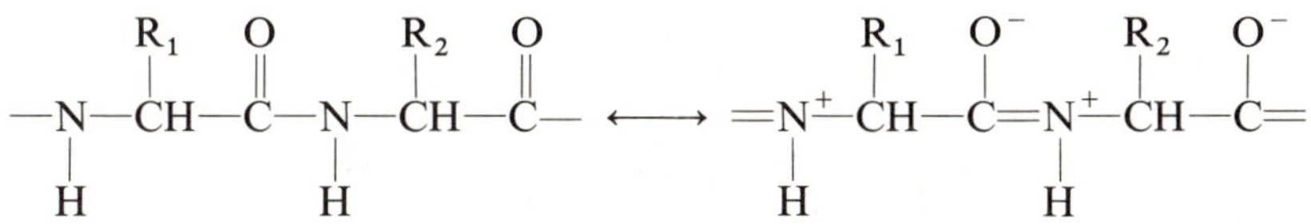

Because of this, the atoms involved in the peptide bond all lie in a single plane:

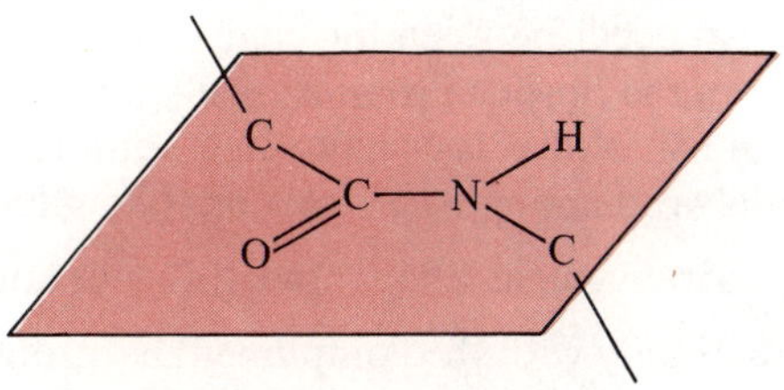

Since the C—C and N—C bonds are true single bonds, rotation can occur at either of these points. As a result, a polypeptide chain can be considered as a series of such planes rotated at various angles to one another:

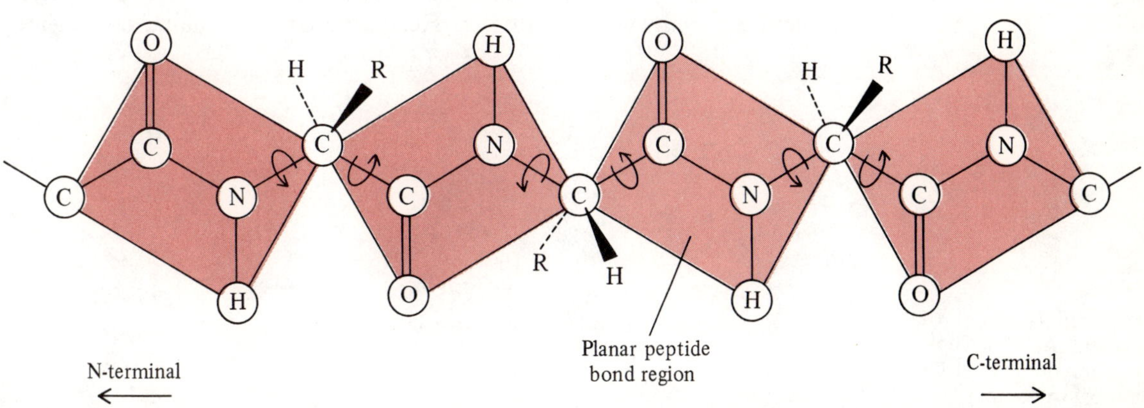

However, even this type of rotation is severely restricted by the problem of interactions between the R-groups in any amino acids other than glycine.

The result of all this is that there are a limited number of possible stable conformations for a polypeptide chain. In fact, there are three different fundamental conformations that between them are found to account for the structural basis of most proteins. Much of our information on the exact nature of these conformations has come from study of the insoluble fibrous structural proteins, such as the keratins of hair, collagen of connective tissue, fibroin of silk, and so on.

Stretched Conformations

There are several stretched conformations, varying only to a small extent in the angles of rotation of the planes of the amino acid units involved. By far the most common is the **β-conformation**, which is essentially the result you might expect if you held both ends of a polypeptide chain and pulled (gently). In this case, the extended chain assumes a form approximately as shown in Figure 3-3.

When several stretched chains line up next to one another, extensive hydrogen bonding can occur between the chains. These types of interaction are most easily seen in the fibrous proteins of silk and collagen.

In the silk fibroin structure, the chains in β-conformation are arranged so that they alternately run in opposite directions (from N—to C—terminal). This "antiparallel" arrangement allows the >C=O and the >N—H of one chain to line up exactly with the complimentary group in a neighboring chain. The resulting pattern of hydrogen bonding is shown in Figure 3-4. A similar result is possible when the chains are lined up in a "parallel" arrangement. In this case, the angles of the hydrogen bonds are somewhat different.

Because of the bond angles involved in either case, and the planar nature of the peptide bond, these interactions produce a sheet that has the appearance of a piece of paper that has been folded repeatedly to form a "pleated" structure as shown in Figure 3-5. Hence the common name of **β-pleated**

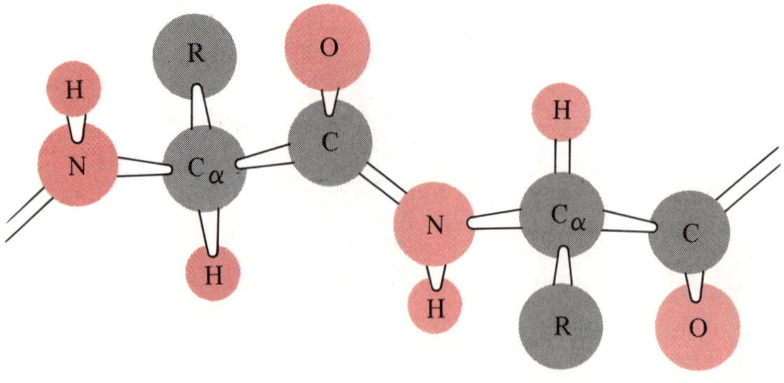

Figure 3-3
A dipeptide unit in a β-conformation. The peptide chain is stretched almost to its limit.

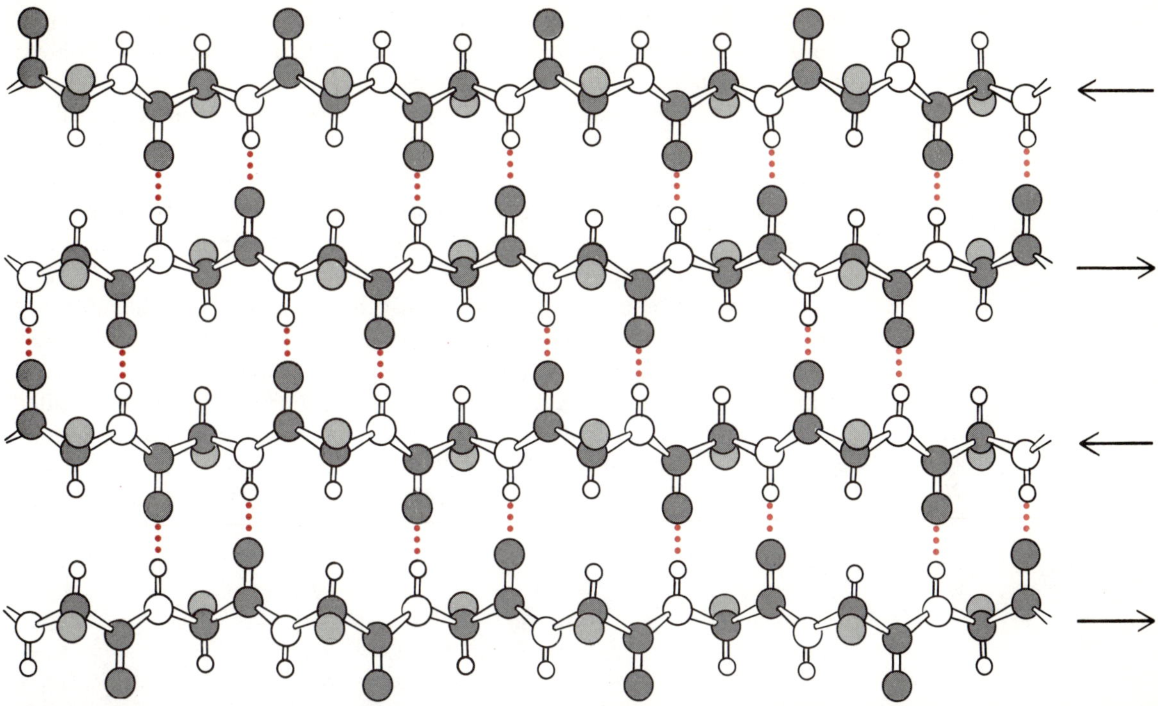

Figure 3-4
A pleated sheet formed by an antiparallel alignment of polypeptide chains in the β-conformation. The triple dots indicate the hydrogen bonds between adjacent chains. In this drawing, the side-chains are above and below the plane of the paper.

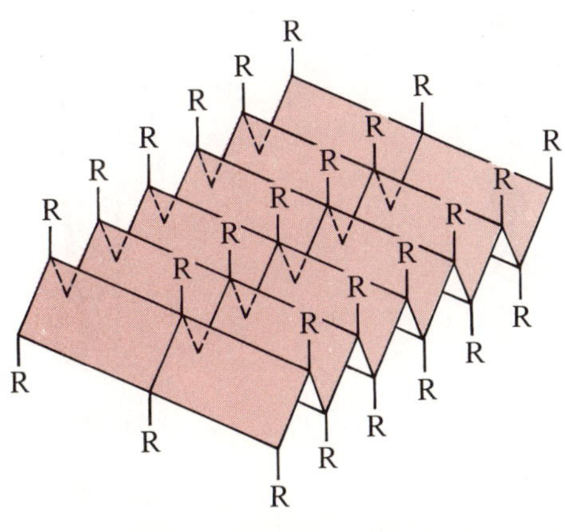

Figure 3-5
A simplified drawing of the pleated sheet structure typical of a protein such as silk fibroin. The three rows of R-groups are the side-chains of the three polypeptide chains involved.

sheet for this type of structure. We will see shortly that these same interactions are also important in maintaining structures of more complex globular (soluble) proteins.

Alternatively, the chains in a stretched conformation may be loosely twisted together to form long rope-like fibers. The basic structure of **collagen**, shown in Figure 3-6, is typical of this type of association. In this case, three strands are loosely coiled together to form a structure called tropocollagen. These molecules associate further to form the collagen fibrils which make up much of the protein of connective tissue.

It may be noted that there are some other constraints on the nature of these two fibrous proteins. In order that the three strands of collagen can come close enough together to hydrogen bond effectively, every third residue in the chain must be a glycine. In the case of silk fibroin, the formed pleated sheets pack down upon one another in a fashion not unlike a stack of empty egg cartons. To permit this, the side chains of the amino acids must be small; consequently, glycine and alanine are the major constituents of this protein.

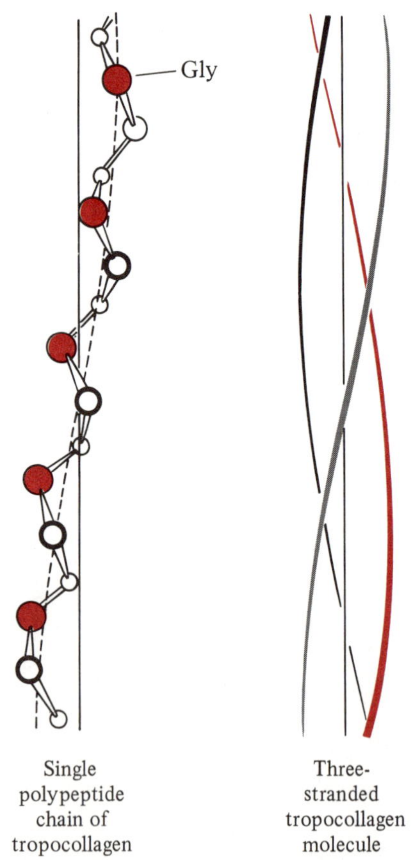

Figure 3-6
A simplified representation of the conformation of the polypeptide chains making up a triple-stranded tropocollagen molecule.

The α-Helix

The second type of conformation for polypeptide chains is one in which the chain is rather tightly coiled upon itself, with hydrogen bonding between adjacent turns of the helix. While several such helical conformations are possible, by far the most common is the **α-helix**.

This is a right-handed helix, with 3.6 amino acid residues per complete turn and the dimension shown in Figure 3-7. The coil is stabilized by the extensive hydrogen bonding shown diagrammatically in Figure 3-8.

A number of fibrous proteins, notably the alpha-keratins of hair, are composed entirely of strands in alpha-helix conformation. Within the globular proteins, the content of alpha-helix varies widely; we will look at some examples shortly.

The ability of a section of a polypeptide to exist in the alpha-helix form is governed in large measure by the amino acid sequence. A series of large bulky residues, or side-chains carrying the same electrical charge, will tend to interfere with the stability of the helix. One amino acid, proline, simply cannot be accommodated. Because of the rigid angular relationships between the carboxyl and the nitrogen in proline, it is not possible for the necessary rotation to take place. In addition, proline has no available hydrogen to form hydrogen bonds. Presence of a proline in a polypeptide will assure a section that is not in helical form.

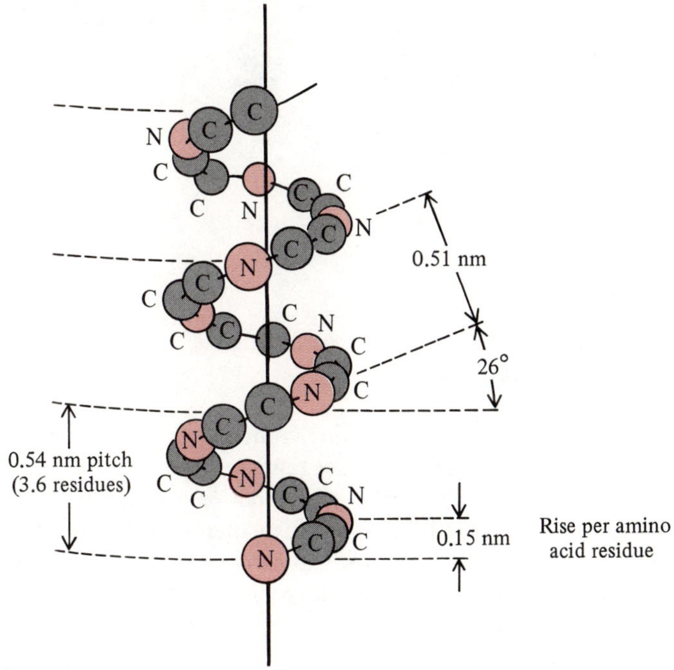

Figure 3-7
A schematic drawing of the backbone of an α-helix showing the average dimensions. [From L. Pauling and R. B. Corey, *Proc. Int. Wool Text. Res. Conf.* 1955, B249 (1956). (CSIRO, Melbourne)]

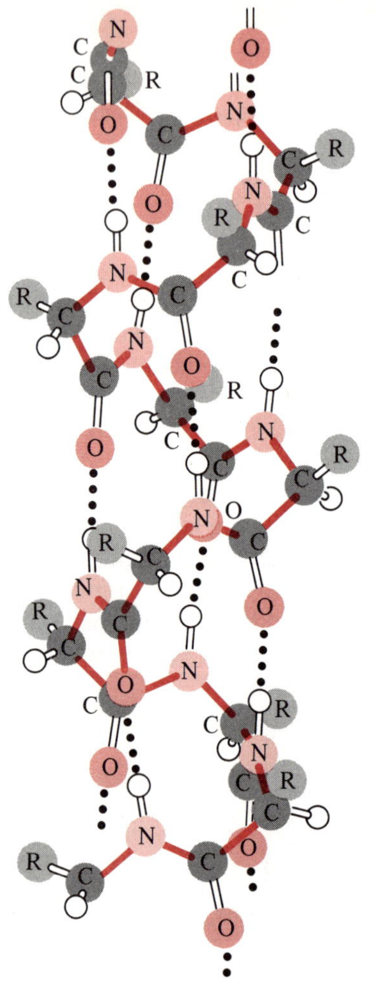

Figure 3-8
A drawing of a typical ball-and-stick model of an α-helix, showing the hydrogen bonding occurring within the chain between every fourth amino acid residue. (Reprinted from Linus Pauling, *The Chemical Bond*. Copyright © 1967 by Cornell University. Used by permission of the publisher, Cornell University Press.)

Random Conformations

The third major type of polypeptide conformation is usually described as **random**. This simply reflects the fact that no clear picture of organized structure is apparent. However, the angles between the planes of peptide bonds usually stay within the limits previously mentioned. Also, the random chains in most globular proteins appear to have rather fixed positions. They are undoubtedly held in conformations specific to the protein in question by the hydrogen, hydrophobic, and electrostatic bonding previously mentioned.

3.3 Structure of Globular Proteins

In contrast to the fibrous structural proteins, most of the biologically active proteins, such as the enzymes, are rather compact structures tending toward some roughly spherical shape. Thus the term "**globular proteins**."

While individual structures of the globular proteins vary greatly, all involve some specific folding and bending of the chain to yield a specific three-dimensional structure. Within this framework, there may be regions of the polypeptide in alpha-helix and/or β-pleated sheet, with some random configuration. Myoglobin, the oxygen-binding protein of muscle, is an example of a protein based almost entirely on sections of alpha-helix. Its structure is indicated in Figure 3-9. On the other hand, glyceraldehyde 3-phosphate dehydrogenase (an enzyme of carbohydrate metabolism) is a representative of a group of proteins containing substantial amounts of β-sheet structure. In Figure 3-10 you can see both parallel and antiparallel alignments being used to form a backbone for the protein molecule.

In either case, the final folded, bent, coiled, sheeting protein will be held in its specific three-dimensional structure by a combination of hydrogen

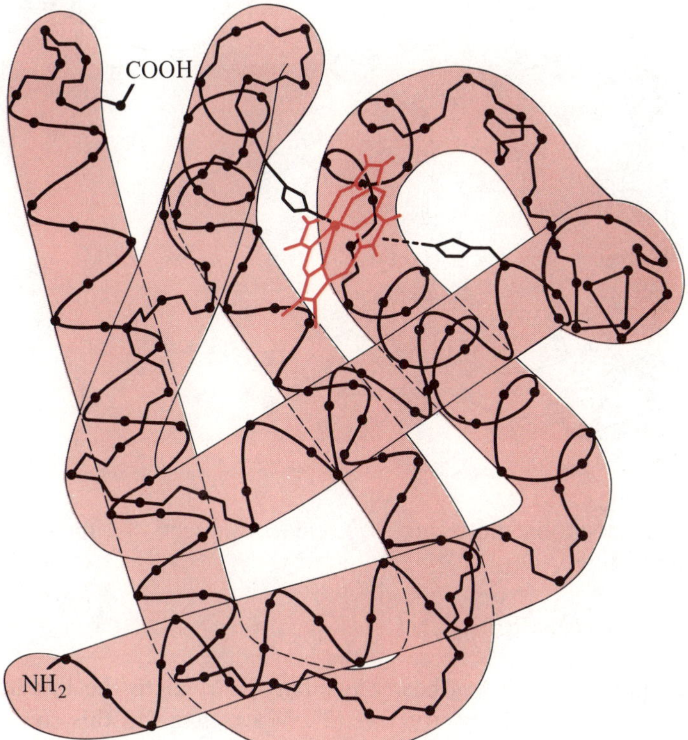

Figure 3-9
A drawing of the three-dimensional structure of myoglobin (an oxygen-binding protein of muscle), showing the extensive coiling of the polypeptide chain into sections of α-helix. The oxygen-binding heme ring, with its central atom of iron, is shown in color. [Based on R. E. Dickerson, in *The Proteins,* 2nd edition, H. Neurath, (Academic Press, New York, 1964).]

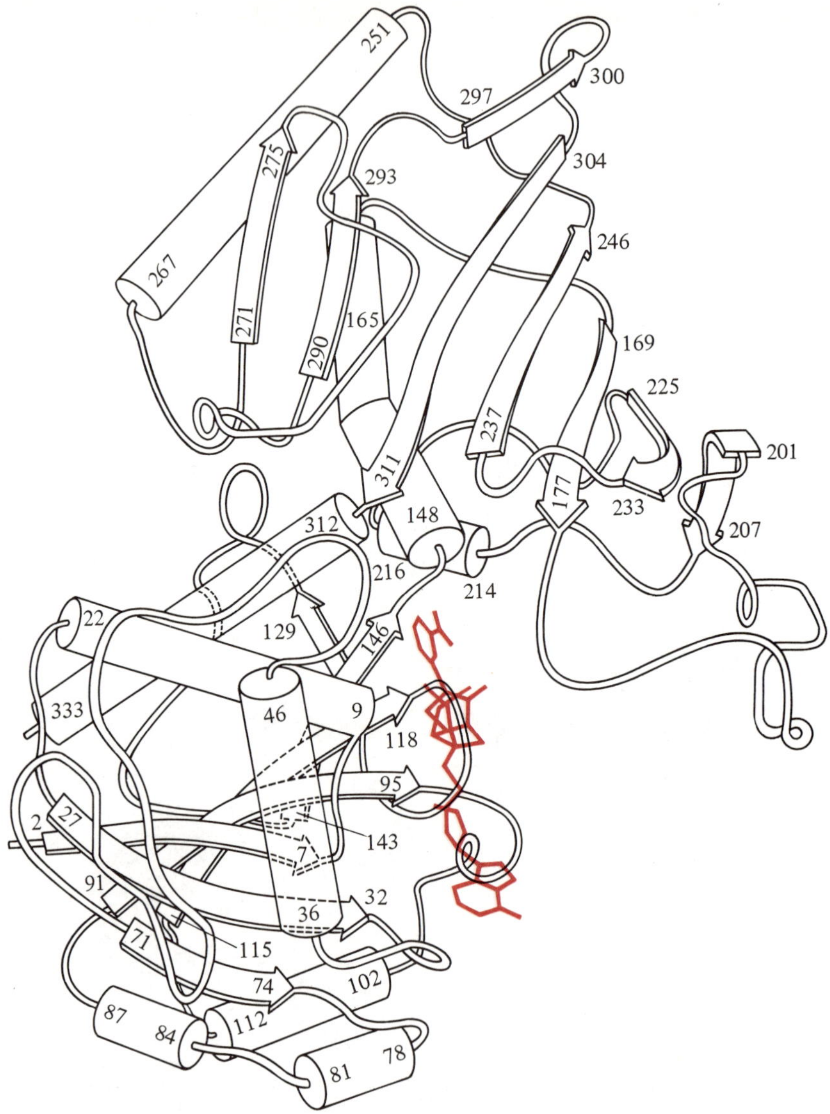

Figure 3-10
The structure of the enzyme *glyceraldehyde 3-phosphate dehydrogenase*. The cylinders represent regions of α-helix and the flat arrows regions of β-conformation. The spaghetti-like strands are regions of "random" conformation. A molecule of the coenzyme NAD is shown in the normal binding position. (From Biesecker, Harris, Thierry, Walker, and Wonacott, *Nature* 266:328, 1977.)

bonding, hydrophobic clustering, electrostatic interactions, and perhaps by disulfide bonding.

A Note on Nomenclature

For many years, it has been customary to refer to the primary (1°), secondary (2°), and tertiary structure (3°) of a protein. In this context, the **primary structure** is synonymous with the amino acid sequence. **Secondary**

structure refers to the folding of a polypeptide chain into α-helix or β-conformation. **Tertiary structure** results from further folding of the chain, stabilized by bonding interactions between distant amino acid residues.

More recently, **quaternary structure** has been used to refer to the aggregation of several polypeptide units to form a native multimeric protein complex.

Amino Acid Sequence and Structure

It is generally considered today that, for most proteins at least, the final structure is inherently determined by the amino acid sequence. As the protein molecule is synthesized, it may be able to enter into a variety of conformations. Those that are the lowest energy forms at that time will predominate, yielding a final product with a single spatial orientation. Experimental support for this concept comes from experiments in which proteins are completely converted to a random configuration by treatment with concentrated urea solution. When the urea is slowly removed, the protein gradually reforms a globular structure. In some cases, a substantial amount of the protein will return to its original form, regaining its biological activity in the process.

This ability to specify structure by specifying the amino acid sequence is very important to cellular function since the only information transmitted from one generation to the next is the data from which to generate a series of amino acid sequences.

Denaturation of Proteins

The conformation of a protein as it is assumed to occur in cells and tissues is termed the **native state**. As the protein is exposed to various environmental conditions, its conformation may change, with attendant alterations in its physical properties and, most significantly, in the ability of the protein to perform its biological role. Such a change in a protein is termed **denaturation**.

Denatured protein molecules often tend to form large aggregates and to precipitate from solution—a process described as **coagulation**.

We cannot describe precisely for any protein the changes that occur as it becomes denatured. Presumably a number of relatively weak secondary bonding interactions are broken and new bonds form in a different pattern. These processes require but little energy; hence proteins may be denatured by rather mild conditions. Just about any condition which will interfere with hydrogen bonding or with electrostatic attractions will increase the tendency of a protein to denature. Prominent among these are increases in temperature, high or low pH ranges, high salt concentrations, detergents, or presence of organic solvents. Proteins are also mechanically denatured when spread in thin films, as in beating an egg white.

For all living cells a certain rate of protein denaturation and replacement is a fact of life itself. For those that exist at the higher ranges of biologically acceptable temperatures, a large percentage of the available energy of the cell must be expended on protein synthesis to replace rapidly denatured cellular components.

Urea Denaturation

The urea molecule is magnificently equipped to take part in hydrogen bonding either through its protons or its unshared electron pairs. Because of its low molecular weight and its ability to hydrogen bond extensively with water molecules, it is extremely soluble in water, readily yielding an 8 molar solution:

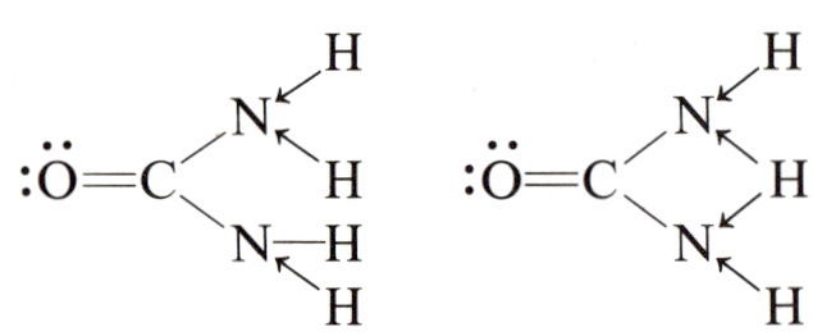

Urea

At such high concentrations, the urea competes very effectively for hydrogen bonding with the >N—H and >C=O groups, as well as all of the eligible side-chain groups. In addition, electrostatic interactions are completely disrupted and hydrophobic interactions rendered less effective:

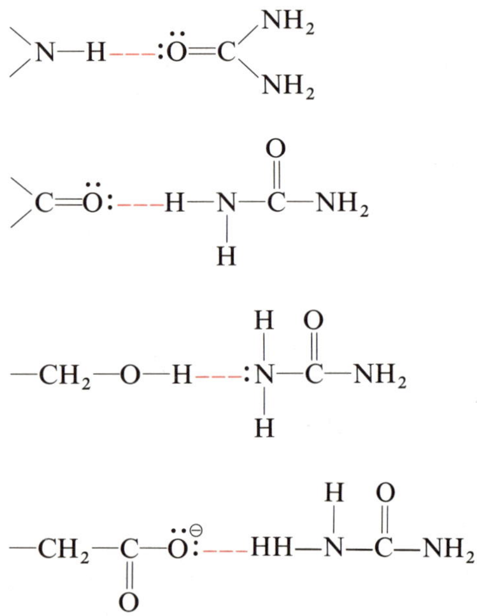

The result of such treatment for globular protein is usually loss of most or all of its secondary, tertiary, and quaternary structure. (Disulfide bonds are, of course, not affected.)

Additional Readings

Scientific American articles

E. O. P. Thompson, "The Insulin Molecule, 192(5), 36 (May 1955).

D. C. Phillips, "The Three-Dimensional Structure of an Enzyme Molecule," 215(5), 78 (November 1966).

R. E. Dickerson, "The Structure and History of An Ancient Protein," 226(4), 58 (April 1972).

A. Light, *Proteins: Structure and Function* (Prentice-Hall, Englewood Cliffs, N. J., 1974). Brief but fascinating discussion of protein structures and functional significance.

R. E. Dickerson and I. Geis, *The Structure and Action of Proteins* (Benjamin, New York, 1969). A concise, exceedingly well-illustrated description of the three-dimensional structures of proteins and the relationship of structure to function.

H. Neurath and R. Hill (Eds.), *The Proteins*, 3rd edition (Academic Press, New York, 1975–77). A three-volume classic covering a wide range of topics related to protein structure and function.

Study Questions

1. Define, or clearly identify, the concepts associated with the following terms:
(a) "sequence" of a protein
(b) a "polypeptide"
(c) "primary," "secondary," "tertiary," and "quaternary" structures of globular proteins
(d) an "alpha-helix"
(e) "hydrophobic" bonding
(f) denaturation

2. List all of the types of bonding involved in protein structures. For each of those that affect 2°, 3°, or 4° structure, give a specific example showing typical side-chains that might be involved.

3. Compare and contrast the structural features characteristic of fibrous and globular proteins.

4. List as many conditions as possible that will lead to rapid denaturation of a protein.

5. Consider an average protein of 1000 amino acid residues that is all in the α-helix conformation. What is the total energy of stabilization of this protein resulting from hydrogen bonding?

6. Hydrogen bonding in an antiparallel pleated sheet is shown in Figure 3-4. Make a drawing to show the nature of hydrogen bonding in a parallel pleated sheet structure.

CHAPTER 4

Enzymes I: The Nature of the Beast

The catalytic units called **enzymes** are the one truly unique feature of the chemical processes of living cells. These protein molecules possess in common with all other catalysts the ability to increase the rates of reactions. But the enzymes have an efficiency in this regard far beyond anything encountered in the laboratory. Rates of some cellular reactions are increased a million-fold over the normal, uncatalyzed, values. In addition, each enzyme has a degree of specificity in its action that permits only desirable reactions to be encouraged and side-reactions thus eliminated or held to a negligible level.

One of the requirements for living systems that we discussed in Chapter 1 was the ability to control the processes occurring in the cell. We will see that enzyme molecules are subject to a variety of control mechanisms, a number of them reflecting the ease with which protein molecules can be modified in structure and conformation.

4.1 The Nature of Catalysis

Before we deal specifically with the action of enzymes, we will review briefly the nature of catalysis. Recall the typical picture of energy changes during a reaction as shown in Figure 4-1. Here we have the reaction

$$A + B \longrightarrow C + D$$

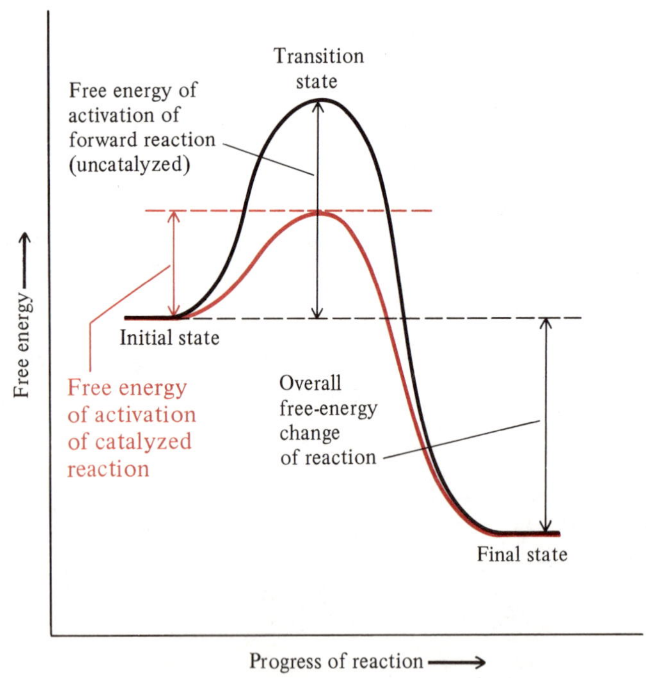

Figure 4-1
Energy diagram for a chemical reaction, catalyzed and uncatalyzed.

The available energy released during this process (the overall change in free energy) is constant no matter what the path from A + B to C + D. On the other hand, the activation energy is dependent on the nature of the activated intermediate state. The basic function of a catalyst is to promote the reaction rate by providing a reaction path with a lower energy of activation.

Enzymes generally accomplish this function by binding the reactants, holding them in positions suitable for reaction, and providing an electronic or other environment that promotes the reaction mechanism involved.

4.2 Structure and Function

We have mentioned above that the process of catalysis by an enzyme involves a binding of the reactant involved (usually called a **substrate**) to the enzyme. Since enzymes are large molecules and substrates are usually small ones, only a relatively small portion of the enzyme molecule will actually be involved in the binding and subsequent catalysis. This region is referred to as the **active site** of the enzyme. If we look a little more closely, we will find some side-chains in the enzyme involved in binding with the substrate, and some involved in promoting the reaction. We can then distinguish between **binding sites** and a **catalytic site** within the active site region of the enzyme. This is illustrated in Figure 4-2 for an example of a hypothetical enzyme catalyzing the hydrolysis of a dipeptide.

Because enzymes often bind substrates at three or more positions, they are capable of stereochemically specific reactions even when the compounds

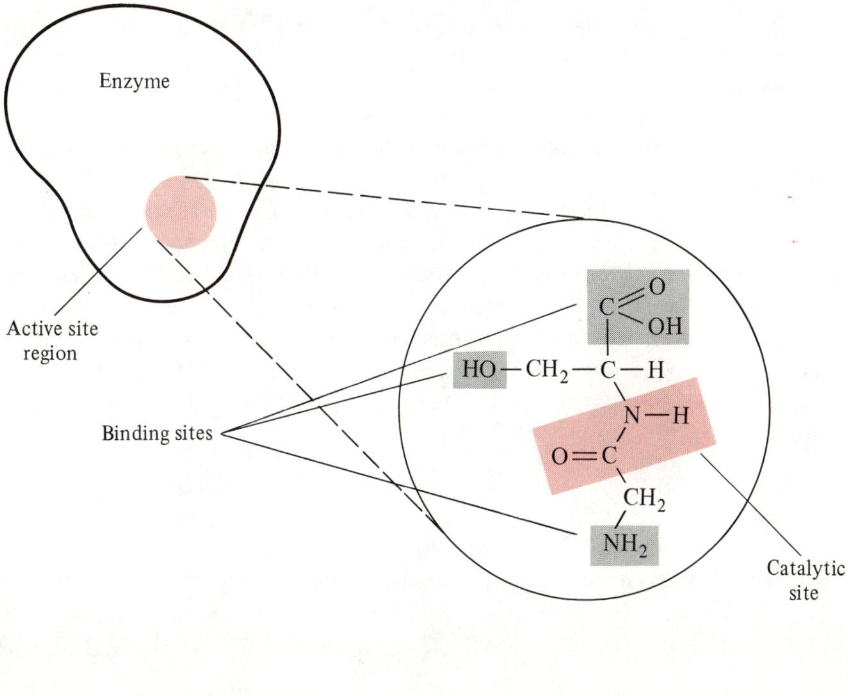

Figure 4-2
A diagrammatic view of the active site of an enzyme. This hypothetical enzyme is one that hydrolyzes certain dipeptides.

involved are themselves symmetrical. An example is the case of aconitase, an enzyme which converts citric acid to isocitric acid:

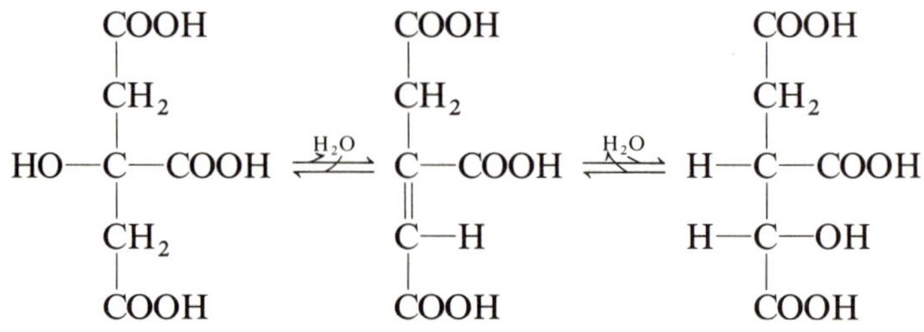

Citric acid   *cis*-Aconitic acid   Isocitric acid

The citric acid molecule can bind to the enzyme only in one way, and thus the transfer of the hydroxyl group will always go to the carbon labeled No. 4 in Figure 4-3 rather than carbon No. 2. Similar binding requirements permit enzymes to deal with or generate only one specific optical isomer of a compound.

In a general sense, if a substrate is to bind productively to the enzyme, there must be a suitable space available, binding groups must match those of the substrate, and the catalytic site must be properly located when the substrate is in the bound form. An example of the necessary spatial relationship between enzyme and substrate can be seen in Figure 4-4. Here we see lysozyme, an enzyme which catalyzes hydrolysis of rather large polysaccharide molecules, with and without its substrate. The catalytic site is located in about the middle of the "valley" of the enzyme.

Catalytic Mechanisms

The variety of reaction mechanisms for which enzymes provide support is too large to go into in any detail here. General acid–base catalysis is probably the most common, often involving a pair of histidine residues to donate and accept protons as required. In addition, one sees nucleophilic catalysis involving the thiol group of cysteine, the hydroxyl group of serine, or the amine group of a lysine residue. Other enzymes appear to function by metal ion or electrostatic catalysis mechanisms. In some cases, as in transfer reactions involving phosphate groups, the entire group may be transferred covalently to a serine or histidine residue on the protein as an intermediate step.

Cofactors, Coenzymes, and Prosthetic Groups

There is considerable overlap and some confusion in the use of the terms **cofactor**, **coenzyme**, and **prosthetic group**. I prefer to use the general

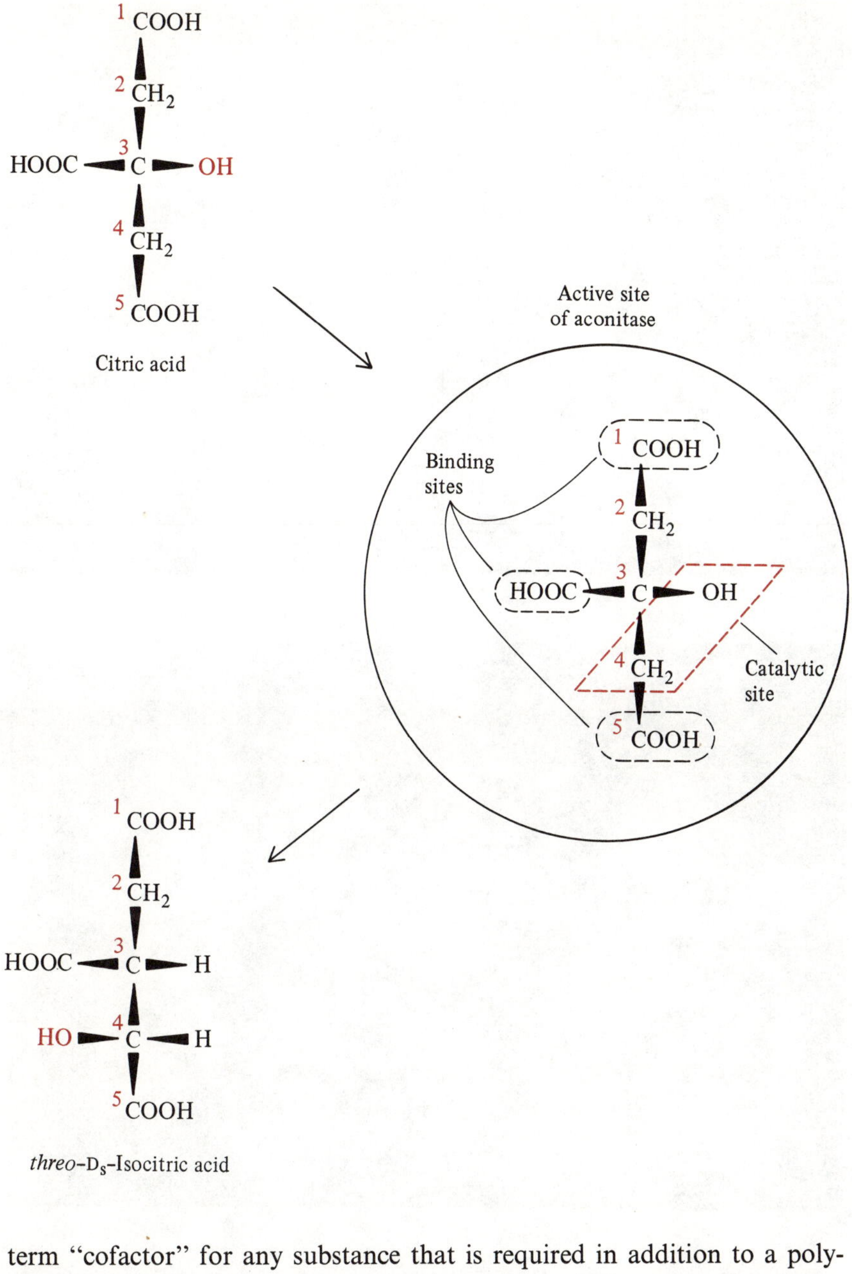

Figure 4-3
The stereospecific action of the enzyme aconitase. The citric acid molecule is able to bind to the enzyme only in this one spatial orientation. Consequently, the hydroxyl group will always be transferred to carbon 4 and not to carbon 2.

term "cofactor" for any substance that is required in addition to a polypeptide chain in order to have an active enzyme. A "coenzyme" is then a subgroup of the cofactors composed of organic molecules actually taking part in the reaction (rather than simply serving to maintain protein structure, for example). The term "prosthetic group" is not used as much today as it

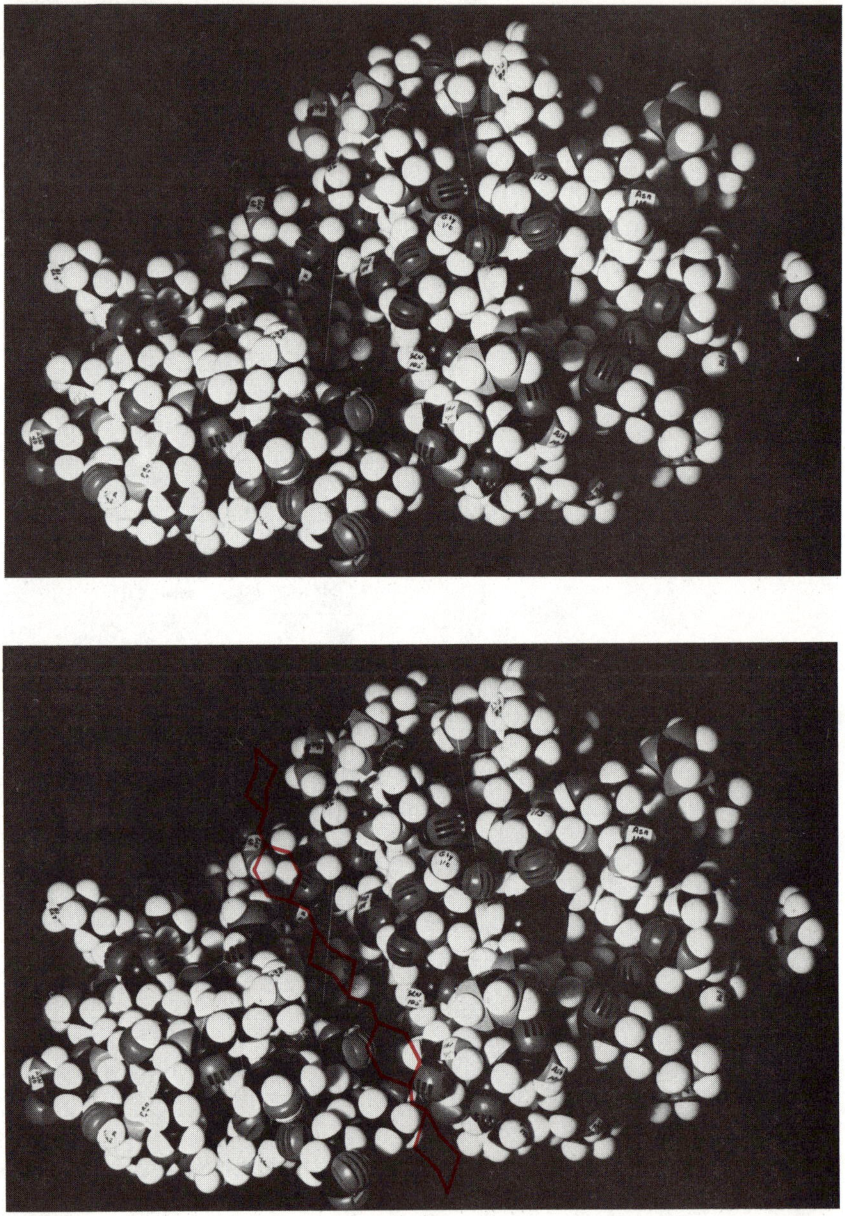

Figure 4-4
A model of the enzyme lysozyme, with and without its polysaccharide substrate. The tertiary structure of the enzyme provides a substantial "valley" into which the substrate can fit. (Model of T4 phage lysozyme courtesy of B. W. Mathews, University of Oregon; photograph by M. Kilgour.)

was some years ago. In practice, it tends to be indistinguishable from the term "coenzyme."

In some cases, coenzymes are loosely associated with the polypeptide chain of the enzyme and are easily removed by dialysis and similar gentle treatment. Others can be removed only after complete denaturation, and a few are covalently bonded to side-chain groups of the polypeptide chain and cannot be removed except by hydrolysis. For those cases where a coenzyme can be reversibly dissociated from the rest of the enzyme molecule, we term the complete enzyme a **holoenzyme** as compared to the **apoenzyme** (the molecule with coenzyme removed).

Coenzymes and B-Group Vitamins

All living cells make use of the same molecules as coenzymes; however, some cells do not have the ability to synthesize completely one or more of these substances. The portions they cannot make are vitamins for these organisms. These compounds are generally water-soluble and thus belong to what are usually called the "B-group" vitamins. A given compound can be a vitamin for one organism and not for the next. Therefore it is preferable to use chemical names such as thiamine, even though they may be trivial names, rather than designations such as Vitamin B_1. The compound thiamine is needed by all living things as part of a coenzyme; it is a vitamin only for a relatively few organisms. It is important here, as in the cases of essential amino acids and essential fatty acids to be encountered later, to distinguish carefully between the metabolic necessity of a compound and the necessity that it be obtained preformed from the diet or the culture medium.

In this connection, it should be realized that the need for the presence in the diet of a compound such as a vitamin implies that the organism concerned has lost the ability to produce one or more enzymes required for the synthesis of the compound from other simple dietary constituents. The fact that genetic deficiencies of this type are widespread among living organisms implies that they do not present much of a hazard to survival, and that through the history of evolution the necessary compounds were readily obtained from the diet in adequate amounts. However, it should be noted that a diet restricted in amount or in variety of foods may well supply inadequate amounts of a variety of substances, including vitamins.

We will examine the roles of the various coenzymes as we encounter them in metabolic reactions. In the meantime, a list of the common B-complex vitamins, with the names of the related coenzymes and some major reactions with which they are associated, is given in Table 4-1.

Inorganic Cofactors

Many of the inorganic substances required by cells, particularly the metal ions, also are needed primarily because of their function as cofactors. This is again consistent with their being needed only in trace amounts, for these are the "trace elements" of animal nutrition, of bacterial culture media, and of fertilizers. Obviously, no organism can synthesize a metal ion; hence

TABLE 4-1
Relationships among water-soluble vitamins, their coenzyme forms, and metabolic reactions in which they participate.

Vitamin	*Coenzyme*	*Types of Reactions*
Thiamine (B_1)	Thiamine pyrophosphate	Decarboxylation of α-keto acids; certain reactions of keto sugars
Riboflavin (B_2)	Flavin mononucleotide Flavin adenine dinucleotide	Several kinds of oxidation–reduction reactions
Pyridoxine (B_6)	Pyridoxal phosphate	Several kinds of reactions involving amino acids, for example, decarboxylation, transamination
Niacin	Nicotinamide adenine dinucleotide Nicotinamide adenine dinucleotide phosphate	Numerous oxidation–reduction reactions
Pantothenic acid	Coenzyme A	Many reactions of fatty acids, particularly those involving the transfer of acetyl groups
Biotin	Enzyme-bound biotin	Certain carbon dioxide fixation reactions
Folic acid	Tetrahydrofolic acid	Various reactions involving single carbon compounds
Cyanocobalamin (B_{12})	Several "cobamide" coenzymes	Carbon chain isomerizations; certain methyl group transfers

these substances, in contrast to vitamins, must be obtained from the environment by all living things.

Typical examples of enzymes requiring metal ions for catalytic activity are: carbonic anhydrase (catalyzing the reaction of CO_2 with H_2O to give H_2CO_3), which requires Zn^{++}; arginase (catalyzing the hydrolysis of arginine to give ornithine and urea), requiring Mn^{++}; and tyrosinase (catalyzing the oxidation of tyrosine), requiring Cu^{++}. Magnesium ion functions as a cofactor for numerous enzymes, particularly those which catalyze reactions of phosphorylated compounds.

Metal-ion cofactors may act in several ways; for example, a cupric ion bound to an enzyme may be reduced to a cuprous ion by a reaction with an organic substrate, and then reoxidized by a second reaction with oxygen. In other cases, a metal-ion cofactor may serve simply as one of the binding points for a substrate or coenzyme. Magnesium ion, for example, is a cofactor for almost every enzyme having as its substrate a phosphate com-

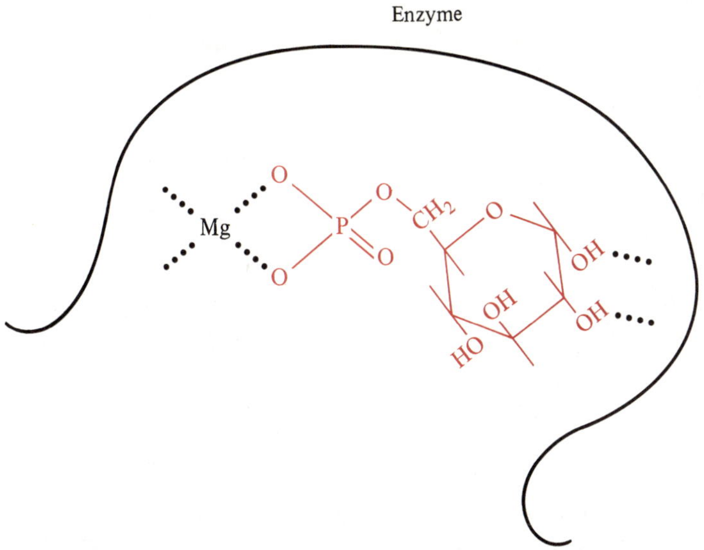

Figure 4-5
A hypothetical example of the action of a magnesium ion in assisting in the binding of a substrate to an enzyme.

pound. Presumably the metal ion acts as a bridge to bind the phosphate group of the compound to the enzyme, as in the diagramatic representation shown in Figure 4-5.

4.3 Factors Affecting the Rates of Enzyme-Catalyzed Reactions

Substrate Concentration

An enzyme is often studied in terms of its effect on the rate of the reaction it catalyzes. Usually we do not determine the time required to reach equilibrium, but rather the initial rate, the rate at which the reaction proceeds in the first few minutes before the accumulation of products makes the reverse reaction significant. If we carry out a series of experiments, identical in all ways except that the initial substrate concentration is varied, we obtain results typified by Figure 4-6. Here the reaction rate must be determined by analysis either for the amount of product formed or for the amount of substrate which reacted in a given time period.

At high substrate concentrations, the reaction rate slowly approaches a constant value in spite of further increases in concentration. To the early biochemists, this was an anomalous finding. In general, in chemical reactions, as the concentrations of the reactants are increased a corresponding increase in reaction rate is observed. In the case of enzymic reactions, we see this only for relatively low substrate concentrations. This apparent anomaly

Figure 4-6
The effects of changing substrate concentration on the rate of an enzyme-catalyzed reaction.

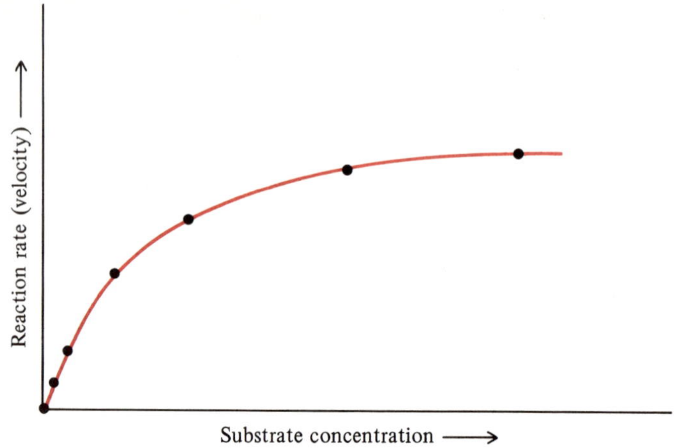

could be explained only by assuming that the enzyme was an active participant in the reaction process, rather than a passive component of the reaction mixture. Furthermore, it was then necessary to assume that the overall reaction occurred not as a single step, but as a multistep process. For example, a hydrolysis reaction might be written simply as:

$$\mathrm{S} + \mathrm{H_2O} \overset{\mathrm{E}}{\rightleftharpoons} \mathrm{P_1} + \mathrm{P_2}$$

This tends to convey the erroneous impression that the effect of the enzyme E on the rate of hydrolysis of the substrate S to the products P_1 and P_2 is simply a reflection of its physical presence. The assumptions given above led to the suggestion that the actual process was of the following nature:

$$\mathrm{S} + \mathrm{E} \rightleftharpoons \mathrm{ES}$$

$$\mathrm{ES} + \mathrm{H_2O} \rightleftharpoons \mathrm{E}\begin{matrix}\diagup \mathrm{P_1}\\ \diagdown \mathrm{P_2}\end{matrix}$$

$$\mathrm{E}\begin{matrix}\diagup \mathrm{P_1}\\ \diagdown \mathrm{P_2}\end{matrix} \rightleftharpoons \mathrm{E} + \mathrm{P_1} + \mathrm{P_2}$$

In this representation of the process, it is clear that the enzyme molecule must combine with the substrate as a prelude to the actual hydrolysis, giving a new compound which is termed the enzyme–substrate intermediate. This enzyme–substrate compound then undergoes hydrolysis of a susceptible bond, yielding yet another substance, the enzyme–product complex. Finally, the products are released from the enzyme, leaving it in its original state.

This picture of the enzyme-catalyzed reaction provides a ready explanation of the observed effects of substrate concentration on the reaction rate. To show this, let us first simplify the equations shown above to a more general form:

$$E + S \rightleftharpoons ES \longrightarrow E + \text{Products}$$

Here consideration is being restricted to the initial stage of the reaction, when the concentration of products is so low that the interaction of the products with the enzyme can be ignored as an effect on the rate. Thus, ES → P is not depicted as a reversible process in the reactions shown above.

Clearly the rate of appearance of products at any instant should be directly proportional to the concentration of the enzyme–substrate compound. The problem reduces then to that of understanding how the concentration of this intermediate changes as the substrate concentration is changed. If we examine the first step alone, the reaction of E and S to give ES, we should expect that an increase in S, at low concentrations, would be accompanied by a shift to a higher concentration of ES and thus a higher rate of reaction. But as S continues to be increased, a maximum concentration of ES should eventually be reached since this value can never exceed the total concentration of the enzyme. The amount of enzyme present therefore determines the maximum concentration of ES. A further increase in S beyond the point at which essentially all of the enzyme is in the form of the ES compound cannot appreciably increase the reaction rate, and a plateau in the graph of rate versus substrate concentration should result.

These qualitative considerations help to explain the experimental observations depicted in Figure 4-6. A quantitative treatment of the kinetics involved is presented in Chapter 5.

Enzyme Concentration

Preceding discussions have emphasized that an enzyme must react chemically with the substrate as the first of a sequence of reactions which ultimately regenerates the enzyme and yields the reaction products. The enzyme is accordingly a reacting molecule, and its concentration should be expected to influence the reaction rate in the same manner as that of the other reactant, the substrate (see p. 53). Therefore we might anticipate that as the enzyme concentration is increased, the rate would also increase to a certain point and then become constant. This should happen theoretically;

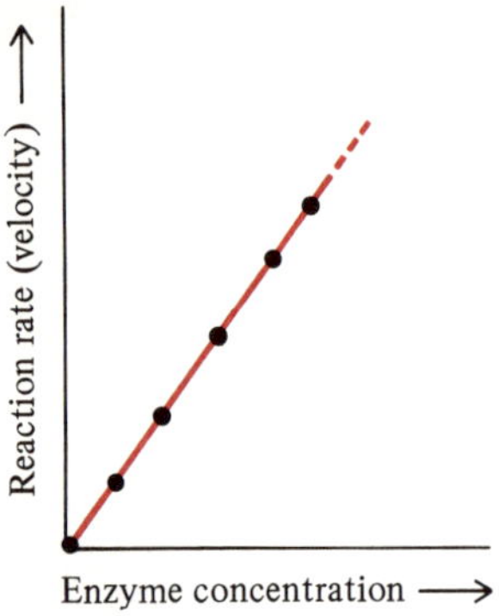

Figure 4-7
The effect of changing enzyme concentration on the rate of an enzyme-catalyzed reaction. The curve should have a shape much like that of Figure 4-6, but is usually limited to the portion shown here by inability to obtain higher concentrations of enzyme

actually it does not—for practical reasons. Here we must recall that the molecular weight of an enzyme is high, often over 100,000. The preparation of 1 molar solution of an enzyme would therefore require the solution of 100,000 grams of enzyme in water to give a final volume of 1 liter—an obvious impossibility. Under normal conditions the enzyme concentration rarely exceeds 0.0001 molar.

Practically, then, we are restricted to working with low concentrations of enzyme. Under these conditions the reaction rate is directly proportional to the concentration of the enzyme, as shown in Figure 4-7.

One further point should be made in connection with this figure. Note that at zero enzyme concentration the rate of reaction is indicated to be zero. It is, of course, not true that reactions do not occur unless enzymes are present. They may be slow, but many occur at a measurable rate. For this reason, in working with enzyme-catalyzed reactions, one always must run a "blank" or "control" experiment in which the enzyme is absent. Any reaction which occurs is then subtracted from that obtained in the presence of the enzyme. The difference between the two values is the effect caused by the enzyme.

The linear relation between enzyme concentration and reaction rate has much practical importance because it allows one to measure the relative concentration of one enzyme in impure protein preparations, or even in the great mixtures of proteins in biological materials. We can express this relation as

$$\text{Rate} = k \times E$$

where E = enzyme concentration and k = a constant characteristic of the enzyme involved.

The reaction rate is the amount of reaction which occurs in a unit of time. It is usually measured and expressed as the amount of product formed in a particular time period:

$$\text{Rate} = \frac{\text{amount of product}}{\text{time of reaction}} = k \times E$$

or

$$E = \frac{\text{amount of product}}{\text{time of reaction} \times k}$$

Since k is a constant for the enzyme at hand, we can compare relative amounts of enzyme simply on the basis of the rate. Two standard units for enzyme activity exist. The first, the older **international unit**, is defined as the amount of enzyme that will cause one micromole of substrate to react every minute. A more recent agreement defines a new unit of enzyme activity known as a **katal**. One katal of enzyme is the amount needed to cause one mole of substrate to react per second. Needless to say, this is usually a very large amount of enzyme and figures are more often expressed in nanokatals.

If a very pure enzyme is available, so that one can measure its precise concentration, it is possible to measure the **turnover number**. This is a measure of how many molecules of substrate react for each enzyme active site in one second. The enzyme must be saturated with substrate and the temperature and pH must be specified. Turnover numbers vary from as low as 1 to as high as 10^6 per second. One of the highest values is that of catalase (which converts H_2O_2 to O_2 and H_2O) with a value of about 10^5 per second.

Temperature

Experiments designed to determine the effect of temperature on the rate of an enzyme-catalyzed reaction often give data of the type presented in graphical form in Figure 4-8. Note that at 0°C the rate is close to zero. As the temperature is raised the reaction rate increases until a maximum is reached. At still higher temperatures the rate decreases very rapidly back toward zero. The temperature at which the maximum rate is observed is termed the **optimum** temperature.

The form of the graph shown in Figure 4-8 suggests at once that two factors are operating—one, predominant at low temperatures, acting to increase the rate, and one predominant at high temperatures, tending to decrease it. The first factor is the typical increase in reaction rate with increased temperature found for most chemical reactions. This is due to the greater energy of the molecules at higher temperatures, enhancing both the

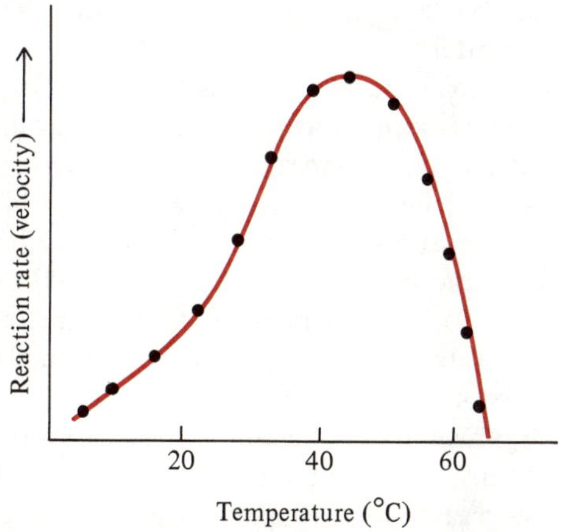

Figure 4-8
Typical effect of temperature on the rate of an enzyme-catalyzed reaction.

inherent reactivity of the molecules and the frequency of their collisions. The temperature coefficient (the Q_{10}) for such a reaction is often of the order of 2; that is, the reaction rate is doubled for each 10°C increase. In the figure shown, this general phenomenon is followed closely until a temperature near 30°C is reached.

The second factor is that of enzyme denaturation. Enzymes are of course proteins, molecules of almost unique complexity whose structures are maintained in major part by relatively weak forces. As the result of this structural complexity the temperature coefficient for the rate of denaturation of an enzyme for a 10°C change in temperature may be of the order of 10 rather than the usual 2. As a few of the secondary bonds responsible for maintaining the native conformation are broken by the increased thermal activity, it becomes increasingly easy for more bonds to rupture, until finally a point is reached at which the entire structure collapses to a random chain.

Because of the high Q_{10} for denaturation, an enzyme may be stable for weeks at 0°C, days at 10°C, hours at 30°C, but only a few seconds at 70°C. You should be able to calculate that the rate of denaturation at 70°C would be one million times that at 10°C, assuming a Q_{10} equal to 10.

The rapidly increasing rate of enzyme denaturation and the accompanying loss of catalytic activity is responsible for the rapid decrease in reaction rate shown in Figure 4-8 for temperatures above 40°C.

The optimum temperature is simply the temperature at which the two opposing factors reach a balance. It has been traditional to assume that enzymes have optimum temperatures near 38°C, the body temperature of humans. However, different enzymes are affected to different degrees by temperature. Small, simple proteins are much less subject to denaturation than large, complex ones. Further, the greater the content of disulfide linkages the more stable an enzyme will be. These bonds are not subject to thermal disruption at the temperatures used in this type of experiment. Consequently they tend to maintain the basic structure of the molecule. Even if secondary bonds are broken, there is a much greater chance for them to reform in the original pattern.

It may be of interest to note that a number of enzymes of microorganisms living in hot springs have been shown to resist denaturation to a greater degree than the usual animal enzymes, thus having higher optimum temperatures under the same conditions. However, for most higher organisms, it does appear that temperatures near 40°C are the most efficient. This is not simply because 40°C is the optimum temperature for the enzymes, but a major factor does seem to be that near this temperature the best balance between reaction rates and denaturation rates is obtained. The denaturation rates, of course, must be held to a level at which they do not seriously strain the capacity of the organism to synthesize new enzyme molecules.

We would like to draw your attention to one further aspect of Figure 4-8—the fact that the reaction rate approaches zero at 0°C. It must be realized that as 0°C is neared, water molecules form crystalline aggregates which interfere with the ability of the substrate and enzyme to interact. Of course, as soon as the solution is frozen solid, collisions between enzyme and substrate molecules become exceedingly few.

The effect of temperature on enzymatic reactions imposes certain restrictions on the capabilities of living organisms. Remember that the activities of organisms, including all physical activities, are based on enzyme-catalyzed reactions. Therefore a change in the temperature of their environment will have marked effects upon the metabolic and physical activities of "cold-blooded" organisms—those unable to control their internal temperatures to an appreciable extent. These creatures expire rather quickly on being exposed to high temperatures because their enzymes and other proteins are rapidly denatured. Pasteurization is a common example of the useful application of this fact, useful to us but not to the bacteria! Similarly, a rattlesnake will soon die if it is kept in the direct rays of a Death Valley summer sun.

On the other hand, cold-blooded organisms become sluggish and even completely dormant as the environmental temperature is lowered toward the freezing point because the rates of all their metabolic processes are greatly reduced. This does permit survival through a frigid winter, the food requirement needed to maintain the lowered activity being met by reserve materials stored in the tissues.

Hydrogen Ion Concentration

The hydrogen ion concentration of an enzyme solution also has a marked effect on the enzyme activity. Low catalytic activities are usually found in quite acidic or basic solutions. These effects are due in major degree to the gross denaturation of the enzyme protein. However, denaturation is not the whole explanation for the striking changes of activity shown by many enzymes with only minor changes in pH, a phenomenon illustrated in Figure 4-9. Rather, this involves a change in the degree of ionization of functional groups of the enzyme. If we assume, for example, that a positively charged group and a negatively charged group are necessary at the active site of a particular enzyme in order for the enzyme–substrate intermediate to form, then any enzyme molecule in which one or the other of these groups is uncharged would be catalytically inactive. Of course, changes in ionization of other groups of the protein may lead to minor alterations in structure with attendant alterations in catalytic activity.

Another common effect of pH variation is upon the substrate molecule. Many of these compounds are also of ionic nature, just like the enzyme. The concentration of any particular one of the various possible ionic forms, as well as the uncharged form, will vary as the pH is altered. Only one of these various forms may be able to combine with the active site of the enzyme.

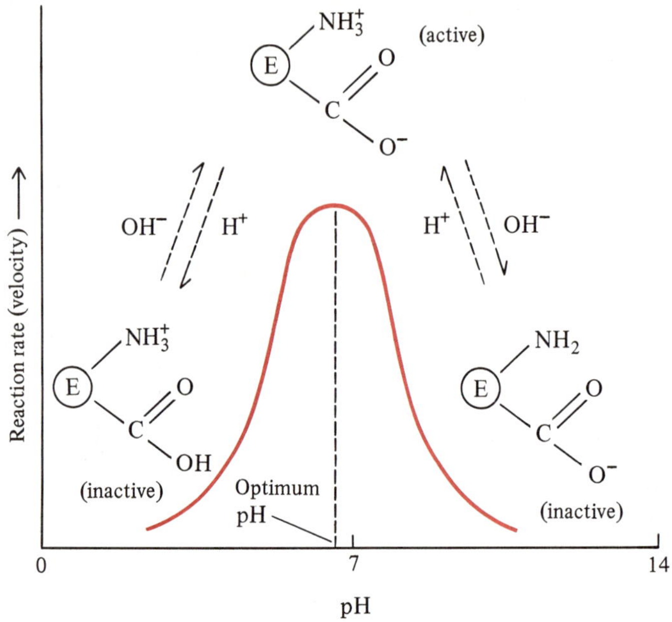

Figure 4-9
One possible case of variation in rate of an enzyme-catalyzed reaction with change in pH. This particular curve would result from a simple case in which one acid group and one base group on the enzyme must be ionized for optimum activity. Because changes in pH can affect many groups on the enzyme molecule, and sometimes on the substrate molecule, a wide variety of curves are possible with different enzymes.

Accordingly, the *effective* substrate concentration is a function of the pH even though the total amount of substrate is constant.

Other conditions being equal, the maximum rate of the reaction will be obtained at a pH which provides for the best balance among these different effects. This is termed the **optimum pH**. This optimum varies widely from one enzyme to another, depending upon the precise structure of the enzyme and the nature of the ionizable groups of both enzyme and substrate. Most enzymes have optima in the pH range of 5 to 8; however, a few are most active in very acidic media, others in quite alkaline solutions. In the living organism, then, the pH will not necessarily be at the optimum for each enzyme. Any change in pH will obviously alter the balance of enzyme activities. If cells are to maintain metabolic control, this necessitates control of pH within rather narrow limits, usually near neutrality, by the selective excretion of acids or bases. In higher organisms, the pH of the extracellular fluids must also be under control. Buffer systems are here the first line of defense against pH change, but ultimately the organism must be able to excrete preferentially the excess of acid or base.

4.4 Specificity of Enzymatic Catalysis

We have noted earlier in this chapter that enzymes have the characteristic of being capable of great selectivity in the nature of the catalysis that they carry out. The first aspect of this is a specificity for the type of reaction that is to be catalyzed. In general, a single enzyme will catalyze only one type of reaction. Thus, an enzyme that catalyzes an oxidation or reduction is most unlikely to catalyze hydrolysis or structural rearrangements.

In addition, enzymes exhibit a greatly varying degree of requirement for specific structural features or spatial arrangements in the molecules that they will accept as substrates. Thus, the enzyme glucose oxidase will catalyze the oxidation only of D-glucose and has no effect upon any other sugar or even on the optical isomer, L-glucose. On the other hand, the phosphomonoesterases (enzymes catalyzing the hydrolysis of monoesters of phosphoric acid) of animal intestines commonly have little regard for the exact nature of the substrate structure, as long as it is a phosphomonoester and is not too sterically hindered. Most enzymes exist somewhere in between these two extremes, acting on a number of compounds, but often with widely varying catalytic efficiency.

4.5 Enzyme Nomenclature

Just as organic chemistry has not yet succeeded in converting completely to a systematic nomenclature after many decades, the much younger field of biochemistry finds itself using a mixture of three distinct forms of nomenclature for enzymes. For the purpose of illustration, these can be called **historic**, **common**, and **systematic** names.

Historic Names

Some enzymes were discovered, or their presence inferred, rather early in the history of biological chemistry. These were given names perhaps based on source or sometimes on classical Greek or Latin reference. The names convey no inherent meaning. Remaining examples would include trypsin, chymotrypsin, pepsin, bromelin, papain, and rennin.

Common Names

A major improvement was made when the practice was adopted of trying to communicate in the name something of the function of the enzyme. Thus for many years it has been the common practice to include the nature of the reaction catalyzed and the most common substrate or group of substrates. To this was added the general agreement that the names of all enzymes would end in "*-ase*." Where the reaction involved is simply hydrolysis, it is not necessary to indicate the nature of the reaction, and the suffix *-ase* is simply added to the name of the substrate.

Thus, as examples of common names:

urease	catalyzes the hydrolysis of urea
sucrase	catalyzes the hydrolysis of sucrose
L-*glutamic dehydrogenase*	catalyzes the dehydrogenation (oxidation) of L-glutamic acid
pyruvate decarboxylase	catalyzes the removal of the carboxyl group of pyruvic acid in the form of CO_2

While common names are still the most popular for many enzymes, this nomenclature is always subject to continuing change. Where systematic names are not too cumbersome for everyday use, they are being employed at a steadily increasing rate.

Systematic (IUB) Nomenclature

The systematic nomenclature devised by the International Union of Biochemistry (IUB) provides a unique and unambiguous name for any enzyme. It also provides for assignment of a specific number to each enzyme, both for convenience in computer mediated retrieval of information and to prevent errors in transfer from one language to another.

In this system, all enzymes are divided into six major classes, reflecting the six fundamental types of reactions that are catalyzed:

1. Oxidoreductases
2. Transferases
3. Hydrolases
4. Lyases
5. Isomerases
6. Ligases (synthetases)

Most of these names are self-explanatory. The **oxidoreductases** catalyze oxidation–reduction reactions of various kinds involving electron transfers from one compound to another. This often involves the apparent addition of an oxygen atom to a compound or perhaps the removal of two hydrogens with their electrons (thus the common name "dehydrogenase").

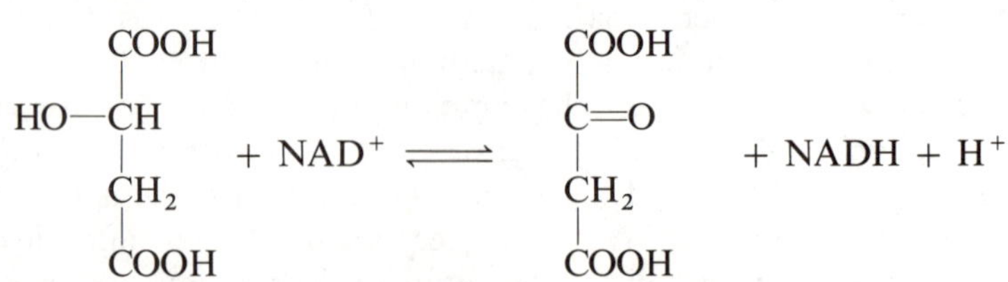

Malic acid Oxaloacetic acid

Transferases transfer some **group** of atoms from one compound to another. Transfer of one or more hydrogens does not fit in this category, since this would be an oxidoreductase:

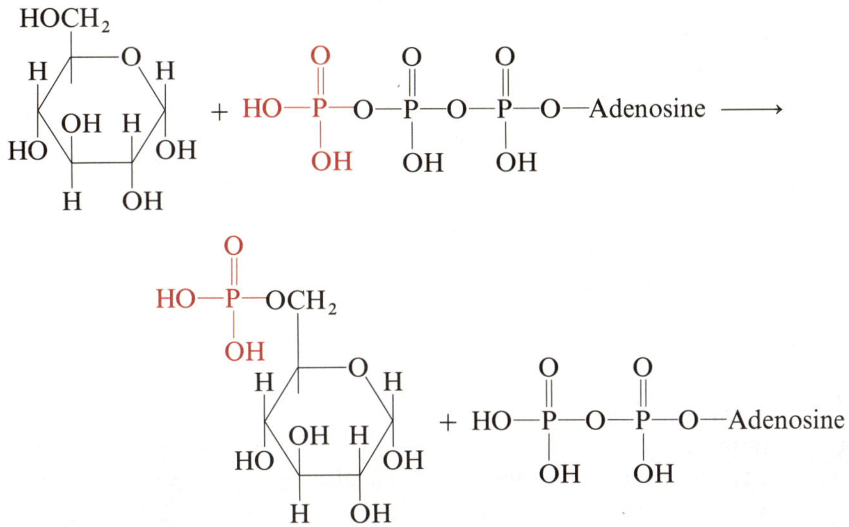

Hydrolases of course catalyze reactions in which some hydrolysis occurs:

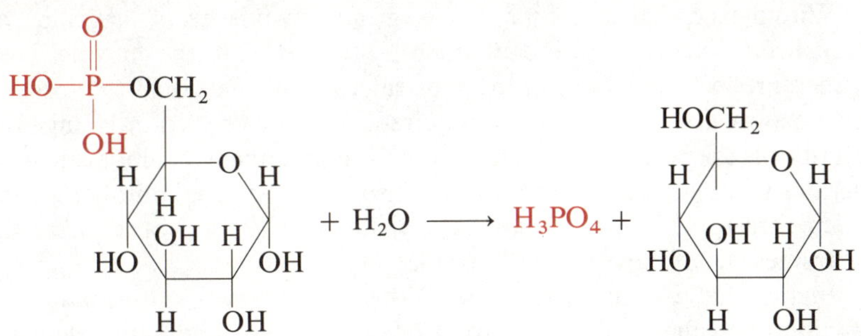

Lyases are enzymes which catalyze the removal of structural portions of their substrates and in the process produce double bonds. Bear in mind that removal of two hydrogens to form a double bond is an oxidoreductase reaction and the addition of water to a double bond is the reverse of a lyase reaction, *not* hydrolysis:

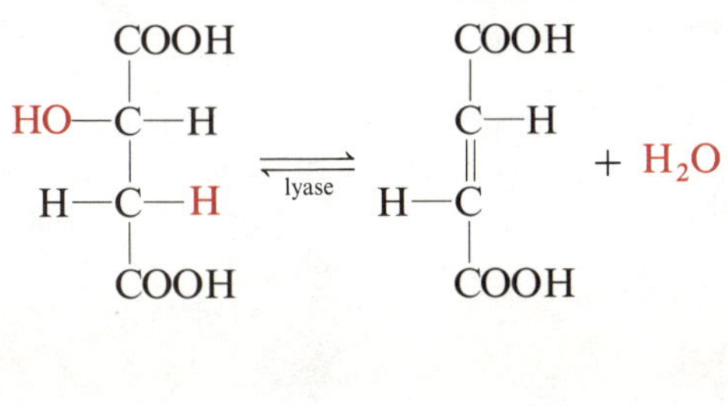

Isomerases catalyze a variety of reactions, all characterized by the fact that they result in some rearrangement *within* a molecule. This may involve formation of structural isomers, optical isomers, and so on, but does not involve permanent changes in any other molecule:

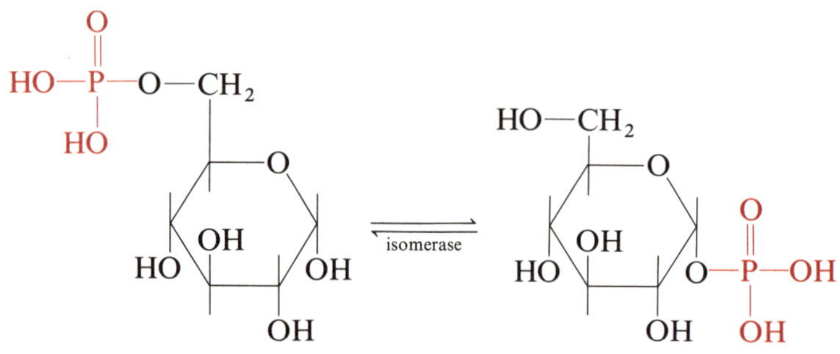

Ligases (sometimes also called **synthetases**) are a special group of enzymes that catalyze the joining together of two molecules by some process which requires the concomitant breakdown of a nucleotide triphosphate or similar source of energy. Since you are probably not familiar with these reactions, we will point out specifically the first several examples we encounter in looking at metabolic reaction pathways.

Within each of these major enzyme groups, two levels of subgroups are established to further separate and identify the nature of the enzyme. These in general reflect the nature of the substrate, specific nature of the reaction, coenzymes or acceptor molecules required, groups removed or changed, or the nature of the products. Assignment to the major and two minor categories generates a series of three numbers; to this is added a serial number for that particular enzyme in the sub-subclass, yielding a four-section number that is characteristic of only that one enzyme.

In practice, systematic names yield a maximum of information but sometimes become rather unwieldy. For example, the L-glutamic dehydrogenase referred to above becomes *L-glutamate: NAD oxidoreductase (deaminating)* (*1.4.1.2*). Use of names such as that is largely restricted to the formal literature.

Additional Readings

M. S. Bender and L. J. Brubacher, *Catalysis and Enzyme Action* (McGraw-Hill, New York, 1973) (203 pp.; paperback). A very readable short introduction to the nature and behavior of enzymes.

P. D. Boyer (Ed.), *The Enzymes* (Academic Press, New York, 1970). A series of advanced articles dealing with many aspects of enzyme structure, function, and control.

A. Fersht, *Enzyme Structure and Mechanism* (Freeman, San Francisco, 1977) (371 pp.; paperback). Chapter 2 with its discussion of chemical catalysis and catalytic mechanisms and Chapter 12, "Structures and mechanisms of selected enzymes," are particularly pertinent to this chapter.

C. J. Gray, *Enzyme-Catalyzed Reactions* (Van Nostrand-Reinhold, New York, 1971). A detailed review based on the types of reactions catalyzed by enzymes.

M. Florkin and E. H. Stotz (Eds.), *Comprehensive Biochemistry*, 3rd edition (Elsevier Scientific Publishing Co., 1973). Volume 13 (1973) contains the full report of the IUB commission on enzyme nomenclature, rate units, and so on.

Study Questions

1. Define a catalyst. How does a catalyst affect (a) the total free energy change, (b) the free energy of activation, and (c) the equilibrium constant for a reaction?

2. What do the terms "active site," "binding site," and "catalytic site" mean when applied to an enzyme?

3. How would you define the terms "cofactor" and "coenzyme"?

4. Name six compounds that are vitamins for humans and that serve within cells as coenzymes or parts of coenzymes.

5. Many disease-causing microorganisms are anaerobes (grow best in the absence of oxygen). We will see in a later chapter that such organisms produce much less energy from their "food" than do aerobic organisms. In light of this, how does an increase in temperature (a "fever") help a human to defend against infection by such microorganisms?

6. In any cell, the observed activity of an enzyme will depend upon
(a) the number of complete enzyme molecules present
(b) the catalytic efficiency of each of these molecules
(c) the concentration of substrate molecules present
(d) the acceptability of each of these substrate molecules to the enzyme.
Indicate which of the above will be affected when the following changes occur:
(i) rate of enzyme synthesis increases
(ii) rate of absorption of substrate into cell decreases
(iii) temperature increases
(iv) pH changes from 6.5 to 7.5
(v) a competitive inhibitor is added
(vi) a noncompetitive inhibitor is added

7. Urease catalyzes the following reaction:

$$H_2N-\overset{\overset{\displaystyle O}{\|}}{C}-NH_2 + H_2O \longrightarrow CO_2 + 2\,NH_3$$

In a specific reaction mixture containing urease, 38 mg of ammonia are produced in 3 minutes. A blank reaction with no enzyme yielded 3 mg of ammonia in the same time. Express the amount of enzyme present in International Units and in katals.

8. Each of the following reactions can be catalyzed by a specific enzyme. In each case, indicate the major class to which that enzyme would belong:
(a) glycylalanine + $H_2O \longrightarrow$ glycine + alanine

(b) 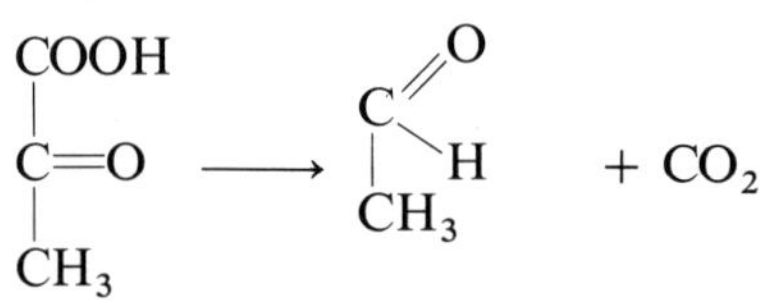

(c) 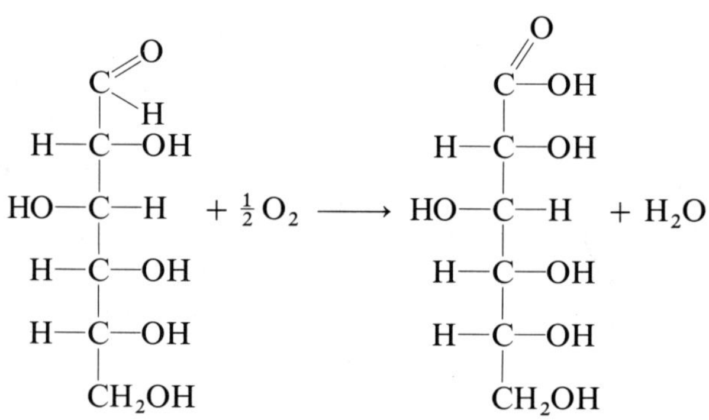

(d) 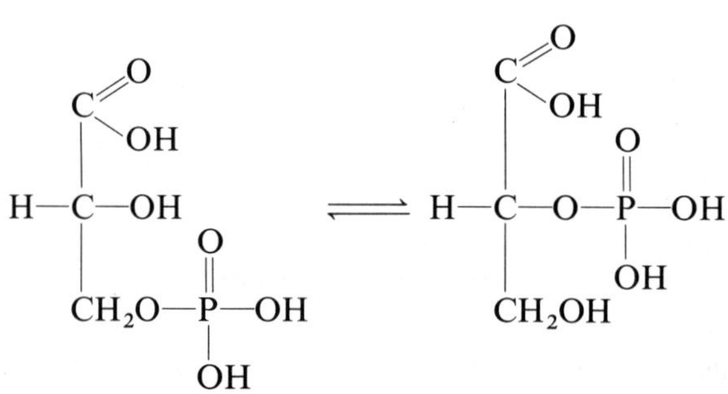

CHAPTER 5

Enzymes II: Kinetics and Inhibition

5.1 Derivation of the Michaelis–Menten Equation

In the previous chapter we discussed in qualitative terms the variation in the rate of an enzyme-catalyzed reaction with variation in substrate concentration. The first satisfactory quantitative treatment of this subject was provided by Michaelis and Menten in 1913 with the development of an equation which is commonly called the Michaelis–Menten equation.

This equation can be derived rather simply with the aid of certain assumptions. The first of these is that the enzymatic process occurs as a stepwise set of reactions, which can be depicted in the early stages (when product concentration is negligible) as

$$E + S \underset{k_2}{\overset{k_1}{\rightleftharpoons}} ES \xrightarrow{k_3} E + P$$

where k_1, k_2, and k_3 are the rate constants for the indicated reactions.

One might expect no problems in describing the kinetics of these reactions in mathematical terms; indeed, there is no theoretical problem, only a practical one. The difficulty arises from our own inability in most cases to determine the concentration of the enzyme–substrate complex, or even the concentration of the enzyme. This was particularly true in Michaelis' day. The success of his derivation lay in replacing these immeasurable quantities with ones that could be measured experimentally.

We can start by writing an expression for the rate of formation of ES complex as

$$v_f = k_1 \times E \times S \qquad (1)$$

Similarly, the rate of disappearance of ES is the sum of the two decomposition reactions and can be written in the form

$$v_d = (k_2 \times ES) + (k_3 \times ES)$$

or (2)

$$v_d = (k_2 + k_3) \times ES$$

For simplicity, we shall omit the brackets generally used to indicate that it is the molar concentration of the substance that is intended. Remember that in these equations the symbols E, S, and ES refer to the **molar concentrations** of the enzyme, substrate, and enzyme–substrate complex, respectively.

If the starting concentration of substrate is very high compared to that of the enzyme, then a "steady state" situation will rapidly result, in which the concentration of ES will remain essentially constant for some period of time. Using equations (1) and (2), we can express this "steady-state" condition as

$$k_1 \times \mathrm{E} \times \mathrm{S} = (k_2 + k_3) \times \mathrm{ES}$$

or (3)

$$\frac{\mathrm{E} \times \mathrm{S}}{\mathrm{ES}} = \frac{k_2 + k_3}{k_1}$$

The expression $(k_2 + k_3)/k_1$ is represented by a single constant called the Michaelis constant (K_m), thus giving us the expression

$$K_m = \frac{\mathrm{E} \times \mathrm{S}}{\mathrm{ES}} \tag{4}$$

The rate of appearance of products v (for *velocity* of the overall reaction) will be given by

$$v = k_3 \times \mathrm{ES} \tag{5}$$

The maximum reaction rate possible, $V_{\max}$, will occur at a substrate concentration sufficiently high so that essentially all of the enzyme is bound to substrate. The maximum concentration of ES which is possible is thus equal to the total enzyme concentration, and accordingly,

$$V_{\max} = k_3 \times \mathrm{E}_{\mathrm{Total}} \tag{6}$$

Dividing equation (5) by equation (6) gives

$$\frac{v}{V_{\max}} = \frac{\mathrm{ES}}{\mathrm{E}_{\mathrm{Total}}} \tag{7}$$

Here we have an equation which allows us to express the immeasurable quantities ES and $\mathrm{E}_{\mathrm{Total}}$ in terms of reaction rates that can be readily measured experimentally.

We now have both equations (4) and (7) containing ES, but one deals with $\mathrm{E}_{\mathrm{Total}}$ and the other with E (the instantaneous concentration of free enzyme). These two are, however, related by the simple expression

$$E = \mathrm{E_{Total}} - \mathrm{ES} \tag{8}$$

Substituting this value for E in equation (4) gives

$$K_m = \frac{(\mathrm{E_{Total}} - \mathrm{ES}) \times \mathrm{S}}{\mathrm{ES}} \tag{9}$$

from which algebraic rearrangement yields

$$\frac{\mathrm{ES}}{\mathrm{E_{Total}}} = \frac{\mathrm{S}}{K_m + \mathrm{S}} \tag{10}$$

Substituting from equation (7) eliminates the ES and $\mathrm{E_{Total}}$ factors and gives

$$\frac{v}{V_{\max}} = \frac{\mathrm{S}}{K_m + \mathrm{S}}$$

This is generally written in the form

$$v = \frac{V_{\max} \times [\mathrm{S}]}{K_m + [\mathrm{S}]} \tag{11}$$

and is referred to as the Michaelis–Menten equation (although the above derivation includes refinements added by Briggs and Haldane).

This equation adequately describes the kinetics of many enzyme-catalyzed reactions. The experimental curve for reaction rate (velocity) versus substrate concentration shown in Figure 4-6 is exactly the result that would be predicted from this equation.

The value of K_m is characteristic of a given enzyme and its substrate. In a given set of experiments, the same value will be obtained no matter what the concentration of enzyme used (as long as the substrate is always in substantially higher concentration). A simple substitution in equation (8) will also show that K_m is equal to the value of substrate concentration at which the velocity is $\frac{1}{2}V_{\max}$.

For some enzymes (but by no means all), k_3 is small compared to k_1 and k_2. Under these circumstances, K_m becomes a measure of the affinity between enzyme and substrate. The lower the value of K_m, the more strongly the enzyme interacts with its substrate, and the higher the proportion of enzyme in the form of ES complex. Thus, a low value of K_m may be considered an indication of a high affinity between enzyme and substrate. This kind of comparison is often made between enzymes utilizing the same substrate. You should be aware, however, that this interpretation is not universally applicable, especially when dealing with enzymes of high catalytic efficiency. Typical values for the Michaelis constant for various enzyme–substrate pairs range from 10^{-2} to 10^{-6} molar.

5.2 Inhibition of Enzymatic Reactions

The need for methods by which to control the reactions that occur within a cell and to be able to vary that control as conditions change has already been mentioned several times. One of the most important tools for such control lies in methods by which individual enzymes may be made into less efficient or more efficient catalysts.

Since enzymes are proteins, they possess a variety of functional groups which are capable of interacting with many different substances. Any such reaction which results in appreciable alteration of the structure of an enzyme is likely to affect its catalytic activity, even though the reaction may not directly involve a group of the active site. In most cases the effect of the reaction is the production of a protein which is a poorer catalyst than the original enzyme. Compounds which adversely affect the rate of enzymic reactions are called **inhibitors** and the process is that of **enzyme inhibition**.

One of the major uses of the Michaelis equation developed above has been in the study of enzyme inhibition, particularly in helping to distinguish which of several types of inhibition is involved in a particular example.

An approach widely used is a **Lineweaver–Burk** or **double-reciprocal** graph obtained from reaction rates at various substrate concentrations in the absence and in the presence of the inhibitor.

The Michaelis equation can be rewritten from

$$v = \frac{V_{\max} \times [\mathrm{S}]}{K_m + [\mathrm{S}]} \quad \text{to} \quad \frac{1}{v} = \frac{K_m + [\mathrm{S}]}{V_{\max} \times [\mathrm{S}]}$$

or

$$\frac{1}{v} = \frac{K_m}{V_{\max}} \times \frac{1}{[\mathrm{S}]} + \frac{1}{V_{\max}}$$

The final equation is of the form $y = mx + b$ if one considers the variables to be $1/v$ and $1/[\mathrm{S}]$. When one plots the reciprocals of the experimental

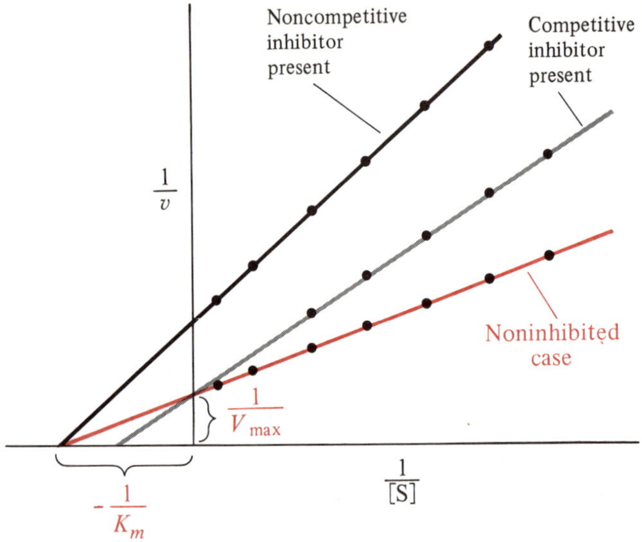

Figure 5-1
A Lineweaver–Burk or double-reciprocal plot of kinetic data obtained with no inhibitor and then in the presence of each of two different types of inhibitors. The values of $1/V_{max}$ and of $-1/K_M$ for the uninhibited case are shown in color. Significance of the data obtained for the inhibited cases is discussed in the text.

velocities against the reciprocals of the corresponding substrate concentrations, that is, $1/v$ versus $1/[S]$, a straight line should result. The slope of the line should be K_m/V_{max} and the $1/v$ intercept should be $1/V_{max}$. While it is not as obvious from the equation in this form, the $1/[S]$ intercept should be $-1/K_m$. A typical graph of this type is illustrated in Figure 5-1. This procedure provides a graphical means for determining the value of the Michaelis constant; it also permits determination of the maximum rate of the reaction for the amount of enzyme used.

If the experiments are repeated with a fixed concentration of an inhibitor present in each of the reaction mixtures, new, slower rates will be obtained. The reciprocals of these velocities are again plotted against the reciprocals of the substrate concentrations. The resulting line will commonly be one of the two major possibilities indicated in Figure 5-1. One of these has the same $1/v$ intercept, that is, the same V_{max}, as in the absence of inhibitor but an apparent increase in the Michaelis constant (a smaller value for its reciprocal). The change in K_m indicates some interference with the binding of substrate to the enzyme; but at very high substrate levels, the same maximum velocity can be obtained as when there is no inhibitor present. This type of inhibition is called **competitive inhibition** for reasons we will see below.

A second possible result is a line with the same $1/[S]$ intercept but a different $1/v$ intercept; that is, the same Michaelis constant but a different, lower, maximum velocity. This is exactly the effect we would see if we used a lower enzyme concentration without any inhibitor. This result is typical of **noncompetitive inhibition**.

Irreversible Loss of Enzyme Function

Certain compounds or ions will bind to a wide variety of protein molecules and, usually, interfere with their biological functions. Most of the molecules we call "poisons" fit in this category. If this binding is essentially irreversible under normal circumstances, such as binding mercuric ions or one of the "nerve gases" to certain enzymes, then the effects on enzyme activity and kinetics are indistinguishable from those of denaturation. For the purposes of our discussions we will restrict the term "inhibition" to interactions that are effectively reversible under cellular conditions.

Competitive Inhibition

As the name implies, this type of enzyme inhibition involves a competition between the normal substrate and the inhibitor molecule for binding space on the enzyme active site. When the inhibitor molecule is bound to the enzyme, the substrate cannot be attached, and vice versa. We can write the situation that results as follows:

$$\mathrm{E + S \rightleftharpoons ES \longrightarrow E + P}$$

$$\mathrm{E + I \rightleftharpoons EI}$$

Just as we can describe a decomposition constant for ES (that is, K_m), we can also develop a similar measure of the binding of the enzyme and inhibitor which is usually represented as K_I.

The net effect of the presence of a competitive inhibitor will depend upon the relative values of K_m and K_I and upon the relative concentrations of the substrate and inhibitor molecules. In the presence of a fixed amount of inhibitor, increasing the concentration of substrate will in theory eventually draw all of the enzyme into ES complex and result in a rate the same as V_{max} in the Lineweaver–Burk plots.

In a physical sense, there are several ways in which competitive inhibition can occur. These are generally called **classical** and **nonclassical** mechanisms.

Classical Competitive Inhibition. The simplest, and earliest known, form of competitive inhibition occurs when the inhibitor closely resembles the structure of the normal substrate, but for some reason cannot undergo the same reaction as the substrate. The greater the structural similarity, the greater the probability that the compound will be likely to fit easily into the active site area of the enzyme and to interact strongly with the binding sites. An example is the very strong inhibition exerted by malonic acid on the enzyme succinate dehydrogenase, which catalyzes the oxidation of succinic acid to fumaric acid:

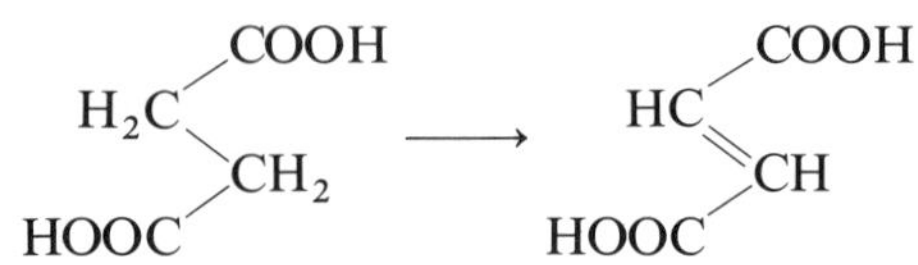

Succinic acid Fumaric acid

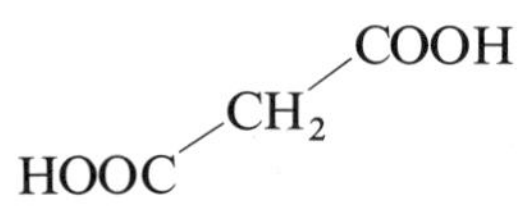

Malonic acid

A more practical example is provided by the class of compounds called "sulfa drugs." These are structural analogs of para-aminobenzoic acid, a compound required by many bacteria but not used, as such, by higher organisms such as man. The sulfa drugs inhibit growth of the bacteria by competing with the para-aminobenzoic acid for the active site of a bacterial enzyme; the reaction involved requires a free carboxyl group:

Para-aminobenzoic acid

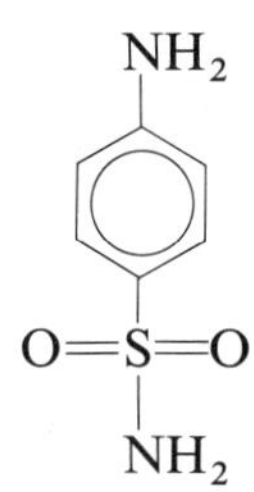

Sulfanilamide

Nonclassical Competitive Inhibition. In cases where some therapeutic purpose is intended, such as might be represented by the sulfa drugs, it is often not possible to use a close structural analog, since this will inhibit all enzymes using the given substrate, when the target may be only one of them.

In certain cases, it has been possible to get around this problem by designing an inhibitor which is specific for the enzyme rather than the active site. This will be competitive inhibition if, while it binds to groups outside the active site, the inhibitor prevents approach of the substrate to its normal binding sites. This situation is indicated symbolically in Figure 5-2.

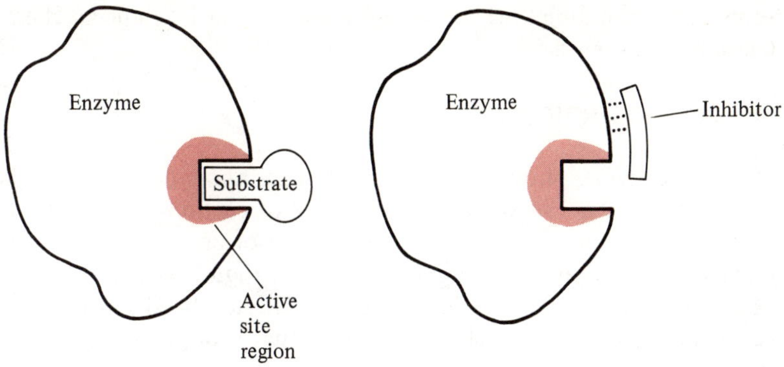

Figure 5-2
Mode of action of a nonclassical competitive inhibitor. Note that the inhibitor binds outside the active site region but blocks access of the substrate.

Noncompetitive Inhibition

Some inhibitors bind to the enzyme in such a way that they affect the function of the catalytic site, but do not interfere with normal binding of the substrate. A hypothetical example appears in Figure 5-3. In this case, there is no competition between substrate and inhibitor, and each can bind to the enzyme whether the other is attached or not:

$$\mathrm{E + S \rightleftharpoons ES \longrightarrow E + P}$$

$$\mathrm{E + I \rightleftharpoons EI}$$

$$\mathrm{ES + I \rightleftharpoons ESI \xrightarrow{\text{slowed}} E + P}$$

$$\mathrm{EI + S \rightleftharpoons EIS \xrightarrow{\text{slowed}} E + P}$$

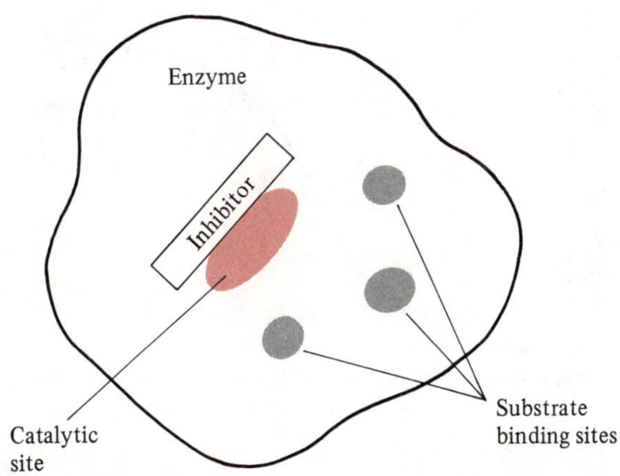

Figure 5-3
Binding of a noncompetitive inhibitor. In this case, the inhibitor molecule is bound near enough to the catalytic site to affect its function, but not in a position to interfere with the normal substrate binding region. Consequently, inhibitor and substrate can be bound at the same time.

In terms of the resulting kinetics there is a decrease in V_{max} but no effect on K_m (as shown in Figure 5-1).

Allosteric Control

Certain enzymes show kinetic data that are anomalous when viewed in the classic Michaelis explanation of enzyme reactions. Typically, they may produce abnormal plots of v versus [S] and/or unusual curved lines in Lineweaver–Burk plots. An example is shown in Figure 5-4.

Further examination of these enzymes led to the discovery of the process called **allosteric control**, a mechanism that is of great importance in the overall control of cellular metabolism.

One fundamental feature of allosteric enzymes is the existence of a second binding site on the enzyme remote from the active site. This is generally called the **effector site** and a compound that can bind to it is then called an **effector** molecule. The second feature is that the enzyme is capable of existing in two different conformations. When the effector molecule is bound, it stabilizes one of these conformations and thus increases the percentage of the total enzyme that is in that form.

This may represent a process of inhibition or of activation of the enzyme, depending upon whether the stabilized conformation is the form most capable of binding substrate and thus being catalytically active. If the effector molecule is serving the function of an inhibitor, it is generally called a **negative effector** and if it promotes the reaction it is called a **positive effector**.

In the simplest possible case, the active site and the effector site would both reside in the single polypeptide chain of a single-unit enzyme. While

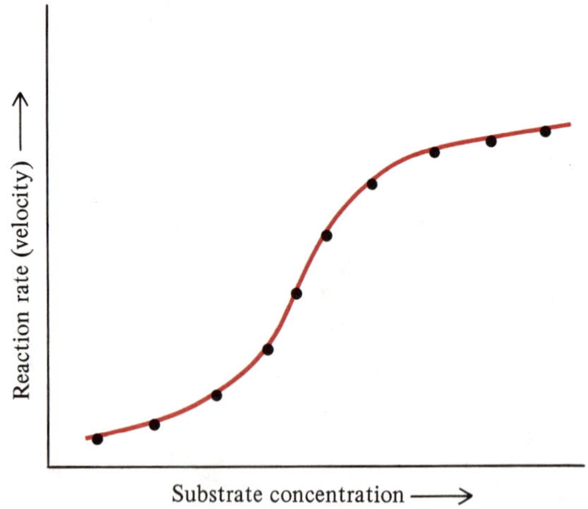

Figure 5-4
Variation in reaction velocity with increased substrate concentration for a typical case of an enzyme that is composed of multiple subunits and is subject to allosteric control.

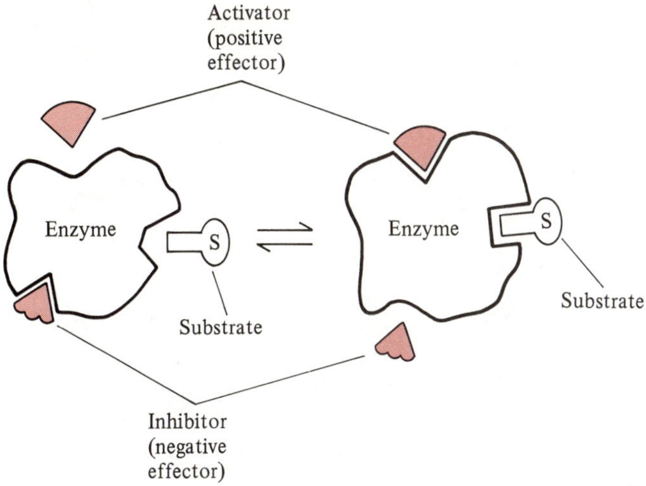

Figure 5-5
The fundamental characteristics of allosteric control. Two forms of the enzyme exist in equilibrium with one another. Effector molecules may bind at sites remote from the catalytic active site and in the process stabilize one or the other of the two forms. Binding of substrate or of a positive effector will promote the reaction; binding of a negative effector will inhibit it.

this case is presently unknown in nature, it serves as a simple illustration of the basic features of allosteric control. Figure 5-5 illustrates diagrammatically the situation with an enzyme having negative and positive effectors.

In most cases, however, the enzyme subject to allosteric control is a multimeric structure with several active sites and several effector sites on the functional enzyme unit. Although the interactions involved are complex, we can simply say here that the effects of binding either an effector or a substrate molecule are magnified by the fact that there is a tendency for the whole unit to exist in one conformation or the other. Thus, a small amount of effector or of substrate can have a large effect on the total number of active sites available, as shown in Figure 5-6.

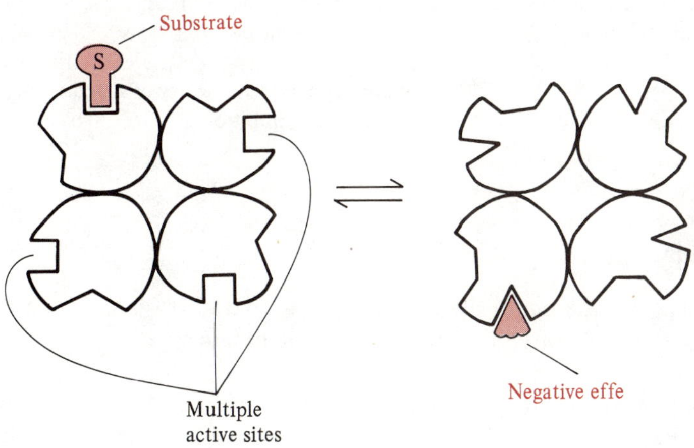

Figure 5-6
Allosteric control in the case of an enzyme composed of multiple subunits. In this situation, binding of substrate or of the negative effector shown results in conformational changes involving a number of active sites. This is the most common situation with allosteric enzymes.

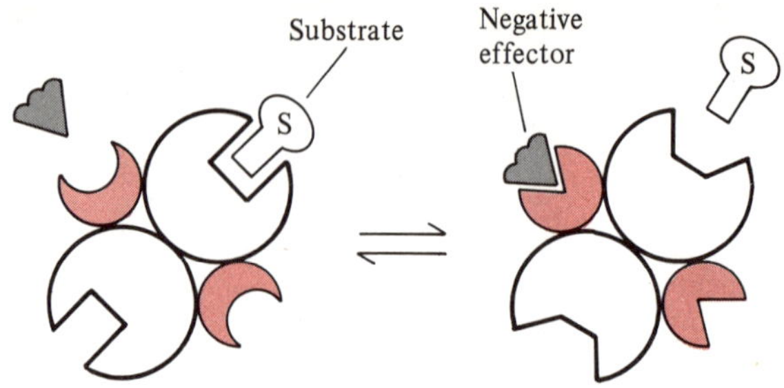

Figure 5-7
The effect of a negative effector (inhibitor) and of substrate molecules on an allosteric enzyme having separate regulatory and catalytic subunits. The regulatory subunits are shown in color.

An indication of the importance of subunit interactions in allosteric enzymes is evident in those cases in which the active sites and effector sites are on separate subunits of a multiunit enzyme complex. We then speak of the "catalytic subunit" and the "regulatory" subunit. An example is illustrated in Figure 5-7. In some cases these can be separated without destroying the enzyme activity; however, the catalytic subunit in these cases no longer shows any response to the presence of effector molecules and displays normal Michaelis kinetics.

One of the major advantages of the allosteric control mechanism is that it permits a reaction to be specifically inhibited (or activated) by a compound that has no structural resemblance to the substrate. This permits a product at the end of a long biosynthetic sequence of reactions to be used to control one of the early steps in the sequence. By analogy to automatic controls often built into electronic circuits, this is usually called **feedback control** of metabolic processes. We will discuss this and similar control mechanisms in Chapter 21.

Additional Readings

All of the references of Chapter 4 also deal with the topics of this chapter. The following additional items are, however, especially pertinent.

Scientific American article

D. E. Koshland, Jr., "Protein Shape and Biological Control," 229(4), 52 (October, 1973).

H. R. Mahler and E. H. Cordes, *Biological Chemistry*, 2nd edition (Harper and Row, New York, 1971). Chapter 6 contains a classic exposition of enzyme kinetics applied to a variety of reaction mechanisms and types of inhibition.

Study Questions

1. What are the major assumptions that have to be made in deriving the Michaelis–Menten formula for enzyme kinetics?

2. The true equilibrium value for $E + S \rightleftharpoons ES$ is sometimes expressed by a dissociation constant K_s. Under what circumstances might this value be essentially equal to K_m?

3. In a mechanistic sense, distinguish between competitive and noncompetitive inhibition of an enzyme.

4. What changes in kinetic parameters would you expect on adding (a) a competitive and (b) a noncompetitive inhibitor to reaction mixtures?

5. What are the essential features that distinguish enzymes that undergo allosteric control?

6. A series of reactions is carried out in which all variables are kept constant except substrate concentration. A similar set is run in the presence of a fixed amount of an inhibitor. The results are shown below:

Substrate Concentration (μmoles/ml)	No Inhibitor Velocity (μmoles/min)	With Inhibitor Velocity (μmoles/min)
50	2.45	0.97
10	2.40	0.95
1	1.70	0.67
0.5	1.25	0.50
0.25	0.825	0.33
0.1	0.42	0.17
0.083	0.36	0.13

(a) Can you give a rapid rough estimate of K_m in moles/liter?
(b) Determine K_m and V_{max} as accurately as possible from the data given.
(c) Is this inhibitor "clearly competitive," "clearly noncompetitive," or "other"?

CHAPTER 6

Biochemical Energetics: An Introduction

The living cell is a highly organized and structured system. To maintain this level of organization requires a constant supply of energy. Many of the chemical reactions and processes of the cell are concerned primarily with maintaining the flow of energy and with its conversion into useful forms. For this reason, it is necessary that we review briefly some aspects of chemical thermodynamics and then look at some of the special mechanisms of energy storage and transfer common to biological systems.

You will recall from general chemistry that a chemical reaction usually involves some change in the total energy of the molecules involved. For our purposes, we will oversimplify the picture and call this energy change ΔH (the "Δ" meaning "the change in"). Not all of this energy is really available to do work. Some of it is involved in the change in the degree of order of the system as a whole—a concept referred to as the **entropy** of the system. The remaining energy *is* available to do work and is called the **free energy change** (ΔG) for the reaction. The relationship can be written as

$$\Delta G = \Delta H - T\Delta S$$

where T is the temperature in degrees Kelvin and ΔS is the change in entropy. In biochemical terms, "work" includes the formation of chemical bonds, transport across membranes, nerve transmission, and so on.

By convention, if a reaction releases energy, the value of ΔG is written as a negative value. If the reaction results in absorption of energy, the ΔG will have a positive value. If we had the reaction

$$\mathrm{A + B \longrightarrow C + D + 10\ kilojoules\ (of\ free\ energy)}$$

the value of ΔG would be -10 kilojoules per mole (kJ/mole). If the reactants start at unit activity and react to form the products at unit activity (1 molar concentrations for our purposes), and if we maintain the temperature at 25°C, then the free energy change is called the **standard free energy change** and is written as ΔG°. For some reactions, the pH of the system is very important in determining the value of ΔG. In most biological systems, we are concerned with reactions at or near a pH of 7. The value of the standard free energy change at pH 7 is usually indicated as $\Delta G^{\circ\prime}$.

The value of ΔG for a reaction occurring in the presence of both reactants and products is determined by the relative concentrations of these entities while the reaction is occurring:

$$\Delta G = \Delta G^\circ + 2.3\, RT \log K_c$$

where R is a constant in appropriate units, T is temperature in degrees Kelvin, and K_c is a ratio of product to reactant concentrations, using the actual concentration values and taking into account the number of molecules that appear in the equation. For the reaction

$$a\mathrm{A} + b\mathrm{B} \rightleftharpoons c\mathrm{C} + d\mathrm{D}$$

the K_c expression becomes

$$K_c = \frac{[\mathrm{C}]^c[\mathrm{D}]^d}{[\mathrm{A}]^a[\mathrm{B}]^b}$$

The real free energy changes for many reactions within the cell will thus differ markedly from the standard free energy change values.

An interesting and useful expression can be obtained by substituting in the expression for K_c the values of the reactants and concentrations at equilibrium; that is,

$$K_c = \frac{[\mathrm{C_{eq}}]^c[\mathrm{D_{eq}}]^d}{[\mathrm{A_{eq}}]^a[\mathrm{B_{eq}}]^b} = K_{\mathrm{eq}}$$

By definition, the free energy change at equilibrium is zero. On substituting these values in the equation

$$\Delta G = \Delta G^\circ + 2.3\ RT \log K_c$$

we obtain

$$0 = \Delta G^\circ + 2.3\ RT \log K_{\mathrm{eq}}$$

or

$$\Delta G^\circ = -2.3\ RT \log K_{\mathrm{eq}}$$

This tells us immediately that whenever ΔG° is negative, the value of log K_{eq} is positive, and thus the equilibrium will lie in the direction of formation of products. The larger the numerical value of ΔG°, the farther the equilibrium will lie in either direction.

6.1 Additive Nature of ΔG Values

Now let us take a look at a reaction sequence starting with a reaction having a positive ΔG°. Consider the reaction

$$\mathrm{E} + \mathrm{F} \longrightarrow \mathrm{G} + \mathrm{H} \qquad \Delta G^\circ = +20\ \mathrm{kJ/mole}$$

Clearly, in this case the equilibrium will lie to the left and, at equilibrium, the reactants E and F will predominate. However, if this is followed by a reaction such as

$$\mathrm{G + H \longrightarrow L + M} \qquad \Delta G^\circ = -48\ \mathrm{kJ/mole}$$

then the equilibrium of the second reaction will lie so far to the right that G and H will be reduced to very low concentrations, forcing a shift of E and F in the direction of G and H. The result would be the overall reaction

$$\mathrm{E + F \longrightarrow L + M} \qquad \Delta G^\circ = -28\ \mathrm{kJ/mole}$$

In a situation such as this, the question of how much of the original reactants will go to products at equilibrium is determined by the overall energy change of the series of reactions, not the individual steps.

If we have the more common situation, where the second reaction involves only one of the products of the first, for example,

$$\mathrm{G + K \longrightarrow Q + M} \qquad \Delta G^\circ = -48\ \mathrm{kJ/mole}$$

the final equilibrium of

$$\mathrm{E + F + K \rightleftharpoons Q + M + H}$$

will usually still lie very much to the right, although calculation of the actual value for ΔG° for this process may be very complex.

6.2 Redox Potentials

In studying energy changes in oxidation–reduction reactions, data are often obtained by looking at the electrode potential generated when a half-cell containing unit concentrations of oxidized and reduced forms of a compound is compared to a standard half-cell. The standard generally used is the hydrogen cell, one in which hydrogen ions at a concentration of 1 molar are in equilibrium with hydrogen gas at a pressure of 1 atmosphere. This half-cell is assigned a voltage of zero.

Table 6-1 lists the redox potentials for a number of compounds that take part in important oxidation–reduction processes in the living cell. In this table, the values have been adjusted to pH 7 instead of 1 molar hydrogen ion concentration. As a result, the hydrogen half-cell has a value of −0.42

TABLE 6-1
Standard redox potentials for some reactions of biological interest.

Reaction	E'_0(volts)*
$\frac{1}{2}O_2 + 2H^+ + 2e^- \rightleftharpoons H_2O$	+0.82
2 cytochrome a(Fe^{+++}) + $2e^-$ $\rightleftharpoons$ 2 cyt a(Fe^{++})	+0.29
2 cytochrome c(Fe^{+++}) + $2e^-$ $\rightleftharpoons$ 2 cyt c(Fe^{++})	+0.25
ubiquinone + $2H^+ + 2e^-$ $\rightleftharpoons$ ubiquinol (CoQ) ($CoQH_2$)	+0.10
2 cytochrome b(Fe^{+++}) + $2e^-$ $\rightleftharpoons$ 2 cyt b(Fe^{++})	+0.03
fumarate + $2H^+ + 2e^-$ $\rightleftharpoons$ succinate	+0.03
flavoprotein–FMN + $2H^+ + 2e^-$ $\rightleftharpoons$ flavoprotein–$FMNH_2$	−0.12
oxaloacetate + $2H^+ + 2e^-$ $\rightleftharpoons$ malate	−0.17
pyruvate + $2H^+ + 2e^-$ $\rightleftharpoons$ lactate	−0.19
acetaldehyde + $2H^+ + 2e^-$ $\rightleftharpoons$ ethanol	−0.20
lipoate + $2H^+ + 2e^-$ $\rightleftharpoons$ dihydrolipoate	−0.29
$NADP^+ + H^+ + 2e^- \rightleftharpoons$ NADPH	−0.32
$NAD^+ + H^+ + 2e^- \rightleftharpoons$ NADH	−0.32
α-ketoglutarate $2H^+ + 2e^-$ $\rightleftharpoons$ isocitrate	−0.38
$2H^+ + 2e^- \rightleftharpoons H_2$	−0.42
acetate + $2H^+ + 2e^-$ $\rightleftharpoons$ acetaldehyde	−0.60
succinate + $2H^+ + 2e^- + CO_2$ $\rightleftharpoons$ α-ketoglutarate	−0.67
acetate + $2H^+ + 2e^- + CO_2$ $\rightleftharpoons$ pyruvate	−0.70

* E'_0 is the standard half-cell potential for the reaction as written at a pH of 7 and a temperature of about 25°C.

volts rather than zero. Note also that the voltages shown are for the reactions going in the direction of reduction; a compound that is a stronger oxidizing agent than hydrogen ion will be above it in the table and will have a more positive value for E'_0. If the reaction as written is reversed, the sign of the voltage must be reversed.

For reactions between the compounds listed, we can obtain values for $\Delta G^{\circ\prime}$, and thus the equilibrium constant K_{eq}, by writing the two half-reactions involved, and then adding together the values of the redox potentials. For example, for the reaction

$$\text{malic acid} + NAD^+ \longrightarrow \text{oxaloacetic acid} + NADH + H^+$$

we write the reactions as

$$NAD^+ + H^+ + 2e^- \longrightarrow NADH \qquad E'_0 = -0.32 \text{ volts}$$

$$\text{malic acid} \longrightarrow \text{oxaloacetic acid} + 2H^+ + 2e^- \qquad E'_0 = +0.17 \text{ volts}$$

$$\text{Overall} \qquad E'_0 = -0.15 \text{ volts}$$

Note that the second reaction is the reverse of that shown in the table, and so the sign of the voltage must be reversed.

Now we can take advantage of the relationship

$$\Delta G^{\circ\prime} = -nF\Delta E_0'$$

where n is the number of electrons being transferred in the balanced equation [(2) in this case] and F is the Faraday constant. Substituting in this example, we would get

$$\Delta G^{\circ\prime} = -2 \times 96.26 \frac{\text{kJ}}{\text{mole·volt}} \times (-0.15 \text{ volts}) = 28.9 \frac{\text{kJ}}{\text{mole}}$$

Obviously, this reaction would be favored in the reverse direction.

6.3 Apply with Caution!

A word of warning: We must be very careful in applying the concepts and calculations of equilibrium thermodynamics to biochemical systems. Most of these concepts are intended to apply to a closed system, and often to a closed system in equilibrium. Cells are only rarely closed systems, depending as they do on a continuous absorption of nutrients and excretion of waste products. And, of course, a cell which has reached equilibrium is a dead one. So we use the values obtained from equilibrium measurements in closed systems only as indications of the nature of the energy changes involved in cellular reactions. Detailed examination of the energy changes really taking place in the cell must be dealt with by theories developed for "steady state" systems, or by the calculations of irreversible thermodynamics.

6.4 Sources of Energy

The cell system will of itself tend to proceed toward equilibrium, as will any isolated system. Therefore the energy required to maintain the non-equilibrium situation must be derived from some other system in the environment of the cell. In the absence of energy transfer into the cell from some external source, the cellular processes will all proceed toward equilibrium and the cell will die.

Green plants and certain microorganisms obtain the necessary energy as the visible radiant energy which we call light. The source of the light is normally the thermonuclear fusion reactions taking place in the sun. As we shall see later, this radiant energy is used by the plant cells for the synthesis of various organic molecules. In effect, nuclear energy is transformed in the sun into radiant energy and the radiant energy is transformed in the green plant into the energy of molecular structure which we can call chemical energy.

All other living cells fill their need for energy through the assimilation from the environment of chemical compounds with their associated chemical energy. For most cells these compounds are organic substances such as glucose, fatty acids, and amino acids, with glucose playing a major role. The ultimate origin of these materials is obviously the green plant. A few microorganisms are exceptional in that they utilize the chemical energy of inorganic substances such as ammonia, sulfur, or ferrous iron.

6.5 Biologically Usable Energy: The Need for Bite-Size Pieces

If we consider for a moment the energy released during the process of complete oxidation of glucose to CO_2 and water, we find ourselves dealing with the following equation:

$$C_6H_{12}O_6 + 6\ O_2 \longrightarrow 6\ CO_2 + 6\ H_2O + 2871\ \text{kJ}$$

(We could equally say that the $\Delta G^\circ = -2871$ kJ/mole.) Of course we are talking here about a process rather than a single reaction. But that is unimportant since you will remember that one of the laws of thermodynamics says that the energy yield will be the same no matter how you get there. Thus, burning the glucose in a flame or metabolizing it in an animal will yield the same amount of total available energy.

However, 2871 kilojoules is a very large amount of energy. Most of the jobs that need to be done in the cell can be handled with something between 8 and 40 kJ/mole. Let us suppose for the moment that we were able to "couple" together the oxidation of glucose and the formation of some compound A–B from A and B. If we assume that the energy needed to form the A–B bond is about 24 kJ/mole, then we would have the expression

$$C_6H_{12}O_6 + 6\ O_2 + A + B \longrightarrow 6\ CO_2 + 6\ H_2O + A\text{–}B + 2847\ \text{kJ}$$

We would not only be wasting the huge majority of the available energy, but the heat produced as a result would probably make life intolerable for the cell.

This is rather a far-fetched example, but it does illustrate one of the primary considerations in cellular energetics: Those processes that are producing the energy for the cell must be carried out in such a way that energy is made available in reasonable amounts and in a form that can be used readily to meet cellular needs. As we will see before long, the actual breakdown of glucose to CO_2 and water in a cell requires some 20 reactions by the most common pathway. Not all of these reactions, of course, release energy; this occurs primarily during the six oxidation reactions that are included. However, if we look at a typical example of one of these oxidations, the conversion of malic acid to oxaloacetic acid, we see that we have still not completely solved the problem. Even this simple oxidation is producing some 188 kJ of energy per mole of substrate:

$$\text{malate} + \tfrac{1}{2}O_2 \longrightarrow \text{oxaloacetate} + H_2O \qquad \Delta G^{\circ\prime} = -188.3 \text{ kJ/mole}$$

We will see in a later chapter the means by which these oxidation reactions are converted into a series of reactions that allow energy to be transferred in amounts on the order of 30 kJ/mole and in forms immediately usable in the cell.

6.6 Coupled Reactions

We have referred above to the possibility of coupling reactions together so that the energy derived from one will carry the other in the desired direction. An example you have possibly encountered in the organic chemistry laboratory would be the formation of a simple ester such as ethyl acetate. If we simply mix ethyl alcohol and acetic acid, the equilibrium is not very favorable for formation of the ester:

$$\text{ethanol} + \text{acetic acid} \rightleftharpoons \text{ethyl acetate} + H_2O \qquad \Delta G^{\circ\prime} = +19.8 \text{ kJ/mole}$$

However, if we couple this reaction with the hydrolysis of acetic anhydride:

$$\text{acetic anhydride} + H_2O \rightleftharpoons 2 \text{ acetic acid} \qquad \Delta G^{\circ\prime} = -91.2 \text{ kJ/mole}$$

we get the overall reaction

$$\text{ethanol} + \text{acetic anhydride} \rightleftharpoons \text{ethyl acetate} + \text{acetic acid}$$

$$\Delta G^{\circ\prime} = -71.4 \text{ kJ/mole}$$

This also illustrates a general feature of compounds of this type that we will see a great deal of in metabolic energy transfers. You will note that the free energy of hydrolysis of the anhydride bond (−91.2 kJ/mole) is much greater than that of the ester bond (−19.8 kJ/mole).

In theory, it would be possible to construct a set of metabolic reactions that would be composed of many coupled pairs, with one reaction specifically supplying the energy for one, two, or three other energy-requiring reactions. However, the practical problems associated with this would be immense. In economic terms, it would be equivalent to a system of strict bartering of goods. For flexibility and any degree of control of a complex system, some centralized monetary system is really required. What we want to look at, then, is the monetary system of the cell.

6.7 The Roles of ATP and Related Compounds

There are within cells a number of compounds that are composed of a nitrogen-containing heterocycle (sometimes called a "nitrogen base") to which is attached a sugar unit that is usually ribose or deoxyribose. These compounds are called **nucleosides**. When one or more phosphate groups are esterified to the sugar portion, the molecule that results is called a **nucleotide**. The nucleotides serve a variety of functions, and we will look at their structures and nomenclature in greater detail later. For the moment we are primarily concerned with those that act as major intermediates in energy transfer in the cell.

Without question the most prominent energy intermediate in the cell is a compound called **adenosine triphosphate** (or **ATP** for short). The structure of this compound is shown in Figure 6-1; note particularly that while the phosphate group actually attached to the ribose is bound by an ester linkage, the other two phosphate groups are attached by anhydride bonding.

Removal of the terminal phosphate group would leave the compound **adenosine diphosphate (ADP)**; removal of the second phosphate would result in a compound usually called **AMP** (for **adenosine monophosphate**). In fact, there are several adenosine monophosphates possible, with the phosphate ester attached at the 2′-, 3′-, or 5′- position of the ribose ring. When the position is not specified, or if the abbreviation AMP is used, it is assumed that the 5′-monophosphate is intended.

Values for the standard free energies of hydrolysis of ATP and related compounds at pH 7 are given in Table 6-2. It should be noted that calculation of a value for the free energy of hydrolysis for a process such as ATP and water going to ADP and inorganic phosphate is a very difficult process if you are trying to arrive at a value that would have significance in a living cellular system. There has been for some years a degree of disagreement

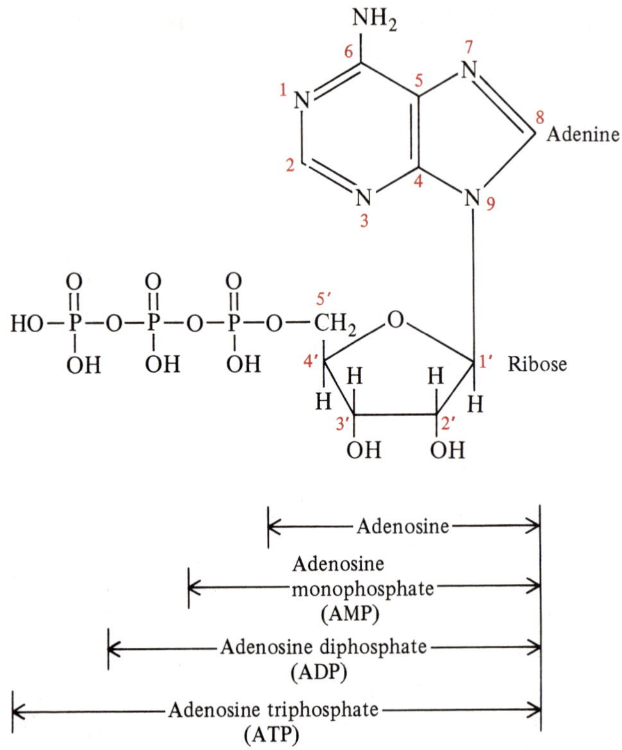

Figure 6-1
Structures of AMP, ADP, and ATP.

over what the value should be. However, for our purposes the exact value is not critical as long as the values for this and related processes are consistent among themselves. The value of -30.6 kJ/mole is one widely used in the biochemical literature.

TABLE 6-2
Standard free energies of hydrolysis of some phosphorylated compounds.

Compound	$\Delta G^{\circ\prime}$ (kJ/mole)
Phosphoenolpyruvate	−61.9
Carbamoyl phosphate	−51.5
Acetyl phosphate	−47.3
Creatine phosphate	−43.1
Arginine phosphate	−38.1
ATP (to AMP + PP_i)	−37.7
Inorganic pyrophosphate	−33.5
ATP (to ADP + P_i)	−30.6
Glucose 1-phosphate	−20.9
Fructose 6-phosphate	−15.9
Glucose 6-phosphate	−13.8
Glycerol phosphate	−9.2

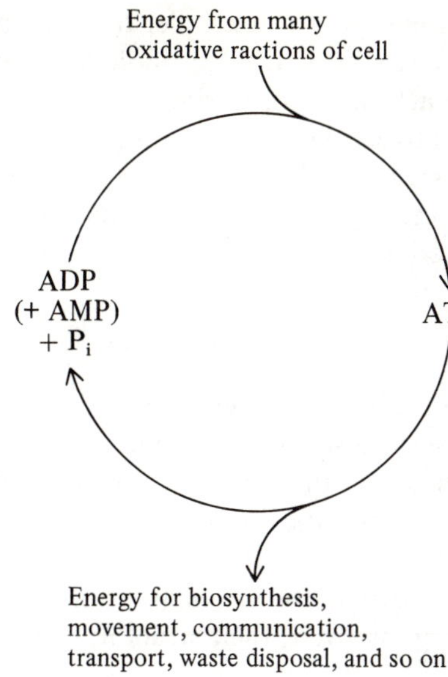

Figure 6-2
The role of ATP as an energy intermediate.

In most cases it is by removal and replacement of the terminal phosphate group that ATP serves as an energy transfer intermediate. The various energy-yielding reactions of the cell result in addition of inorganic phosphate to ADP to form ATP. For example, a single oxidation step, such as the oxidation of malic acid mentioned above, through a process called **oxidative phosphorylation** gives rise to the formation of three molecules of ATP from ADP and inorganic phosphate. The energy thus stored in this phosphate anhydride bond can be used in some coupled mechanism to carry out a variety of energy-requiring reactions or processes. This cyclic process and the central role of ATP is outlined diagrammatically in Figure 6-2.

Why ATP?

While it is not possible to answer the question of "why ATP?" in any real sense, it is worth pointing out some of the features that render this molecule rather well suited to serve as a primary energy transfer intermediate. First, this nucleotide is required in substantial amounts for transmission of genetic information (Chapter 18) and thus the cell must have synthetic pathways by which to make it. This may seem to be reasoning in reverse, but cells are quite conservative in the sense that they are inclined to get along with as few enzymes as possible to conserve the large amounts of energy that must be expended for protein synthesis.

Second, the amount of energy either needed for ATP formation or obtained on hydrolysis to ADP is well located in the range of energies we have talked about as representing the needs for most biosynthetic reaction steps. There is enough energy, for example, to carry out an esterification reaction and still have a few kilojoules per mole left over to give up as heat in order to drive the reaction to completion.

Third, ATP is a relatively stable compound. At physiological pH values, ATP is only relatively slowly hydrolyzed in the absence of enzymes. Under similar circumstances, more reactive compounds such as acetyl phosphate are subject to rapid decomposition.

The Other Nucleoside Triphosphates

While ATP is the central reservoir for chemical energy in most cells, there are a number of very closely related compounds that serve similar but more specialized functions. In their use as energy intermediates, they serve primarily to allow for very specific control of certain reactions or pathways of reaction without disturbing all of the processes in which ATP is involved.

These other nucleoside triphosphates differ only in the structure of the nitrogen heterocycle ring involved; their structures are shown in Figure 6-3.

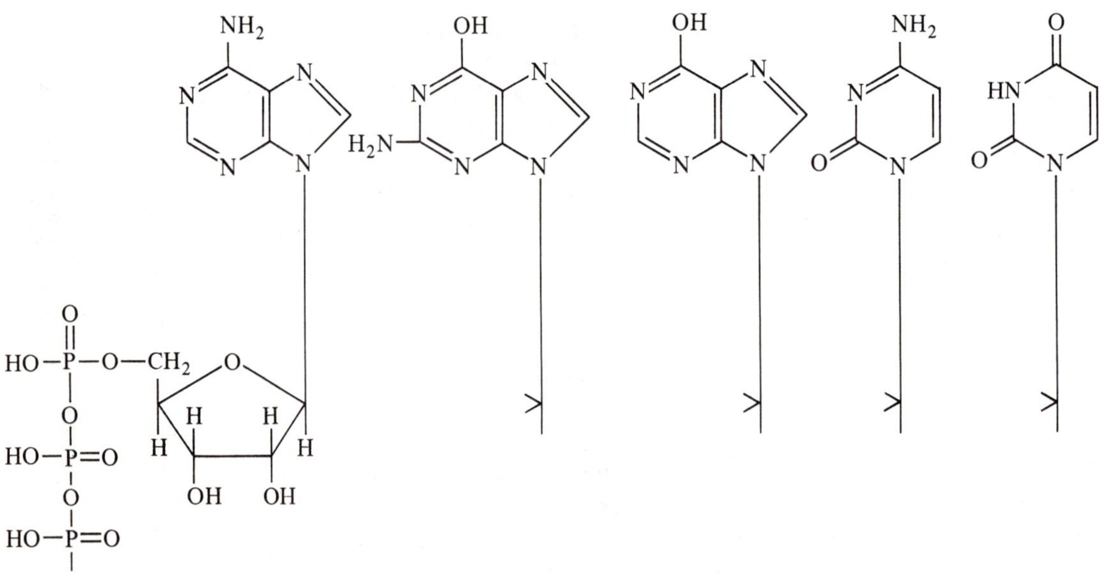

Adenosine triphosphate (ATP) | Guanosine triphosphate (GTP) | Inosine triphosphate (ITP) | Cytidine triphosphate (CTP) | Uridine triphosphate (UTP)

Figure 6-3
The structures of the nucleotide triphosphates involved in energy metabolism.

Each of these triphosphates tends to be concerned with a particular area of metabolism. For example, UTP is much involved with the synthesis of polysaccharides, and the interconversions of various sugars. Similarly, CTP is concerned with a number of reactions involved in the synthesis of lipid compounds such as the triglycerides and glycerol phospholipids.

Typically, utilization of these nucleoside triphosphates other than ATP results in production of the diphosphate and monophosphate compounds. Regeneration to the original triphosphate is at the expense of an ATP molecule. Enzymes capable of catalyzing interconversions of this type are widespread in the tissues of most if not all organisms; for example,

$$\text{UMP} + \text{ATP} \rightleftharpoons \text{UDP} + \text{ADP}$$

or

$$\text{UDP} + \text{ATP} \rightleftharpoons \text{UTP} + \text{ADP}$$

6.8 Oxidation–Reduction Coenzymes—Role in Energy Transfer

While they do not have the extensive application that ATP and its relatives have in the conservation and distribution of cellular energy resources, certain of the oxidation–reduction coenzymes do serve a critical role in transferring energy directly from the oxidative reactions (such as those involved in the total oxidation of glucose) to reduction reactions that are necessary in a number of biosynthetic pathways.

The two compounds most involved in this process are **nicotinamide adenine dinucleotide (NAD)** and **nicotinamide adenine dinucleotide phosphate (NADP)**. The structure of the oxidized and reduced forms of NAD is shown in Figure 6-4. NADP differs only in having another phosphate ester group attached at the 2′-position of the ribose of the AMP portion.

Because the reduced forms of these intermediates are stable to molecular oxygen (unless subjected to enzyme-catalyzed reactions), they are reasonably stable in the cell even under aerobic conditions. Thus, they may accept electrons from a compound such as malic acid; for example,

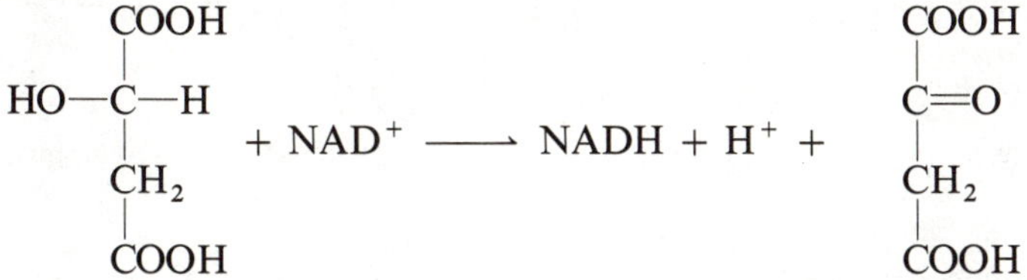

Malic acid Oxaloacetic acid

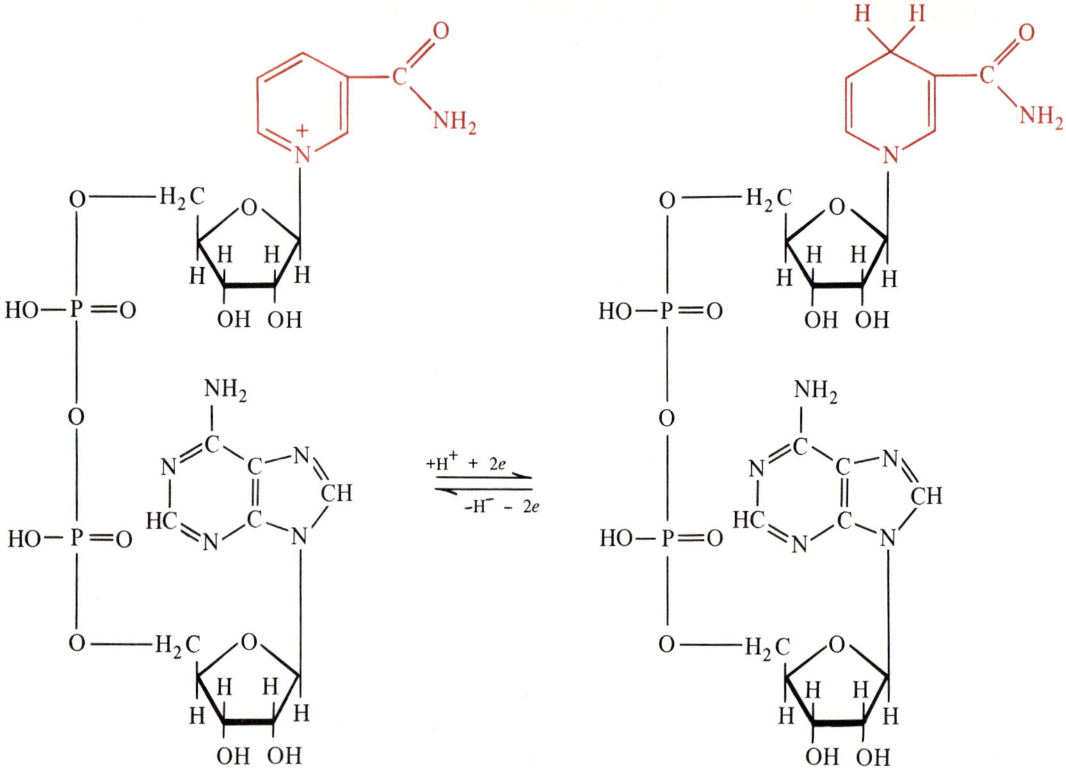

NAD$^+$
(oxidized form)

NADH
(reduced form)

Figure 6-4
Structures of the oxidized and reduced forms of nicotinamide adenine dinucleotide.

and then this NADH may be used to reduce a compound needed in a synthetic pathway, such as

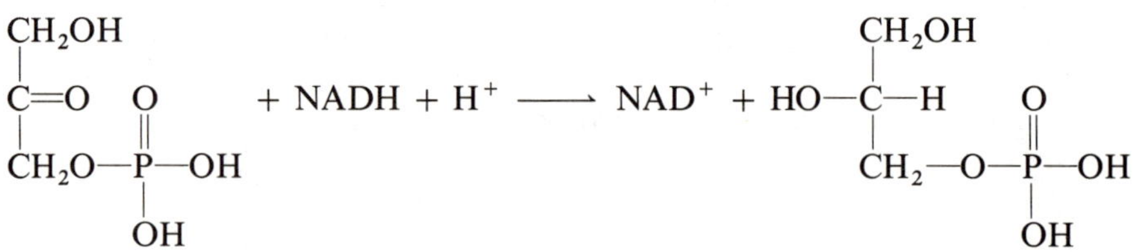

Dihydroxyacetone phosphate

sn-Glycerol-3-phosphate

It should be noted that in these cases the energy being saved and transferred is rather a larger amount than in the case of a single ATP molecule.

The free energy difference for the reaction

$$NADH + H^+ + \tfrac{1}{2} O_2 \longrightarrow NAD^+ + H_2O$$

is on the order of -218 kJ/mole.

Generally, in animal cells for example, the level of NAD^+ is held higher than NADH, and the effect of this is to promote the energy-yielding oxidative reactions of the cell. At the same time, NADPH normally predominates over $NADP^+$ and helps drive the reduction reactions of many biosynthetic pathways.

In plant cells, NADP is intimately involved in the processes of photosynthesis and many related reactions.

The availability of two sets of nucleotides carrying out similar, but not identical, functions again permits an extra degree of flexibility and control of metabolic process. We will look into these questions in more detail in later chapters.

Additional Readings

Scientific American articles

P. K. Stumpf, "ATP" (April 1953).

A. L. Lehninger, "How Cells Transform Energy", 205 (3), 62 (September 1961).

I. M. Klotz, *Energy Changes in Biochemical Reactions* (Academic Press, New York, 1967). An excellent short monograph at an introductory level dealing primarily with the application of thermodynamic relationships to biochemical systems.

W. M. Becker, *Energy and the Living Cell* (J. B. Lippincott, Philadelphia, Pa., 1977). A very lucid description of the role of energy exchanges throughout biochemistry. Chapters 1 and 2 are particularly appropriate at this point.

J. G. Morris, *A Biologist's Physical Chemistry*, 2nd edition (Addison-Wesley, Reading, Mass., 1974).

Study Questions

1. Correct the following statement: "While a reaction is occurring, the actual ΔG will be a function of the concentrations of reactants and products, of the temperature, and of the rate at which the reaction is proceeding."

2. The reaction of glucose with oxygen to yield CO_2 and water has a standard free energy change of about -2871 kJ/mole. At the same time, a dish of pure glucose may be left out in the air for a long time without noticeable reaction taking place. Can you explain this situation? (Hint: You may want to review Figure 4-1.)

3. Given that the value of R is 8.32 J/mole-°K and that °K = °C + 273, find the equilibrium constant at 25°C and pH 7 for the reaction

$$\text{oxaloacetate} + \text{NADH} + \text{H}^+ \rightleftharpoons \text{malate} + \text{NAD}^+$$

4. Using the data of Table 6-2, obtain a value for ΔG° for the reaction

$$\text{glycerol} + \text{ATP} \longrightarrow \text{glycerol phosphate} + \text{ADP}$$

[Hint: This is the sum of two coupled reactions: (i) ATP + $H_2O \longrightarrow$ ADP + P_i and (ii) glycerol + $P_i \longrightarrow$ glycerol phosphate + H_2O.]

5. Creatine phosphate is used in some organisms as an immediate reserve source of phosphate groups:

$$\text{creatine} + \text{ATP} \rightleftharpoons \text{creatine phosphate} + \text{ADP}$$

Given the data of Table 6-2, with the value of R from Question 3 above, and given a steady state ratio of ATP/ADP of about 100 in a resting muscle cell at 25°C and pH 7, what will be the ratio of creatine phosphate to free creatine in this cell?

6. Clearly, the reactions shown in Questions 4 and 5 show the possibility of transferring a phosphate group from creatine phosphate to glycerol through the intermediate formation and use of ATP. What would be the value of (a) $\Delta G^{\circ\prime}$ and (b) K_{eq} for the overall process

$$\text{creatine phosphate} + \text{glycerol} \longrightarrow \text{creatine} + \text{glycerol phosphate?}$$

CHAPTER 7

Carbohydrates

Since we are going to start our examination of the metabolic activities of cells with a look at reactions involving carbohydrates, we will pause briefly to review some key aspects of the structure and nomenclature of some of the carbohydrates of biological importance.

The term **carbohydrate** was originally derived from the fact that the large bulk of the compounds being described fit the empirical formula $C_n(H_2O)_n$. This is now considered to be excessively restrictive, and a more useful definition might be "polyhydroxyaldehydes and polyhydroxyketones and their derivatives." This would include deoxy sugars, amino sugars, and even sugar alcohols and acids.

Carbohydrates occur in a variety of forms, including simple sugars (incapable of being hydrolyzed to simpler compounds) and a variety of derivatives and polymers, such as starch and glycogen.

7.1 Monosaccharides

Compounds possessing a free (or "potentially free") aldehyde or ketone group and two or more hydroxyl groups are called **monosaccharides** or **simple sugars**.

The simplest compounds which meet these requirements are glyceraldehyde and dihydroxyacetone. Having three carbon atoms, these are known as **trioses**. Sugars having four carbon atoms are then **tetroses**; five carbons, **pentoses**; six carbons, **hexoses**, and so on. They may be further differentiated on the basis of the functional group involved into **aldo-** or **keto-** sugars. Thus, for example, glyceraldehyde is an **aldotriose** and dihydroxyacetone is a **ketotriose**:

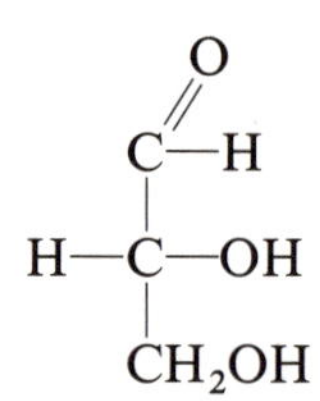

Glyceraldehyde

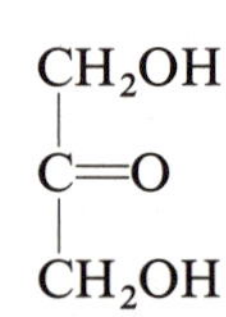

Dihydroxyacetone

When the three-dimensional structure of the glyceraldehyde molecule is arranged and projected according to the Fischer system (aldehyde on top and primary alcohol at bottom; hydrogens and hydroxyl groups on other carbons projecting up out of the plane of the paper), we find that two different enantiomorphic structures are possible:

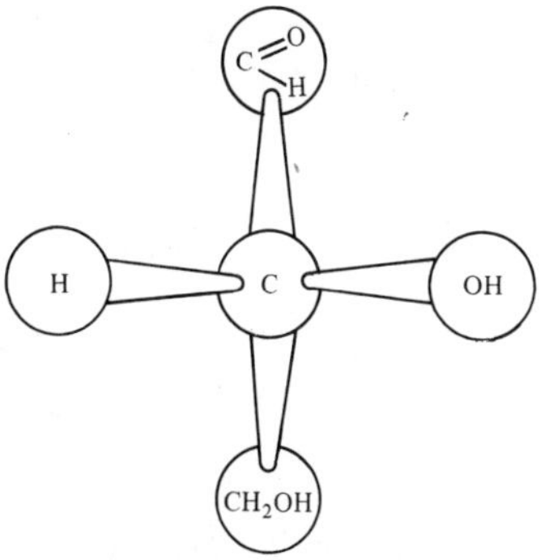

D-glyceraldehyde

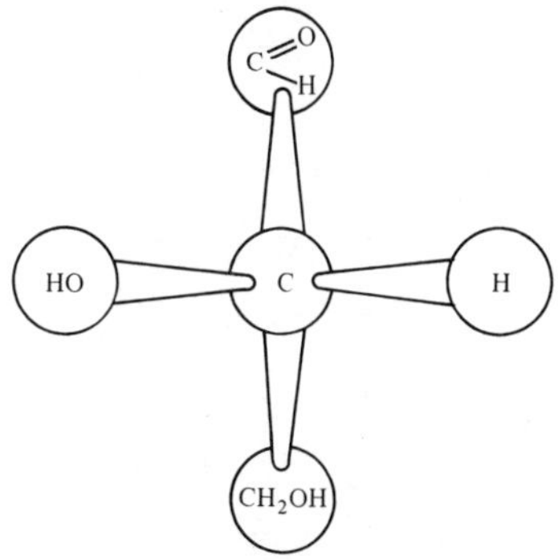

L-glyceraldehyde

The form in which the hydroxyl group next to the primary hydroxyl is projected to the right of the carbon chain is designated as the D-form and the form in which it is projected to the left is the L-form. These D- and L-designations have nothing to do with the direction in which the compound rotates the plane of polarized light. If it is desired to indicate the direction of rotation, the name may be written D-(+)-glyceraldehyde.

Whatever else may be inserted between the aldehyde and the alcohol next to the primary alcohol, the assignment to the D- or L-series of compounds still depends on whether the hydroxyl of the asymmetric carbon atom farthest from the aldehyde (that is, next to the primary hydroxyl) is projected to the right or the left. For example:

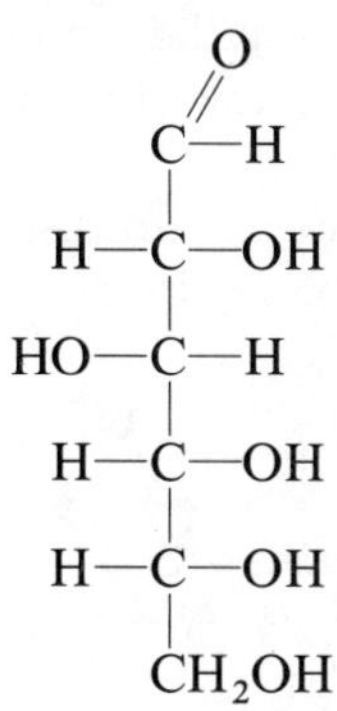

D-glucose

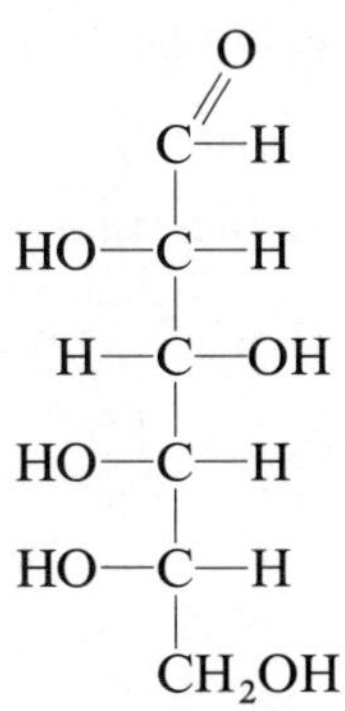

L-glucose

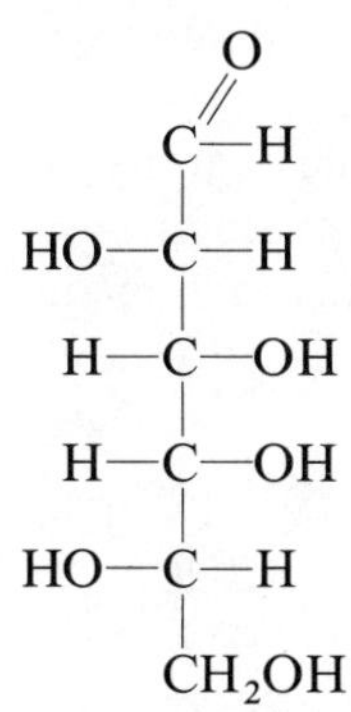

L-galactose

When we go from glyceraldehyde to a four-carbon aldose, we get the possibility of four stereoisomers because, as shown by the asterisks, there are two chiral carbon atoms:

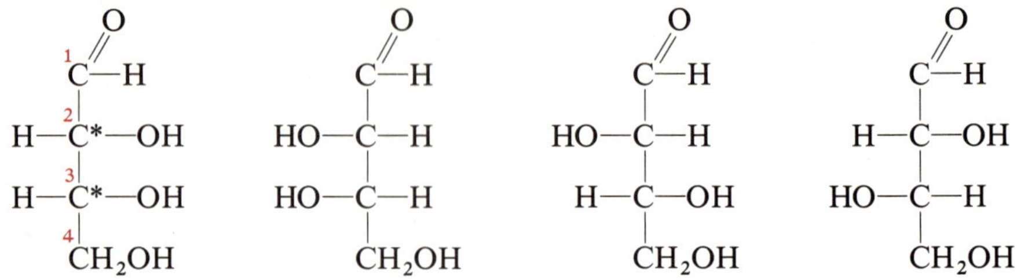

D-erythrose L-erythrose D-threose L-threose

For aldopentoses, there are eight isomers; the D-forms are shown below (for each of these there is a mirror image L-form):

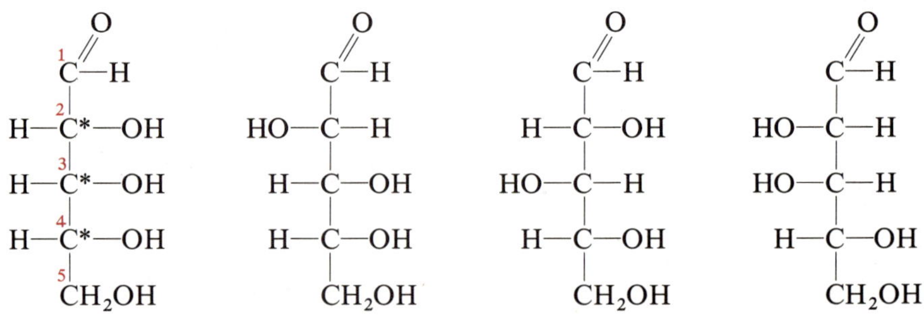

D-ribose D-arabinose D-xylose D-lyxose

Both D-ribose and D-xylose occur widely in nature, while lyxose is most uncommon. Arabinose is a special case since L-arabinose is the common form rather than the D-isomer. This is the only L-series sugar known to occur in any appreciable quantity in nature.

For the aldohexoses, there are 16 possible stereoisomeric forms, of which only D-glucose, D-mannose, and D-galactose are commonly found in nature:

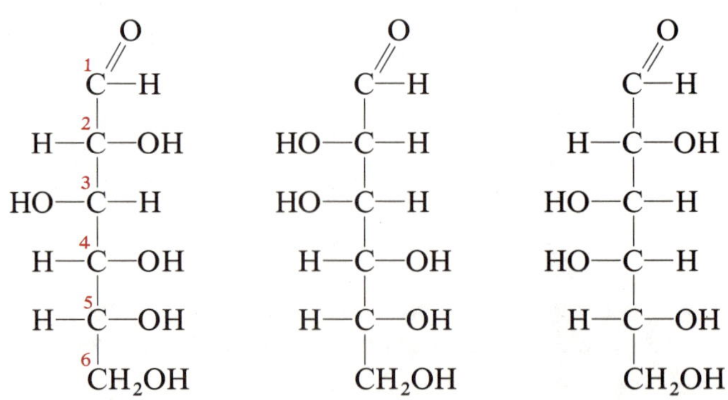

D-glucose D-mannose D-galactose

For the ketoses, a series of sugars can similarly be formed, all of which will be related to dihydroxyacetone:

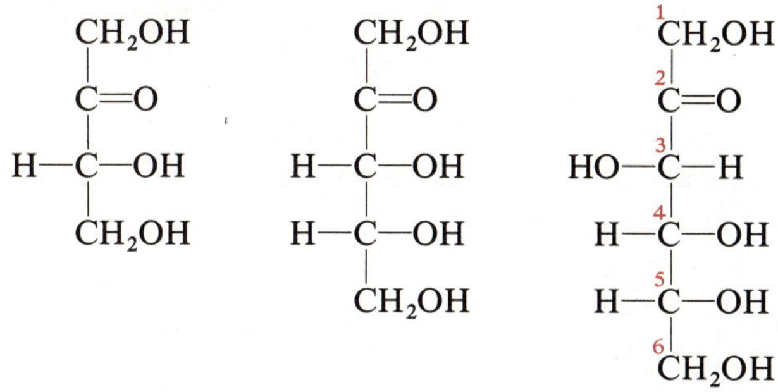

A D-ketotetrose (D-erythrulose) A D-ketopentose (D-ribulose) A D-ketohexose (D-fructose)

A number of the keto sugars, and their derivatives, play important roles in cellular metabolism. D-fructose is of particularly common occurrence; the other ketohexoses are found only to a limited extent. Derivatives of the ketopentoses ribulose and xylulose are important intermediates in carbohydrate metabolism related to photosynthesis. Note that the ending *-ulose* designates a keto sugar.

Cyclic Forms of Sugars

The Fischer projection formulas for the sugars obscure the fact that the aldehyde group and the alcohol group of carbon 5 can readily approach each other. For example, Figure 7-1 is a more realistic representation of one

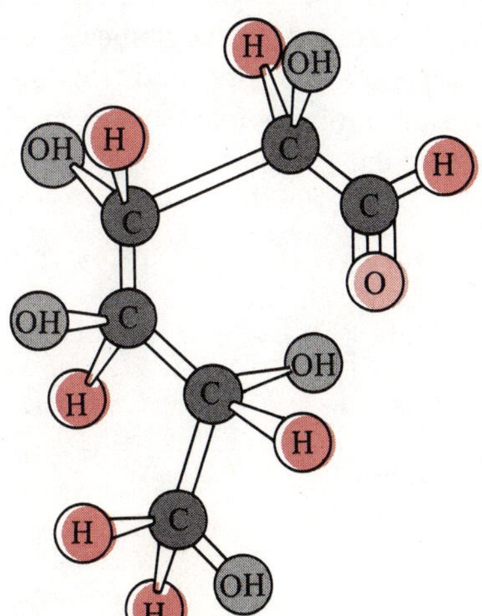

Figure 7-1
One possible conformation for the glucose molecule. Note the close proximity of the hydroxyl group of carbon 5 to the carbonyl group.

possible conformation of a glucose molecule which takes into account the normal bond angles. It is also possible for the molecule to twist in such a way that the alcohol group of carbon 4 is brought close to the aldehyde group. Similarly, the keto group of a keto sugar can approach the alcohol group of either carbon 6 or carbon 5.

The close proximity of an aldehyde (or ketone) group and an alcohol group in the sugars facilitates their reaction to give cyclic hemiacetals (or hemiketals). Of course this can occur with the alcohol group of either carbon 4 or of carbon 5:

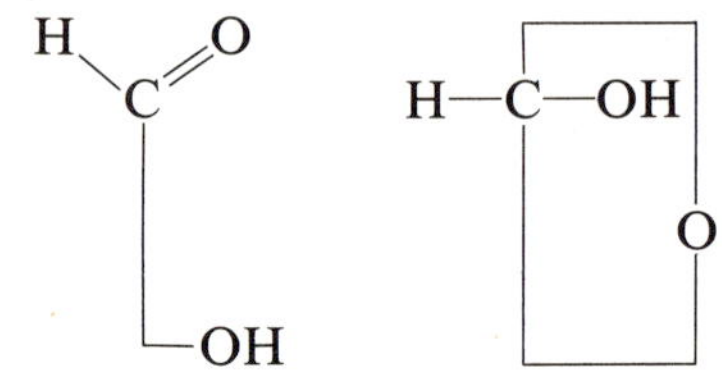

Aldehyde Cyclic hemiacetal

This reaction does not involve the loss or gain of any atoms by the sugar molecule; hence the aldehyde and cyclic forms are isomers. The sugar is said to exist in either the aldehyde or hemiacetal form. The cyclic forms are actually the most important forms of the sugar molecules. Glucose, for example, exists in the crystalline state or in water solution almost entirely as one or another hemiacetal, not as the free aldehyde. The sugars display many reactions typical of aldehydes only because hemiacetal formation is an equilibrium process, so that small amounts of free aldehyde are always present in solution (about 0.003% in the case of glucose).

The formation of the cyclic forms introduces two new structural problems into the chemistry of the sugars. There is first the question of the size of the ring. As we indicated above, a cyclic hemiacetal can be formed which involves either carbon 4 or carbon 5. In the first case a ring consisting of four carbon atoms and one oxygen atom is created; this is called a **furanose** ring. In the second case the ring contains five carbon atoms in addition to the oxygen; this is a **pyranose** ring.

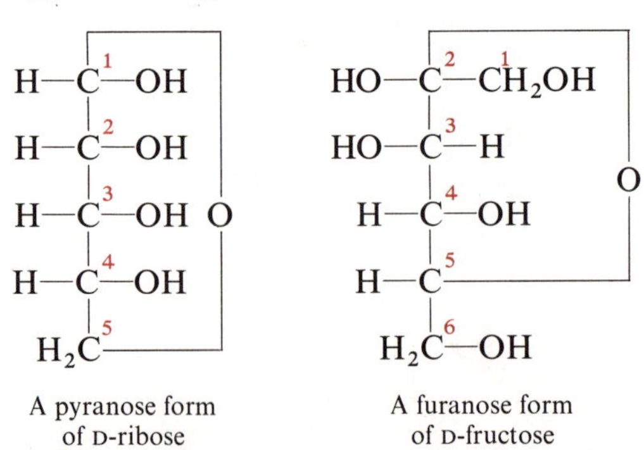

A pyranose form of D-ribose (α-D-ribopyranose) A furanose form of D-fructose (β-D-fructofuranose)

The second structual problem originating with the cyclic forms of the sugars is the question of the stereochemistry of the hemiacetal or hemiketal carbon. Examination of the structures given above should reveal that a new chiral center exists at the position of the original carbonyl: C_1 for ribose and C_2 for fructose. We find, therefore, that each ring form of a sugar can exist as either of two isomeric structures. These are termed the α- and β-isomers.

In depicting the structure of an α-isomer, we place the hydroxyl group of the hemiacetal or hemiketal carbon on the same side of the carbon chain as the hydroxyl group which determines whether the sugar is D or L. In practical terms, if the sugar structure is given in the Fischer method, the hydroxyl will be written to the right for a D-series sugar. Similarly, a β-isomer of a D-series sugar would be shown with the hydroxyl group of the critical carbon projecting to the left. The ribose molecule shown above is therefore an α-isomer; the fructose compound is a β-form.

Haworth Projections

The Fischer projection, while very useful to indicate differences in configuration at various chiral centers, does not represent very accurately the ring forms of the sugars. For this purpose we have a second conventional picture of the molecule which is known as the Haworth projection. In this system, the basic structures of the two ring forms are

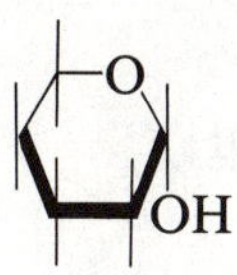

Pyranose

and

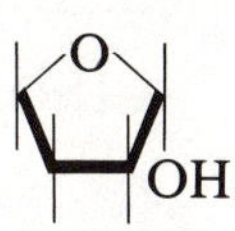

Furanose

When the rings are viewed in this manner, the attached groups will be found to lie either above or below the plane of the ring. Instead of trying to twist the Fischer structure around mentally, or actually constructing a molecular model, we may follow certain "rules" which permit us to obtain the correct Haworth form.

The first rule for translation of Fischer projections to Haworth forms is that any group to the *right* of the carbon chain is written *down*; those to the *left* are written *up* (that is, above the plane of the ring). Thus, for example:

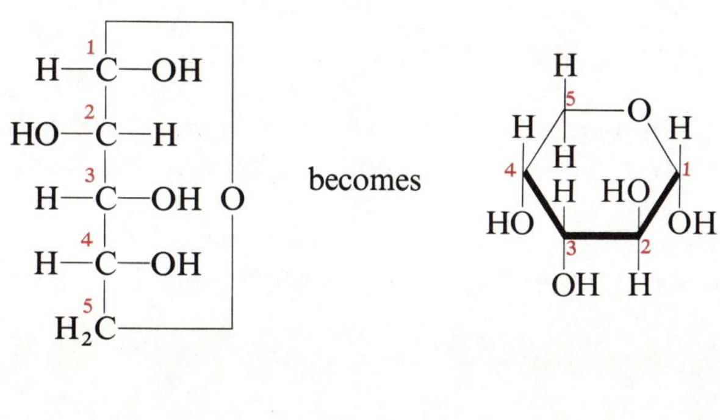

When there are more carbon atoms in the sugar than are involved in ring formation, it becomes necessary to determine whether they should be written above or below the plane of the ring. Here, the rule is that if the ring is to the *right*, the extra carbon or carbons will be *up* (that is, above the plane). For example:

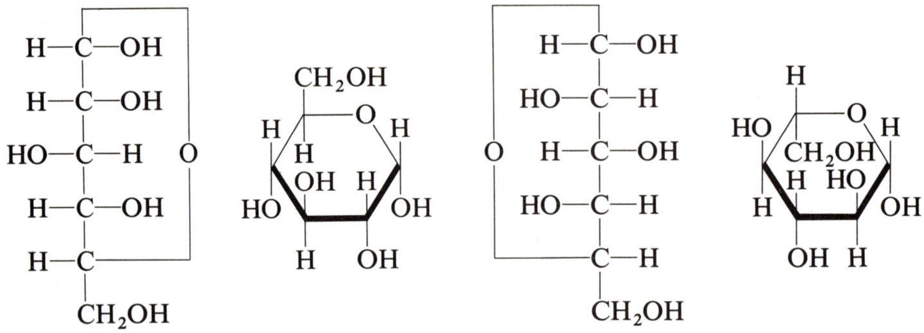

α-D-glucopyranose β-L-glucopyranose

In the case of a ketohexose in the furanose form, the position of the 6-carbon will be determined as above; the 1-carbon will be up or down depending on whether the sugar is in the α- or β-form:

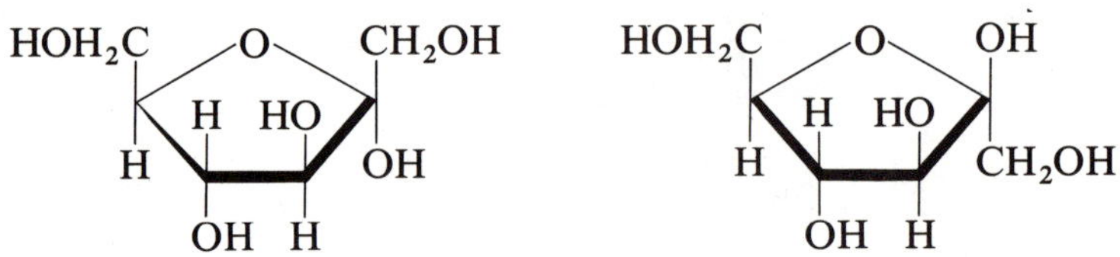

α-D-fructofuranose β-D-fructofuranose

These Haworth projections are the most common representations for the ring forms of sugars. When using them we must bear in mind that they do not accurately represent the real conformations of the molecules. The true state is usually a mixture of several conformations, depending on the actual molecules. Some examples are shown in Figure 7-2.

Sugar Derivatives

In defining the carbohydrates as a class, we agreed to include a number of sugar derivatives. The more common of these are the following:

Deoxy Sugars. These are sugars in which one of the hydroxyl groups is replaced by a hydrogen. The most common example is 2-deoxy-D-ribose

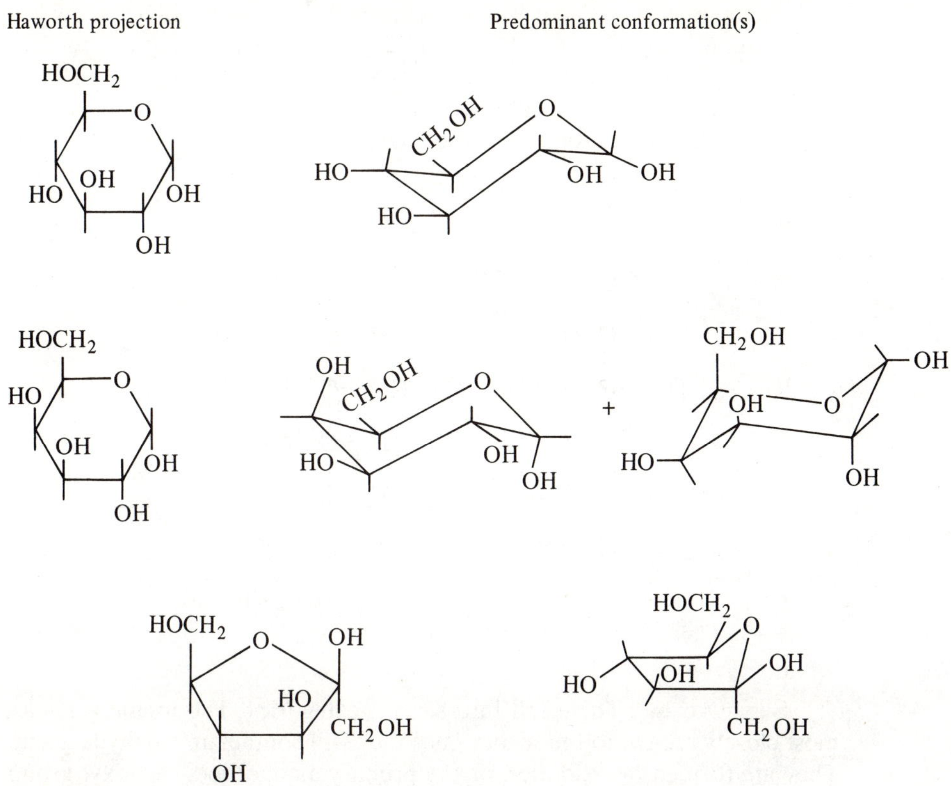

Figure 7-2
Examples of the actual predominant conformations of representative sugar molecules compared to their standard Haworth representations.

(or D-2-deoxyribose). This sugar is one of the major components of the nucleic acids of cell nuclei:

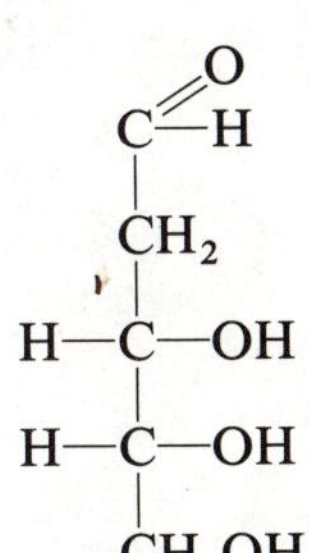

D-2-deoxyribose

Amino Sugars. These are sugars in which a hydroxyl has been replaced by an amino group, generally at the 2-position. Hexosamines are most common; glucosamine, galactosamine, and mannosamine have been found in a wide variety of biological materials. Frequently these compounds occur with an acetyl group attached to the nitrogen:

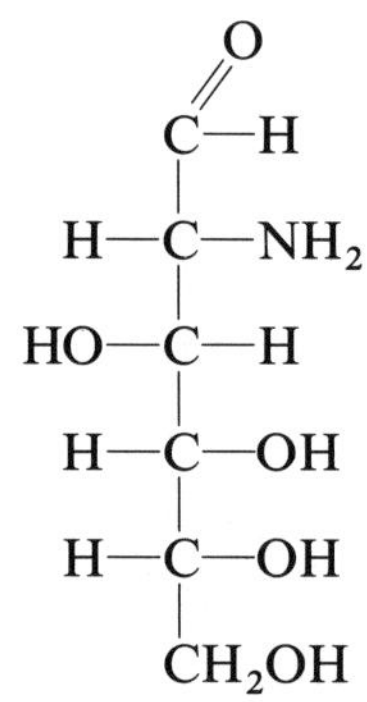

D-glucosamine or
D-2-deoxy-2-aminoglucose

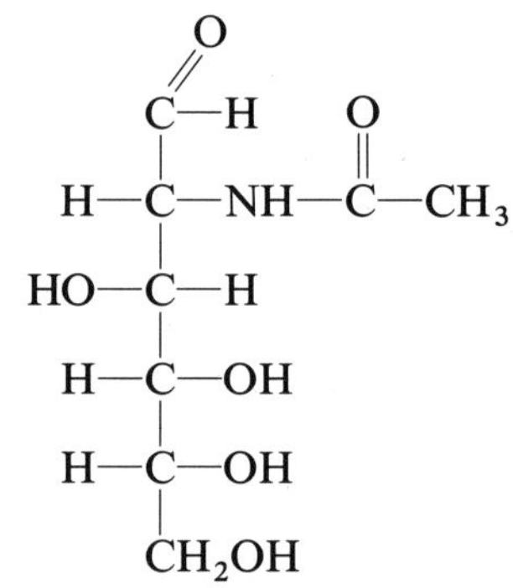

N-acetyglucosamine

Sugar Acids. These fall into several categories. The **uronic acids** are most closely related to the sugars since they still contain an aldehyde group. They are formed by oxidation of the primary alcohol to a carboxyl group.

The **glyconic acids** are formed by oxidation of the aldehyde group of the sugar.

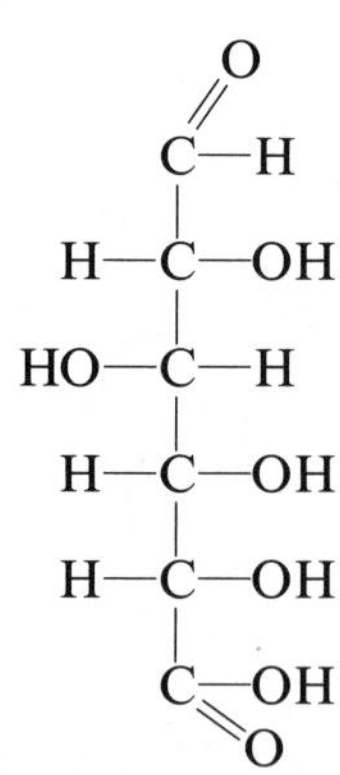

D-glucuronic acid
(a uronic acid)

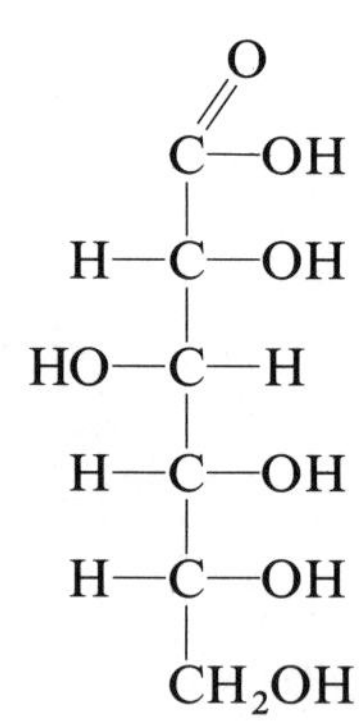

D-gluconic acid
(a glyconic acid)

7.2 Glycosides

Since all of the sugars in the ring forms are hemiacetals, it is possible for them to react with another alcohol to form acetals. These are given the general class name of **glycosides**; when a particular sugar is involved, a more specific name is applied; for example, glucoside, galactoside, fructoside, and so on. Further, it is of course possible to have α- and β-forms and pyranose or furanose ring structures. When all of these are specified, we get, for example (on reaction with methanol):

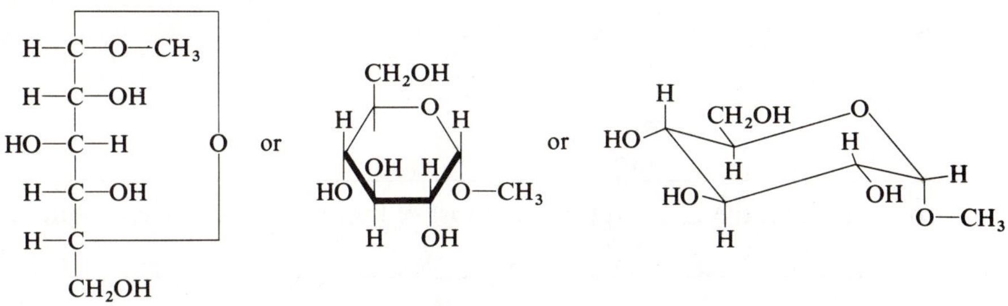

Methyl-α-D-glucopyranoside

In contrast with hemiacetals, in which the α- and β-forms are always in equilibrium with each other through the free aldehyde or ketone, α- and β-glycosides are individually stable with respect to such transformations.

7.3 Disaccharides

When the alcohol used in formation of a glycoside is part of another sugar molecule, we obtain a disaccharide as a product. Some common examples are as follows:

Maltose (two glucoses; α-1, 4-glucoside linkage):

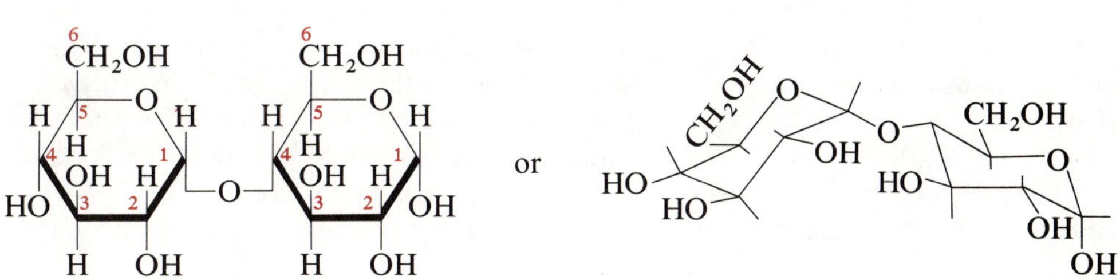

Maltose is obtained as an intermediate during the digestion of starch to glucose.

Note that because of the free hemiacetal form of one of the sugar units, maltose in solution will consist of an equilibrium of three forms—the α-, β-, and *aldehydo*-forms:

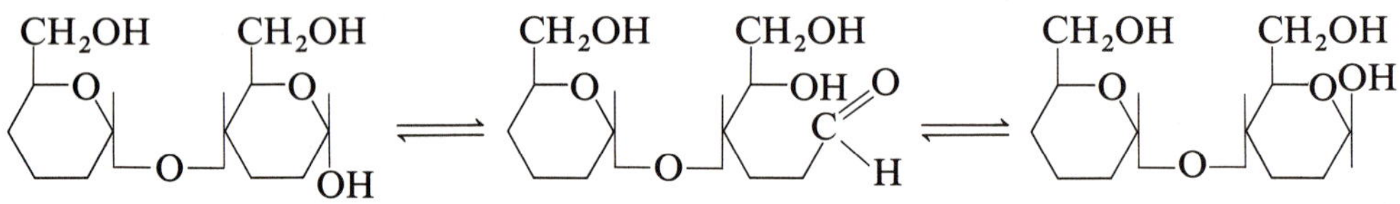

α-Maltose *aldehydo*-Maltose β-Maltose

For simplicity, the H's and OH's of most of the carbons have been omitted.

"Reducing" and "Nonreducing" Carbohydrates

Any aldehyde is capable of being oxidized, and thus bringing about the reduction of some other substance. Any of the carbohydrates that carry a free aldehyde, or any form in equilibrium with a free aldehyde, will fit this category. This includes all of the hemiacetal forms.

In addition, under any alkaline conditions, an equilibrium is established that brings about the conversion of ketoses in part to aldoses. For example, fructose in alkaline conditions isomerizes into a mixture with glucose and mannose. While the fructose predominates, this type of reaction allows any ketose under alkaline conditions to act as though it were an aldose for the purposes of taking part in an oxidation/reduction reaction.

A variety of agents may be used to carry out the oxidation (and be themselves reduced). The most common is cupric ion, which is the active ingredient in Fehling's, Benedict's, and Barfoed's reagents for the detection of reducing sugars. The reaction is

$$\text{reducing sugar} + \underset{\text{Blue}}{2\,Cu^{++}} \longrightarrow \text{oxidized sugar} + 2\,Cu^{+}$$

$$2\,Cu^{+} + 2\,OH^{-} \longrightarrow \underset{\text{Yellow}}{2\,CuOH} \longrightarrow \underset{\text{Red}}{Cu_2O} + H_2O$$

Any carbohydrate capable of giving this reaction without having to be hydrolyzed first is designated as a "reducing" carbohydrate: others are then "nonreducing." Obviously, all free monosaccharides must be reducing carbohydrates.

Lactose (galactose plus glucose; β-1,4-galactoside linkage):

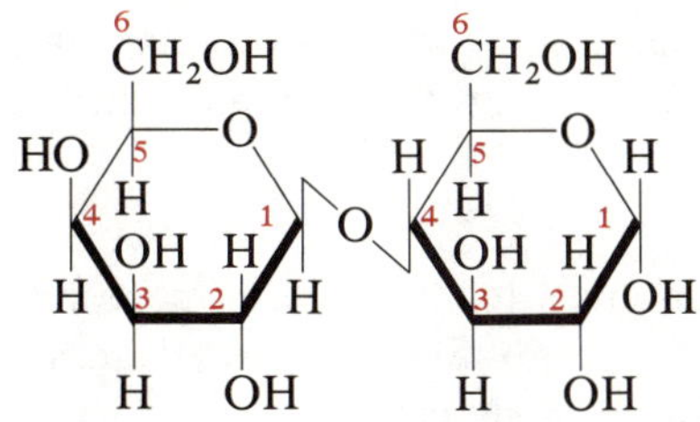

This molecule, like maltose, has one free hemiacetal group (carbon 1 of the glucose portion of the molecule). Consequently, lactose also is readily oxidized by cupric ion and is therefore designated as a reducing disaccharide.

The occurrence of lactose in nature, in contrast to most carbohydrates, is almost completely restricted to the animal kingdom. Here it is found as the major carbohydrate component of mammalian milks.

Sucrose (glucose plus fructose; α-D-glucopyranosyl-β-D-fructofuranoside):

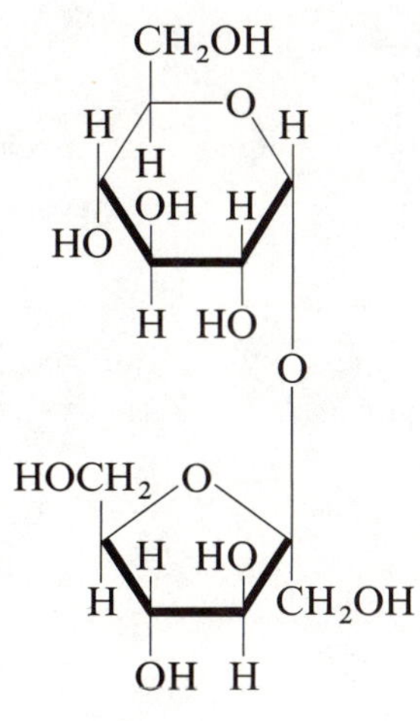

Sucrose, our common "table sugar," is very widely distributed in plants. In the case of sugar cane, sugar beets, and sugar maple, the concentration is sufficiently high to make its isolation commercially feasible.

Since both of the carbonyl groups are involved in formation of the glycoside bond, neither will be in equilibrium with its free form. Consequently, sucrose is a nonreducing disaccharide.

7.4 Polysaccharides

These are polymeric structures containing many units. Those in which the units are all the same sugar are called homopolysaccharides and those composed of two or more different sugars are called heteropolysaccharides.

Starch

This is the name given to the material comprising the major storage form of carbohydrate in plants. Starch usually occurs in the form of compact insoluble grains inside the plant cells.

The starch grains are actually composed of two different polysaccharides having quite different properties:

1. **Amylose.** A polysaccharide in which glucose units are joined in **α-1,4-glucoside linkages** to form long slender chains

These long chains can approach one another very readily and can then be held together by hydrogen bonding between the hydroxyl groups. The large structures which result are quite stable and compact, and are insoluble in water unless the temperature is raised high enough to break most of the hydrogen bonds (for example, by boiling):

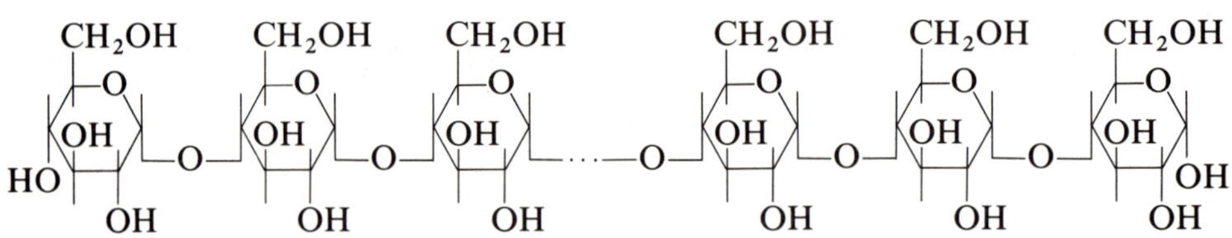

2. **Amylopectin.** This polysaccharide possesses the same basic chain of **α-1,4-glucoside** linkages as does amylose, but has in addition a number of side chains attached by **α-1,6-glucoside** linkages, as shown:

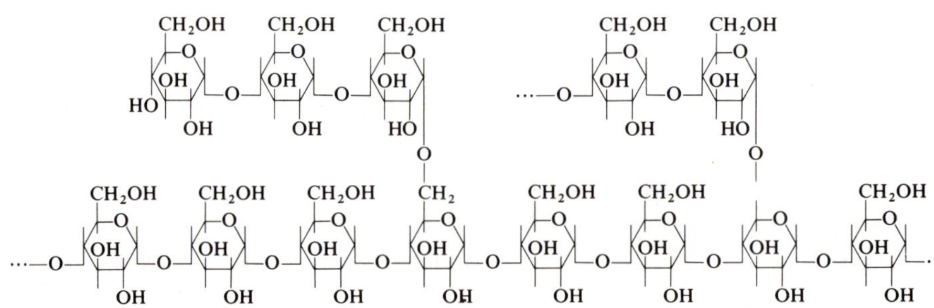

This branching decreases the ability of the chains to get together with one another and increases the binding of water molecules to the chains. Thus, the amylopectins are reasonably soluble in warm water.

Glycogen

This is the major carbohydrate reserve in animals. In most mammals, deposits of glycogen are maintained especially in the liver and in the skeletal muscles.

The structure of glycogen is essentially the same as that of the amylopectins, except that there is much more extensive branching. This confers an increased solubility, to the degree that glycogen is quite readily put into suspension even in cold water.

Cellulose

This is one of the major structural components of plants. It is similar in structure to amylose, except that all of the glucose units are held together by **β-1,4-glucoside** linkages:

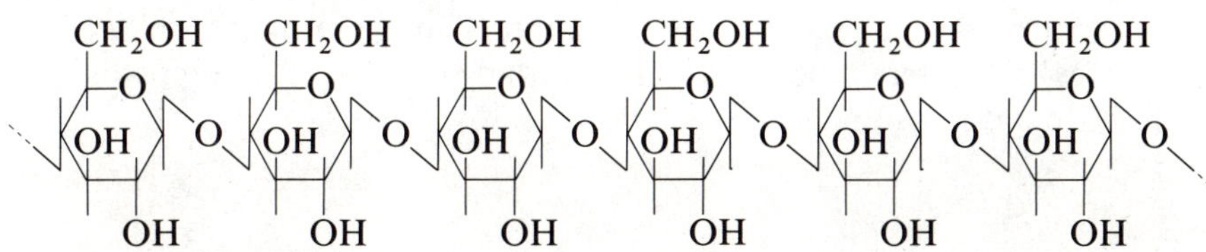

For the reasons mentioned in connection with the amyloses, this yields an insoluble macromolecule. The cellulose chains are rather longer and more efficiently hydrogen-bonded to one another, and the aggregated molecules are thus very large and extremely insoluble. However, these are just the properties that make them so useful as fibers for paper, cloth, and so on.

Animals that do their own digesting are unable to make any use of cellulose as a food, since none of them possesses an enzyme capable of catalyzing the cleavage of β-glucoside bonds. Ruminants and a variety of wood-eating insects (including termites) are able to use cellulose only because the microorganisms of their digestive tracts do possess such enzymes and use them to the mutual benefit of microorganism and animal.

Mucopolysaccharides

This term is used to denote those heteropolysaccharides which contain amino sugars or their derivatives.

The members of this class may be separated into several subgroups. First, the **neutral** mucopolysaccharides, such as the immunologically important blood-group substances, which contain only an *N*-acetlyhexosamine and a hexose.

A second group is distinguished by the presence of a hexuronic acid (such as D-glucuronic acid), along with an *N*-acetylhexosamine. Typical of this group are the **hyaluronic acids** which are found as components of connective tissue. These are straight-chain polymers with D-glucuronic acid and *N*-acetylglucosamine alternating in the chain.

The third group is a rather complex one in which a polysaccharide may contain hexoses, amino sugars (free and acetylated), hexuronic acids, and sugar sulfates. Examples from this group are **chondroitin sulfate** of cartilage (*N*-acetylgalactosamine, glucuronic acid, and sulfate esters of those sugars), and **heparin**, the natural anticoagulant of blood (glucosamine *N*-sulfate, glucuronic acid, and their sulfate esters).

Glycosidic linkages involving uronic acids or amino sugars are unusually resistant to chemical hydrolysis. Consequently, the polysaccharides containing these units tend to be extremely stable and are often found where strength and chemical resistance are most important; for example, in skin, connective tissues, insect exoskeletons, umbilical cord, and so on.

Others

All of the polysaccharides mentioned above are merely selected examples of different types of structures. Many other similar compounds are to be found in nature.

Additional Readings

R. D. Guthrie, *Guthrie and Honeyman's Introduction to the Chemistry of Carbohydrates*, 4th edition (Clarendon Press, London and New York, 1974).

D. A. Rees, *Polysaccharide Shapes* (Chapman and Hall, 1977). A very short description of various types of polysaccharides and their structures.

R. Barker, *Organic Chemistry of Biological Compounds* (Prentice-Hall, Englewood Cliffs, N.J., 1971). Includes an interesting review of some selected aspects of carbohydrate chemistry.

W. Pigman and D. Horton, *The Carbohydrates*, 2nd edition (Academic Press, New York and London, 1970–72) (4 volumes). The standard reference in the field of carbohydrates.

Study Questions

1. Draw structures for all of the possible D-aldohexoses.

2. Draw appropriate structures for all of the forms of D-glucose that will exist at equilibrium in water solution at neutral pH.

3. An "epimer" is a compound that differs from another only in the configuration about *one* chiral center, although there are other chiral centers common to the two molecules. Draw structures for all of the epimers of α-D-glucopyranose and name those for which you can obtain the structures from this chapter.

4. For many biochemical purposes, the α or β nature of the glycosidic bond, the ring form, and the nature of the sugar involved in contributing the glycosidic carbon are all important considerations. For each of the following structures, indicate the nature of the glycosidic bond by specifying an α-glucopyranoside, a β-fructofuranoside, a β-ribofuranoside, or whatever:

(a)

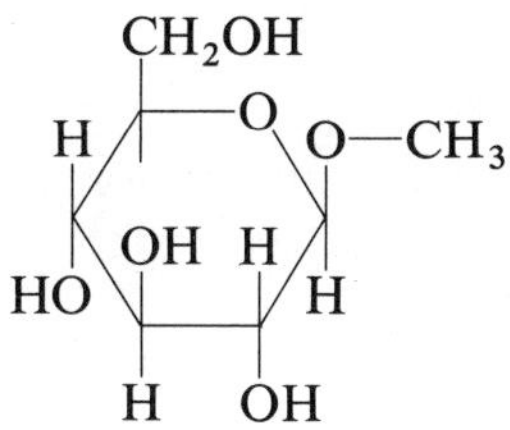

(b)

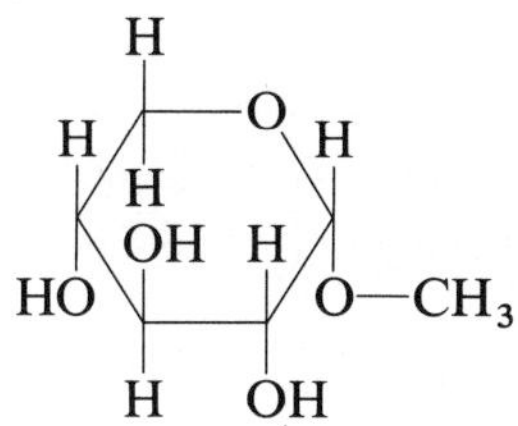

(c)

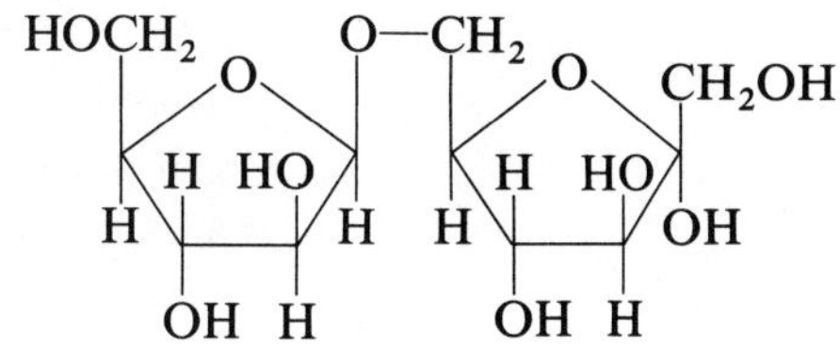

(d)

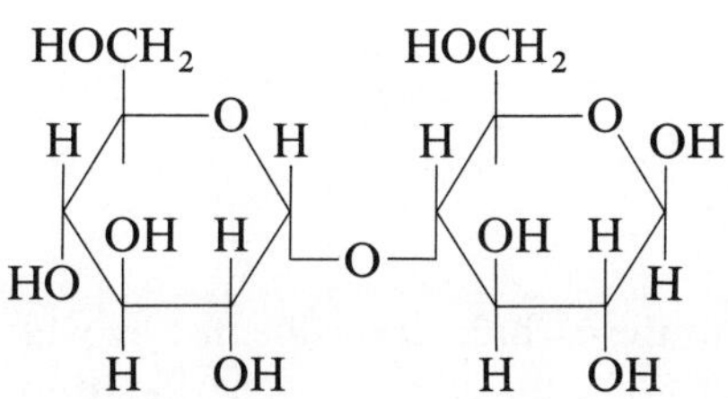

(e)

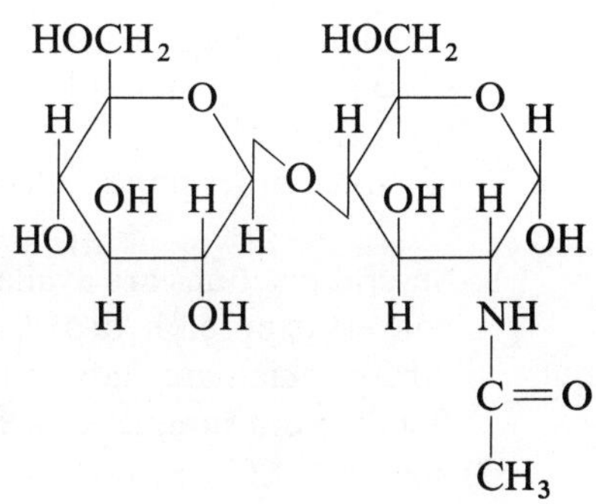

5. Name, as accurately as you can from the structure given, each of the following:

(a)

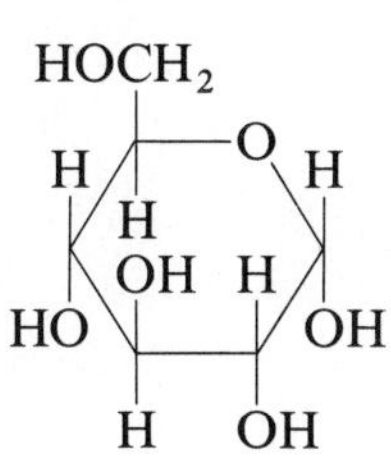

(b)

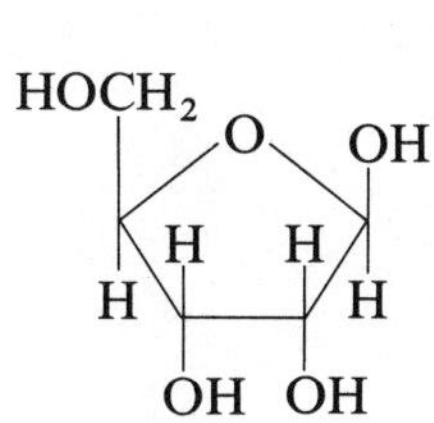

(c)

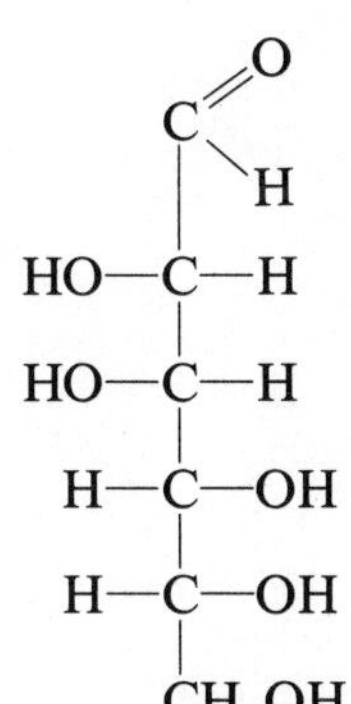

(d)

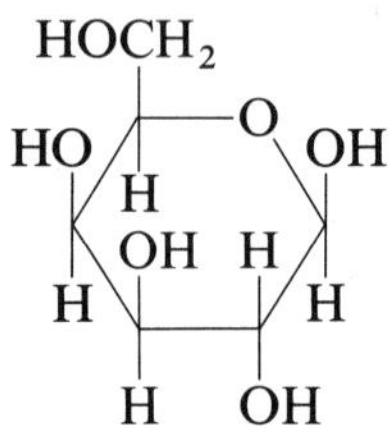

(e)

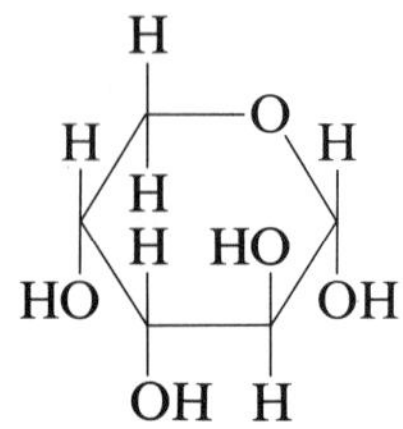

(f)

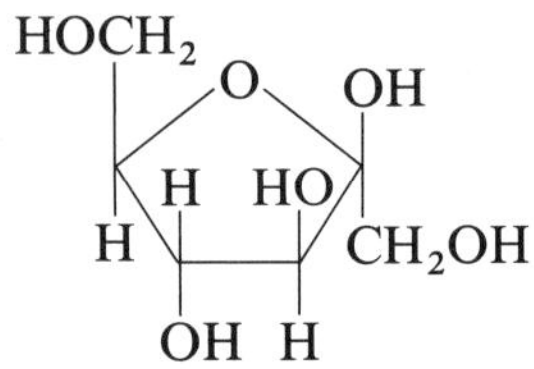

6. Which of the compounds of Questions 4 and 5 are capable of reducing an alkaline copper reagent (that is, are "reducing carbohydrates")?

7. Draw a portion of structure sufficient to show the essential features of the following:
(a) maltose
(b) sucrose
(c) an amylose molecule
(d) a glycogen molecule
(e) cellulose
(f) some other polysaccharide

8. Several reactions are available to the carbohydrate chemist by which to convert exposed hydroxyl groups of a di- or polysaccharide to methyl ethers, which are stable to conditions used to hydrolyze the glycosidic bonds. For example, subjected to these procedures maltose yields equal amounts of 2,3,5,6-tetra-*o*-methylglucose and 2,3,4,6-tetra-*o*-methylglucose. A trisaccharide named melezitose (which is sometimes a problem to honey producers) yields two moles of 2,3,4,6-tetra-*o*- methylglucose and one mole of 1,4,6-tri-*o*-methylfructose.
(a) Draw one possible structure for melezitose.
(b) What are the remaining points of uncertainty in the structure?

CHAPTER 8

Metabolism of Carbohydrates—Glycolysis

We have mentioned in the preceding chapter the wide range of carbohydrate compounds that serve in nature a variety of functions from structural components through reserve energy supplies to the more specialized functions such as the blood-group substances of higher animals. However, from the viewpoint of the cell, everything starts with a limited number of monosaccharides, since these are the only forms that are admitted into the cells. These monosaccharides are then central to the production of whatever di- and polysaccharides the cell needs for storage or structural purposes. In addition, they provide, during the process of oxidation, much of the energy needed by living cells. Further, they are a major source of carbon compounds that can serve as intermediates in the synthesis of amino acids, lipids, nucleotides, and so on. They even serve as critical intermediates in the photosynthetic processes that convert carbon dioxide and water to the various organic compounds of plant cells.

The reactions involved in carbohydrate metabolism, especially the processes that are involved in the complete oxidation of glucose and related compounds, have significance beyond simply serving as a means of generating useful energy for the cell. They also provide a basic framework or pattern of interrelated reactions to which all of the other metabolic processes of the cell can be related. If we are to build a picture of the body of knowledge that is cellular metabolism, it is well to start with the skeleton of carbohydrate metabolism.

Because we will in many cases be considering for the first time certain types of biochemical reactions that will recur throughout metabolism as a whole, we will examine a number of these reactions in detail in the context of carbohydrate metabolism. This will permit us to concentrate primarily on novel reactions when looking at other areas of metabolic activities.

8.1 Absorption of Sugars

The structure of cell membranes and the passage of materials, the sugars included, through these membranes are matters of great complexity and of intense study at the present time. We will not deal with the matter at this time except to make a few very general observations.

For one thing, the nature of a cell membrane is by no means continuous. That is, the structure at any one point on the surface is unlike that at most of the rest of the surface. These different regions can display a number of properties, including a remarkable ability to permit or encourage the passage of certain compounds while rejecting others, or to allow passage in only one direction. In fact, we can distinguish three different ways in which materials get across the membrane. The first is by simple diffusion; pentoses, for example, can diffuse into and out of cells at rates appropriate for the normal laws of diffusion processes.

The second process is one sometimes called "facilitated diffusion"; here diffusion takes place at rates in excess of what would be expected on the basis of diffusion based on random motion of the molecules, and so on. One can think of this as possibly being due to the action of certain areas of the membrane that "catalyze" the motion of the compound through the membrane.

The third process is one which allows absorption across a membrane to be accomplished even in the face of an opposing concentration gradient. Obviously, energy must be expended to carry out this operation. The term **active transport** is often used to describe such energy-requiring membrane processes.

In both of these cases, there is evidence that the absorption is mediated by specific protein units in the membrane which serve functions not unlike those of enzymes. Because of this, these absorption proteins are often referred to as **permeases**. While the picture of monosaccharide absorption is by no means clear, it would appear that a separate "permease" is required for each one that is entering by a mechanism other than simple diffusion.

In this regard, it is worth noting that a major function of the hormone **insulin** is the facilitation of glucose absorption into cells. Although the mechanisms involved are not known, it may serve as a necessary part of the specific "glucose permease" areas of cell membranes. This would be consistent with the finding that the cellular absorption of fructose is not affected by insulin.

8.2 Phosphorylation of Sugars

As soon as the individual hexoses (and some of their derivatives) reach the interior of the cell, they encounter enzyme systems which catalyze the formation of phosphate esters. As a result, free sugars do not exist in living cells to any appreciable extent.

The formation of a phosphate ester of a sugar such as glucose cannot be accomplished by the simple reaction of the sugar with phosphate ion, because the equilibrium is exceedingly unfavorable. The synthesis of the sugar phosphates in biological systems actually requires the donation of a phosphate group by ATP. The reaction involves the cleavage of a phosphate anhydride with the formation of a phosphate ester of an alcohol group of the sugar:

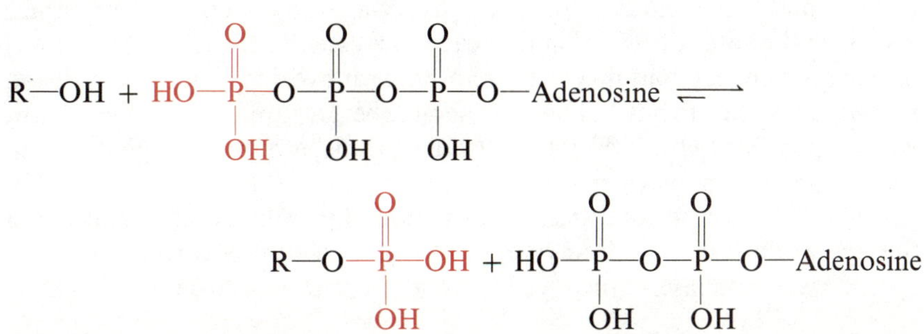

The equilibrium for this reaction lies far toward the formation of the ester. In effect, the chemical energy of ATP is utilized to overcome the unfavorable equilibrium for the direct reaction of the alcohol and the phosphate ion. Note that this is accomplished by making possible a different route to the ester. This is the first of many examples we will encounter in which the energy of ATP is utilized in the formation of compounds otherwise difficult to synthesize.

Since the reaction involves transfer of a phosphate group, the enzyme catalyzing it would properly be called a **phosphotransferase**. However, very early in the history of enzymology, such enzymes were called **kinases**, and this trivial name has persisted in common use. The most common enzyme involved, present both in animal cells and in yeasts, is one called **hexokinase**. The equation for the hexokinase reaction with glucose is as follows:

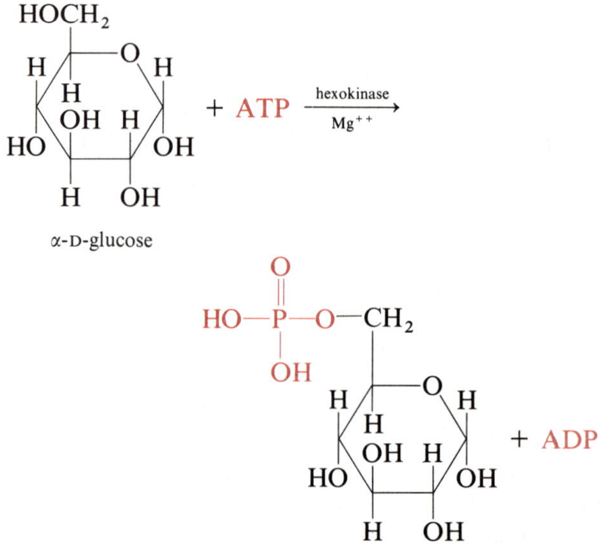

α-D-glucose

α-D-glucose 6-phosphate

$\Delta G^{\circ\prime} = -18$ kJ/mole

Note the requirement for magnesium ion as a cofactor in this reaction. This is the first of many examples we will see of the involvement of this ion in reactions that include phosphorylated compounds. You will recognize also from the value of $\Delta G^{\circ\prime}$ that the reaction is essentially irreversible. This is true of all such reactions in which a phosphate anhydride bond is cleaved and a phosphate ester formed. Thus, we should be prepared for the fact that it is never possible to form ATP from ADP by transfer of the phosphate group from a phosphomonoester.

Hexokinase also catalyzes the formation of mannose 6-phosphate from the appropriate hexose. Phosphorylation of galactose is carried out by a special galactokinase, which yields as a product galactose 1-phosphate.

Similarly, fructose 1-phosphate is the product of action of the enzyme fructokinase. We will deal with the question of interconversion of one hexose into another and the integration of each into carbohydrate metabolism in Chapter 12.

Sugar phosphates cannot be released directly from living cells. Highly ionized substances in general cannot pass through cell membranes. In order for sugars to be returned to the environment, as exemplified by the release of glucose to the blood from liver cells, the phosphate group must be removed. This is accomplished by hydrolysis of the ester linkage in the presence of an enzyme of a type called a **phosphomonoesterase** (or, commonly, just a **phosphatase**). Many cells, including muscle cells of animals, lack glucose 6-phosphatase and therefore cannot liberate glucose. Naturally, the release of glucose to the environment is useless for all cells except those of specialized tissues (such as the liver of an animal) which are responsible for maintaining a supply of glucose to other tissues.

The products of phosphatase action are the alcohol and inorganic phosphate ion, as is shown for the case of glucose 6-phosphatase in the following equation:

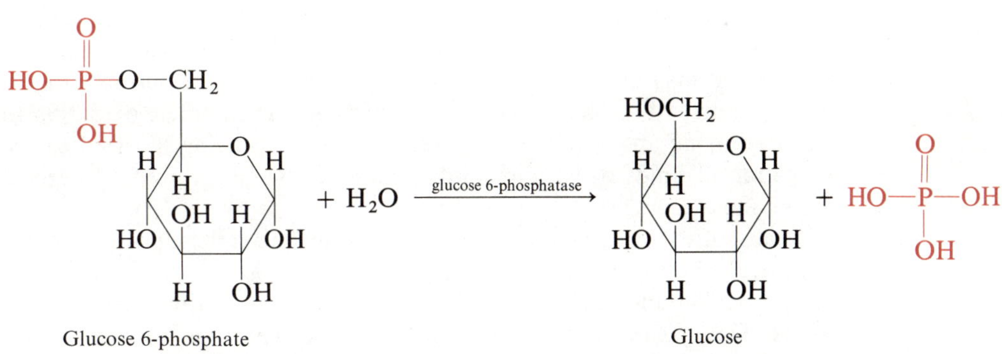

For simplicity, the phosphate groups here are indicated as being un-ionized. Actually, however, you should remember that under neutral conditions these groups will be highly, though not completely, ionized.

8.3 The Many Fates of Glucose 6-Phosphate

Some idea of the complex system that is carbohydrate metabolism, and a beginning look at our strategy in examining it, can be obtained by looking at the competing processes that face glucose 6-phosphate. As outlined in Figure 8-1, a single molecule of glucose 6-phosphate may go in the direction of complete oxidation, with the production of cellular energy, or in the

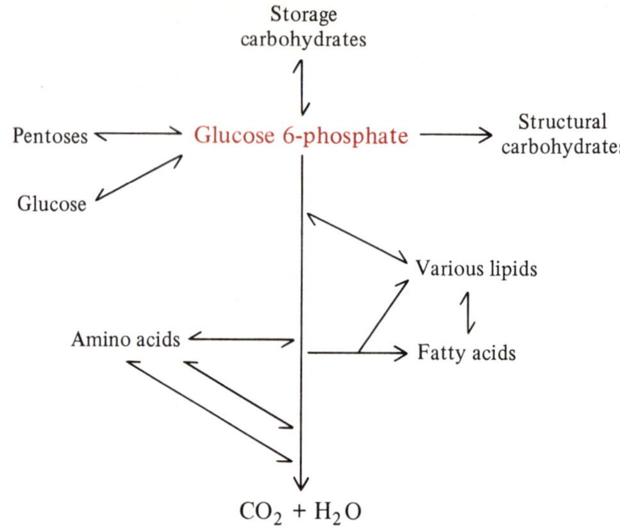

Figure 8-1
Some of the possible metabolic fates of glucose 6-phosphate.

direction of storage within the cell as a reserve polysaccharide. It may alternatively serve as a source of necessary pentose units, structural polysaccharides, or (in the case of liver cells) of extracellular glucose. If it is in a plant cell, it may be deeply involved in the carbon cycle of photosynthesis.

Of course, if it does proceed in the direction of complete oxidation to carbon dioxide and water, it may very well end up being diverted to the formation of lipids of various kinds, amino acids, or a variety of derived compounds.

Naturally, any one molecule can follow only one of these possible paths. In general, some of the glucose 6-phosphate molecules present in the cell at any particular time will proceed down one path, some down each of the others. The quantitative distribution of the molecules among the various metabolic paths will be determined to a major extent by the activities of the various enzymes. These in turn will be governed by the physiological state of the cell.

Under conditions wherein glucose molecules are available to the cell in large quantities and there is little demand for ATP generation for synthetic or physical activities, many of the glucose 6-phosphate molecules will be converted to storage polysaccharide—glycogen in the animal. At the same time a greater percentage of the intermediates of the oxidation of the glucose 6-phosphate will be diverted from further oxidation to the synthesis of lipids. As some of us have reason to know all too well, animals are capable of storing almost unlimited amounts of lipids, while storage of glycogen, except by liver cells and to a lesser extent by muscle cells, is very limited.

On the other hand, when there is a continuing demand for relatively large amounts of ATP for various purposes, the balance of activities shifts so that a greater percentage of the available glucose units undergoes oxidation. At the same time, storage polysaccharides are converted back to glucose 6-phosphate and lipids are transformed in larger quantities into compounds which can enter oxidative pathways.

We will describe each of the metabolic processes involved in the formation and utilization of glucose 6-phosphate in some detail in separate sections. You must bear in mind, however, that the whole metabolic pattern diagrammed in Figure 8-1 represents a single unified system in which each reaction is intimately related to every other reaction, not only those indicated here but every other metabolic process of the cell.

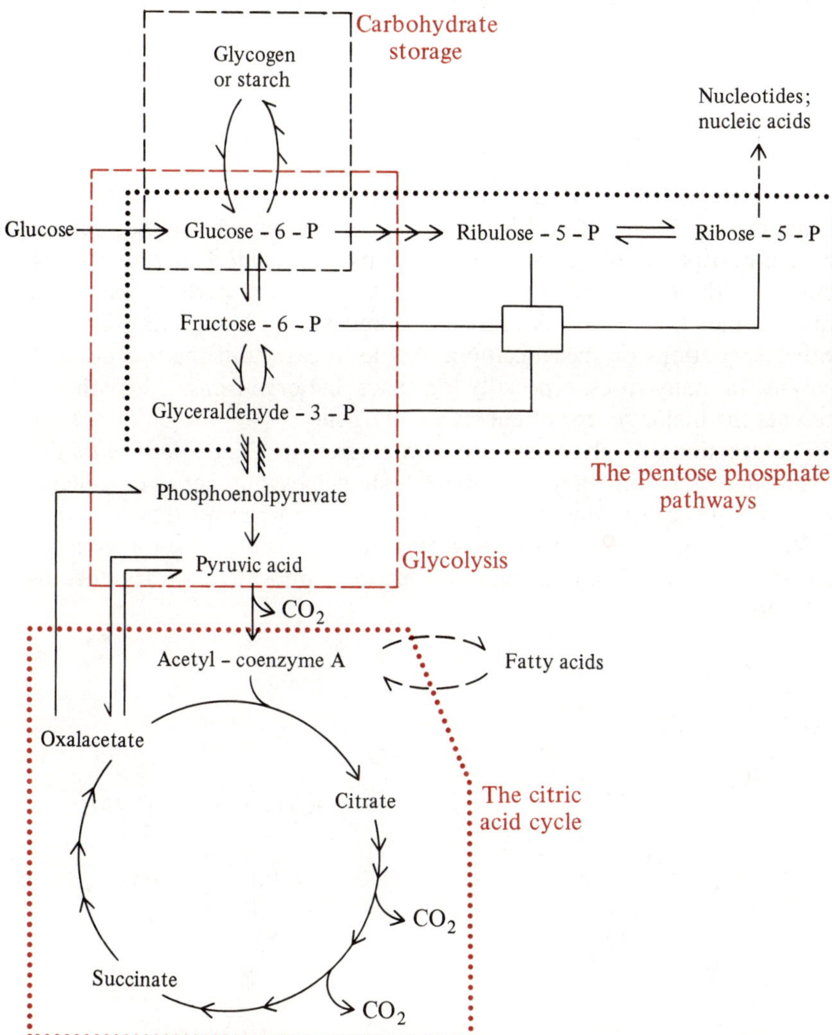

Figure 8-2
A summary of some of the processes of carbohydrate metabolism. The major named pathways are enclosed by broken lines.

The Plan

Figure 8-2 is a diagrammatic representation of some of the processes of carbohydrate metabolism. Because of the number of reactions involved, the complexity of the interactions, and cellular localization and structural organization of some parts of this system, it is not practical to try to consider it all at once. Thus we tend to break it up into a series of metabolic pathways and cycles that accomplish some defined part of the whole. The subdivisions that we will use are enclosed in broken lines in Figure 8-2.

As we complete our study of each of these pathways, however, we will want to stop and remind ourselves of where that particular segment fits in the overall scheme of carbohydrate metabolism.

8.4 Glycolysis

$$\text{Glucose 6-phosphate} \longrightarrow 2 \text{ pyruvic acid}$$

The word **glycolysis** means simply "sugar splitting," and reflects the fact that the overall result of this process is the splitting of one molecule of glucose 6-phosphate into two molecules of pyruvic acid. This sequence of reactions, with some variations we will see shortly, is perhaps the most ubiquitous metabolic pathway in nature. Almost every living cell, whatever its other limitations or specialization, is able to carry out the reactions of glycolysis. In many cases, especially for anaerobic organisms, this pathway represents the major source of energy for the cell.

The reactions of glycolysis are summarized in Figure 8-3, with the common names of the enzymes involved shown by number. The general class to which each enzyme belongs is also shown in parentheses.

We will look at certain significant aspects of each of these reactions. The first step is the isomerization of glucose 6-phosphate to fructose 6-phosphate:

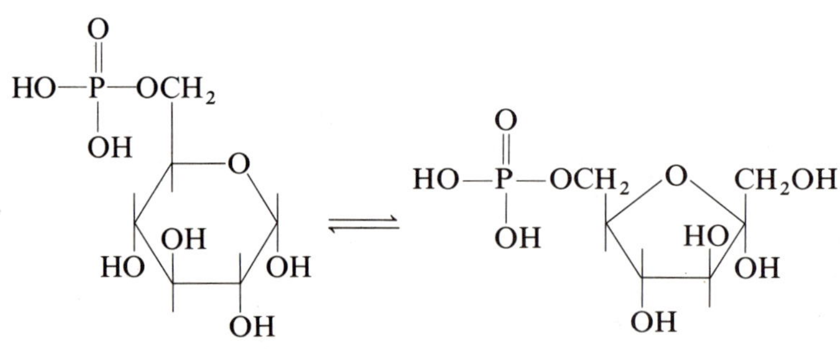

Glucose 6-phosphate

Fructose 6-phosphate

$\Delta G^{\circ\prime} = +1.7$ kJ/mole

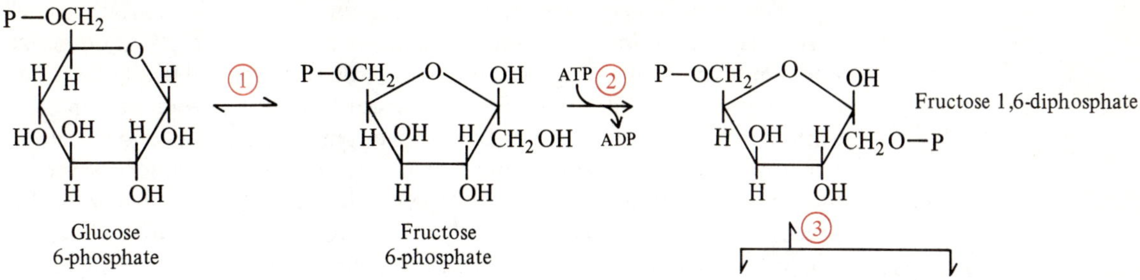

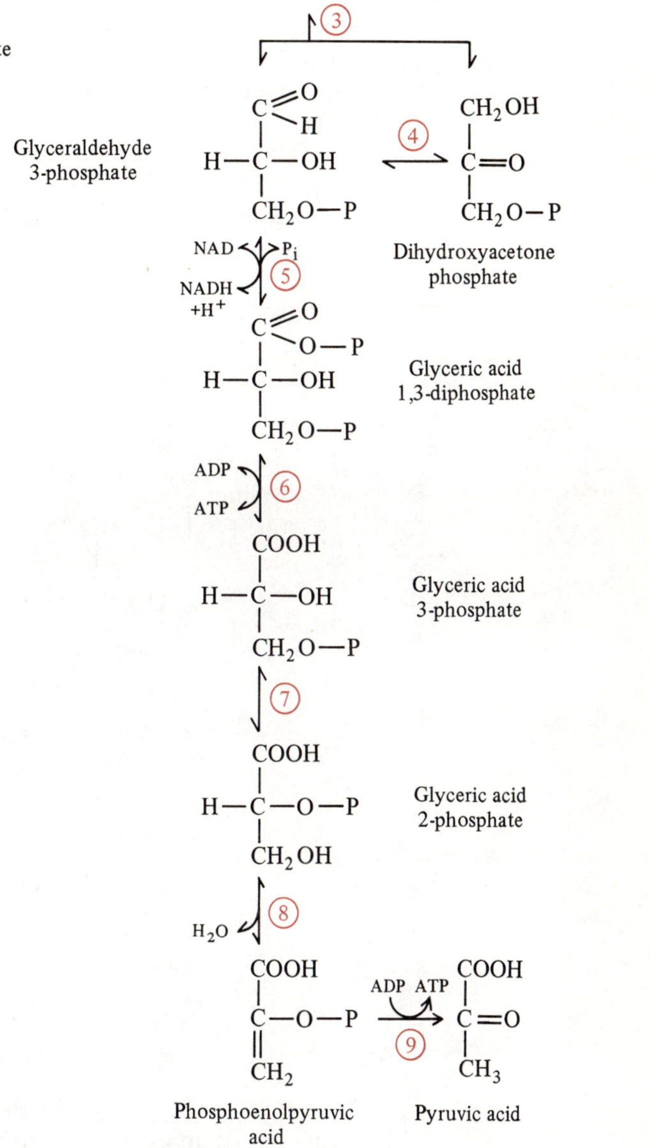

Enzymes

1. Phosphohexose isomerase [Isomerase]
2. Phosphofructokinase [Transferase]
3. Aldolase [Lyase]
4. Triosephosphate isomerase [Isomerase]
5. Glyceraldehyde 3-phosphate dehydrogenase [Oxidoreductase]
6. Phosphoglycerate kinase [Transferase]
7. Phosphoglyceromutase [Isomerase]
8. Enolase [Lyase]
9. Pyruvate kinase [Transferase]

Figure 8-3
The glycolytic pathway from glucose 6-phosphate to pyruvate.

You may remember being told when you first studied carbohydrates that under conditions of basic pH values any free sugar will begin to isomerize into a mixture of three sugars, composed of two aldoses and one ketose. The two aldoses will differ only in configuration at the second carbon. For example, glucose in base yields a mixture of glucose, fructose, and mannose. Fructose predominates in this mixture at equilibrium and mannose is present only in small amounts. Because of the much sweeter taste of fructose, this process is used widely commercially in the production of "corn sweetener" and similar syrups. In making these products, corn starch is hydrolyzed to glucose and then treated with a base, usually lime, to produce the sweeter syrup.

One of the great things about enzymes is that often they can not only do the job faster, but can direct the process and eliminate unwanted reactions. In this case the enzyme specifically reacts only with glucose 6-phosphate and fructose 6-phosphate; no mannose derivatives or other sugars are formed.

The next reaction is another transphosphorylation reaction involving ATP and a kinase:

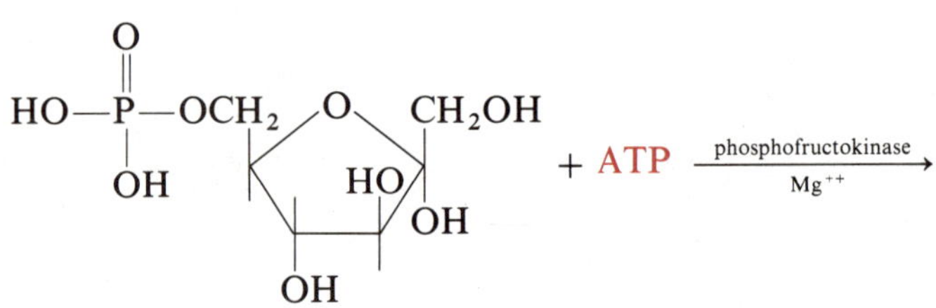

Fructose 6-phosphate

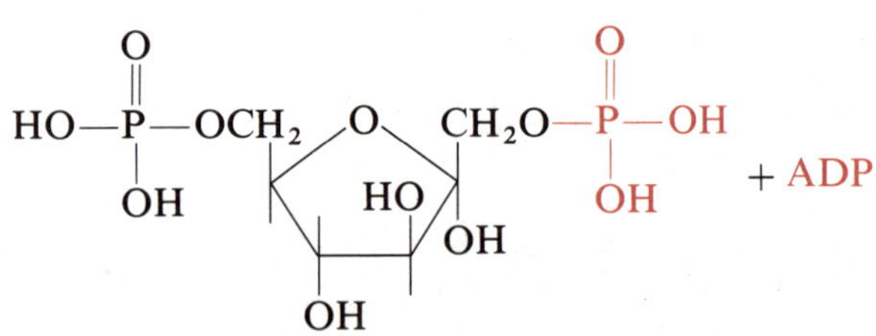

Fructose 1,6-diphosphate

$\Delta G^{\circ\prime} = -14.2$ kJ/mole

Note that this reaction, like the hexokinase reaction, is a typical case of formation of an ester by reaction with an anhydride and is not effectively reversible. To convert the fructose 1,6-diphosphate back to fructose 6-

phosphate requires the action of a very specific phosphatase:

$$\text{fructose 1,6-diphosphate} + H_2O \longrightarrow \text{fructose 6-phosphate} + P_i{}^1$$

In the next step in glycolysis, the fructose 1,6-diphosphate undergoes cleavage into two molecules that are **triosephosphates**. The reaction is the reverse of an aldol condensation, which involves the addition of an activated methylene or methyl group across a carbon–oxygen double bond. Accordingly, the enzyme is called **aldolase**. Properly speaking, the enzyme is fructose diphosphate aldolase, since other aldolases exist; however, by custom use of just "aldolase" implies this enzyme:

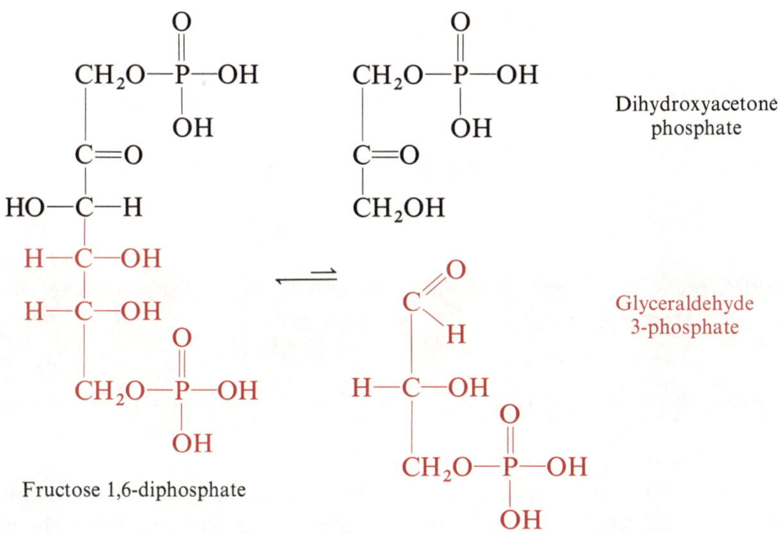

$$\Delta G^{\circ\prime} = +24 \text{ kJ/mole}$$

Aldol Condensation

Because we will be seeing a number of aldol condensations as we go along, it is worth taking time out to recall the basic organic chemistry involved. You will remember that a carbonyl group (or for that matter any electron-attracting group) will cause electrons

1. Because of the difficulty in representing a molecule of phosphoric acid in its proper ionization state at neutral pH, we will use the abbreviation P_i to represent "inorganic phosphate" and the symbol P–P_i to represent "inorganic pyrophosphate."

to be withdrawn from the carbon or carbons next to it. This in turn causes hydrogens attached to those carbons to become more acidic. Whether one calls this "electron delocalization," "methylene group activation," or whatever matters little; the result is a polarization of the carbon–hydrogen bond. If the methylene group is doubly "activated," the effect is even more pronounced. A typical example is the well-known malonic ester synthesis:

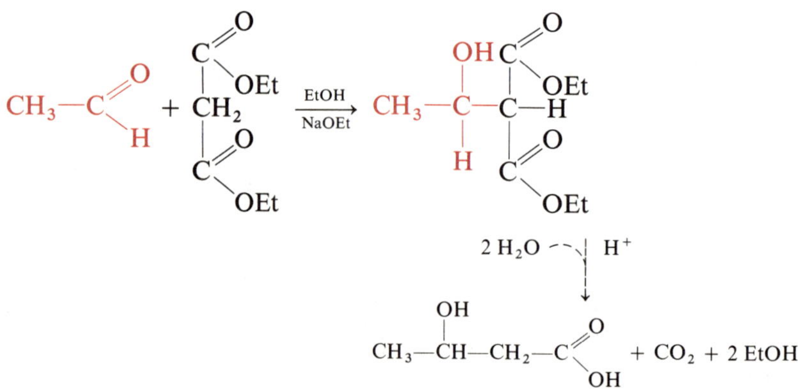

In the proper solvent, it is actually possible to form the sodium salt of the malonic ester

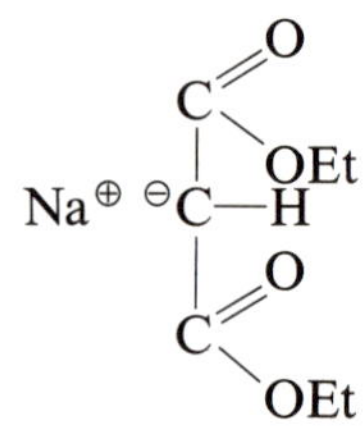

The two triosephosphates that result from the aldolase reaction are isomeric, and can be interconverted by a reaction very similar to the isomerization of glucose 6-phosphate and fructose 6-phosphate:

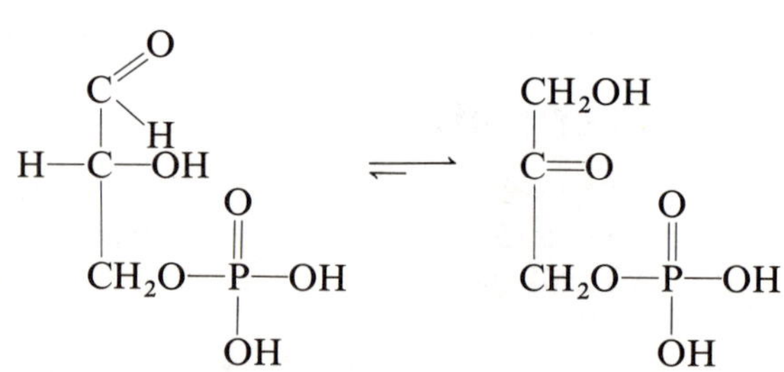

Glyceraldehyde 3-phosphate — Dihydroxyacetone phosphate

$\Delta G^{\circ\prime} = -7.7$ kJ/mole

As you will have noted from the arrows we have drawn, the equilibrium lies substantially to the left in the aldolase reaction and slightly in the direction of dihydroxyacetone phosphate in the isomerization reaction. The formation of glyceraldehyde 3-phosphate would seem to be a poor bet. Yet it is just this compound that we want to follow the rest of the way to pyruvate. We will see shortly that this is simply an example of what we were saying in Chapter 6, that an unfavorable equilibrium can be overcome by having one of the products take part in a reaction with a very favorable equilibrium.

Let's take a moment to take stock of the situation. Starting with glucose 6-phosphate, we have managed to rearrange the molecule and split it into two three-carbon pieces. Because these pieces are interchangeable, we can say that we have converted one glucose 6-phosphate into two glyceraldehyde 3-phosphates. We must bear in mind, of course, that starting with the glucose coming into the cell we have invested 2 ATPs worth of energy. Eventually, the cell will have to get this back and then some in order to stay alive.

The Oxidation Step

The next reaction is the first oxidation–reduction reaction in this pathway of carbohydrate metabolism. It is also a rather unusual one, in that several things happen to the substrate molecule, although only a single enzyme, **glyceraldehyde 3-phosphate dehydrogenase**, is involved. The aldehyde group is not only oxidized to the level of the acid, but also converted to a phosphate anhydride, using inorganic phosphate. In effect, part of the energy released by the oxidation of the aldehyde is used to reduce the NAD and part to form the anhydride bond:

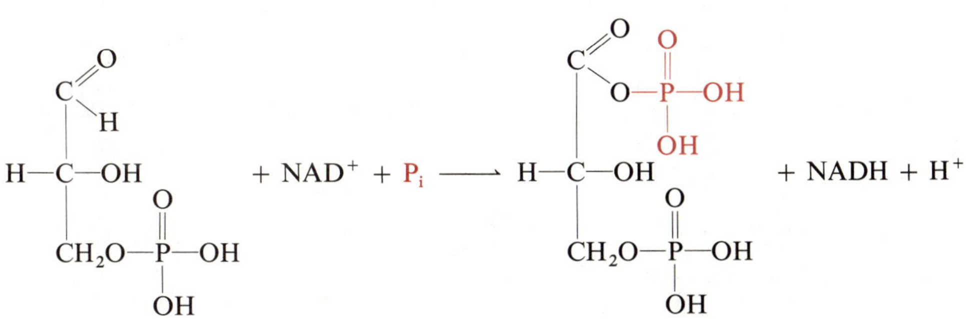

Glyceraldehyde 3-phosphate

Glyceric acid 1,3-diphosphate

$$\Delta G^{\circ\prime} = +6.3 \text{ kJ/mole}$$

As an aid in understanding complex enzymatic reactions in general, and as a means of seeing how the energy of oxidation can be coupled both to reduction and to bond formation, it is worthwhile to look a little more closely at this reaction process. Remember that there is a single enzyme involved, and that it contains within its active site a cysteine side-chain with a free –SH group and a strongly bound molecule of NAD. There are four distinct stages in the overall reaction:

1. Formation of a thiohemiacetal by reaction of the aldehyde with the sulfhydryl group of the enzyme. This represents one of the few known cases where the first enzyme–substrate complex formed is held together by a covalent bond:

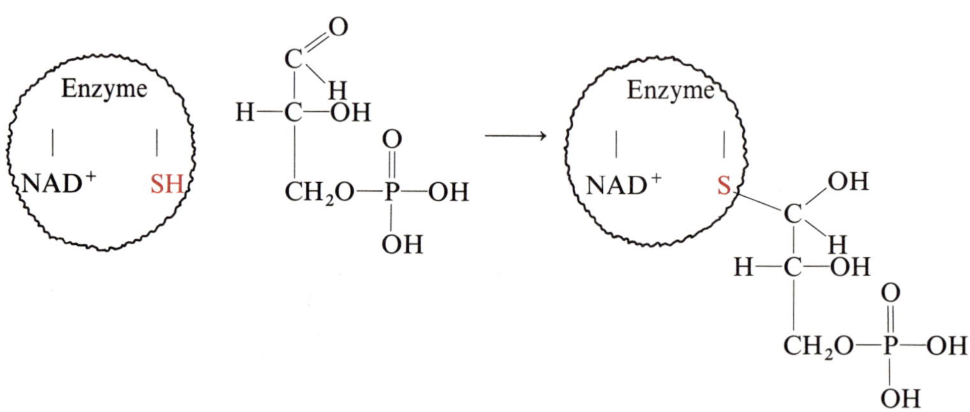

2. Oxidation of the thiohemiacetal to the thioester at the expense of reduction of the tightly bound NAD^+:

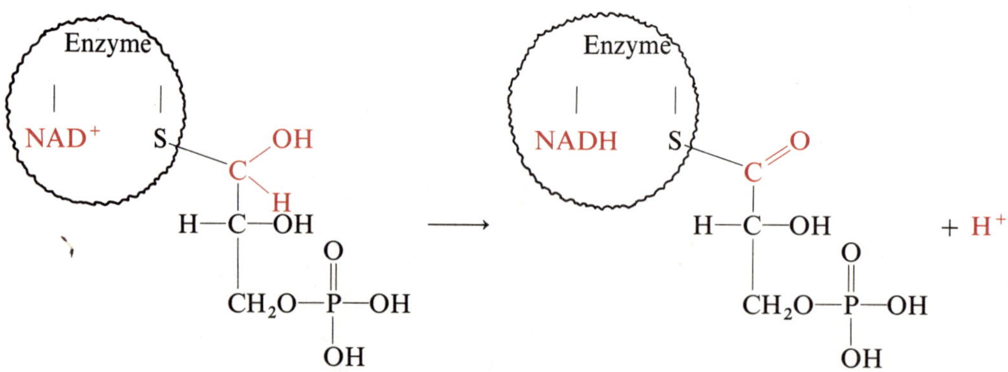

3. Phosphorolysis of the thioester, yielding the phosphate anhydride and releasing the free –SH group of the enzyme:

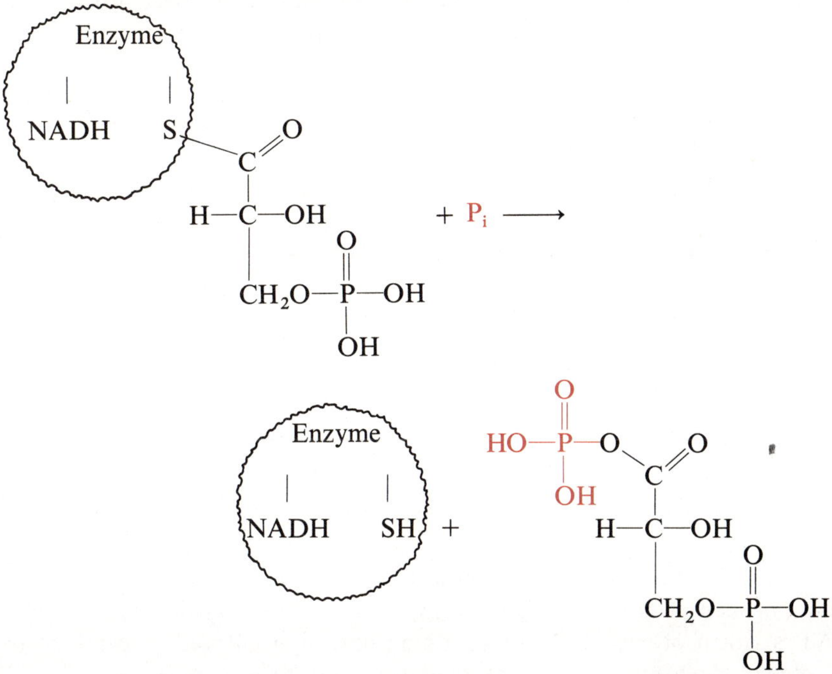

Glyceric acid
1,3-diphosphate

4. Reaction of the bound NADH with external NAD^+, resulting in regeneration of the original enzyme:

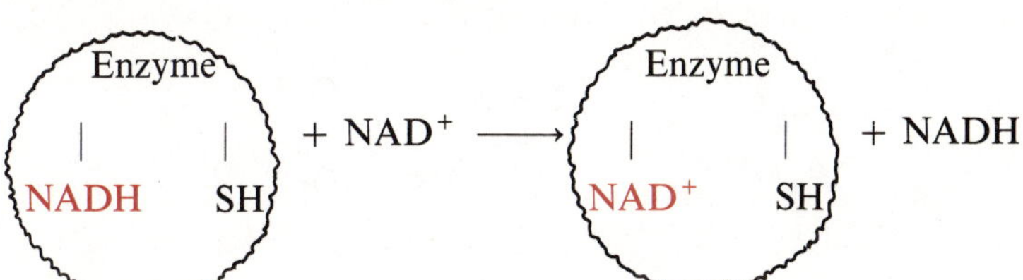

For the moment, we will just leave the reduced NAD lying there. Eventually, of course, we must be able to re-oxidize it in order to keep the system going, but we will look at that problem a little later.

The major product of the oxidation step, **glyceric acid 1,3-diphosphate**, has a phosphate ester on the 3-position, but an anhydride on the 1-position. And one of the things that anhydrides do with some ease is to participate in exchange reactions. In this case, the phosphate from the glyceric acid anhydride is transferred to ADP to form ATP. Reactions of this type are sometimes called **substrate-level phosphorylations** of ADP to distinguish them from the reactions by which ATP is formed during the electron transport processes of respiratory oxidation and of photosynthesis:

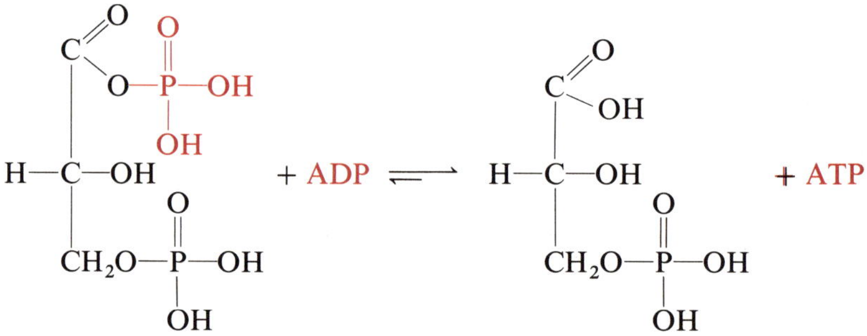

Glyceric acid 1,3-diphosphate

Glyceric acid 3-phosphate

$\Delta G^{\circ\prime} = -18.8$ kJ/mole

Another time out is necessary at this point to mention that the cell has now broken even in terms of energy. Starting with glucose, it invested 2 ATPs worth of energy. Now, assuming both molecules of glyceraldehyde 3-phosphate went through the oxidation, we have 2 molecules of ATP reformed. From here on anything the cell makes is profit.

Phosphate groups have a habit of migrating to next-door hydroxyl groups, especially in acid conditions. In this case, the enzyme **phosphoglyceromutase** does the job and moves the phosphate from the end to the 2-position of the glyceric acid:

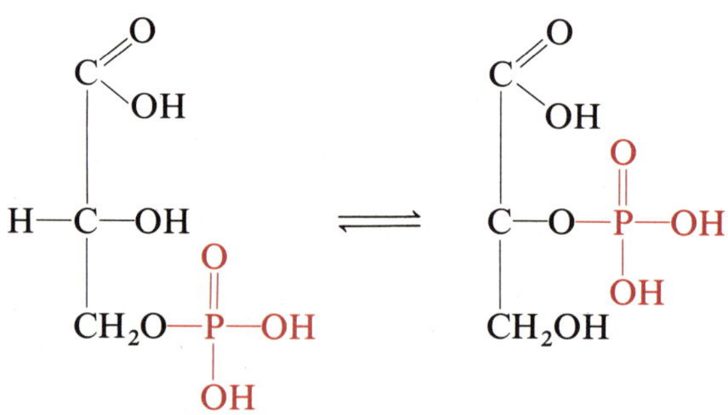

Glyceric acid 3-phosphate

Glyceric acid 2-phosphate

$\Delta G^{\circ\prime} = +4.4$ kJ/mole

This sets the stage for a simple dehydration reaction and a most unusual product. Dehydration reactions are exceedingly difficult in the organic laboratory, but present no problems for enzymes. We will see a number of them, usually fully reversible and often very specific with regard to structural or optical isomerism of the products. The product of this particular reaction is unusual in that it has the phosphate group attached to a hydroxyl group that is actually part of the *enol* form of a keto acid:

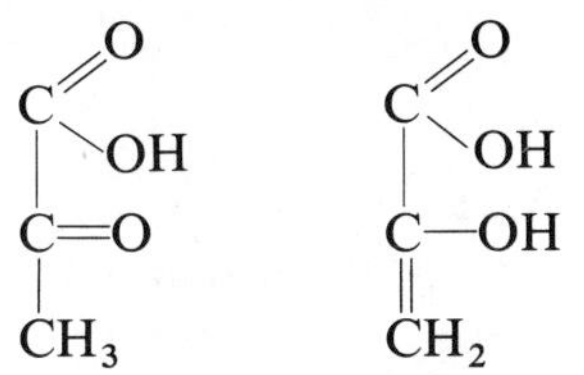

Pyruvic acid *enol*-Pyruvic acid

The reaction at hand converts glyceric acid 2-phosphate to phosphoenol-pyruvate.

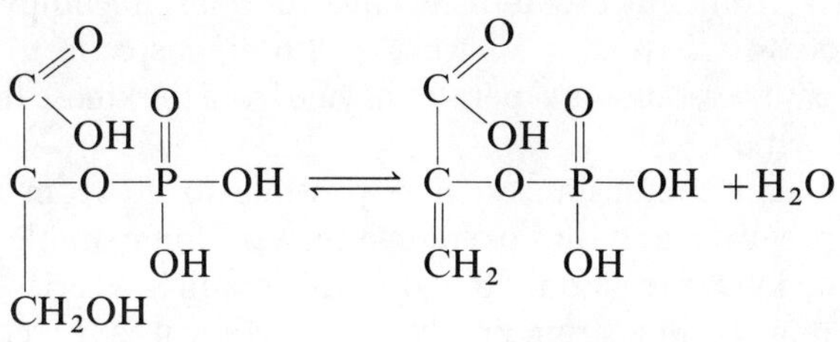

Glyceric acid 2-phosphate Phosphoenolpyruvate

$\Delta G^{\circ\prime} = +1.7$ kJ/mole

You will recall that there is always an equilibrium mixture of *keto* and *enol* forms of any ketone in solution; but usually the equilibrium lies far on the side of the *keto* form. This case is no exception. One can say that the concentration of the *enol* form at equilibrium is negligible, that the energy of the *enol* form is much higher than the *keto* form, or that the free energy change in going from the *enol* to the *keto* form is a relatively large negative number. All of these really convey the same fact. In addition, enolic hydroxyl groups experience the powerful electron withdrawing influence of the double bond, and consequently are just about as acidic as phenols. Therefore, the *enol*-phosphate resembles a true anhydride and has a relatively high tendency to transfer a phosphate group to ADP. The reaction is as follows:

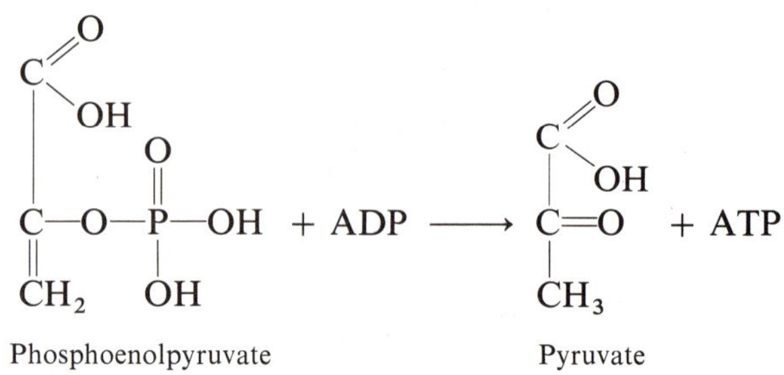

Because of the two factors mentioned above, the standard free energy for this reaction as written is $\Delta G^{\circ\prime} = -27.6$ kJ/mole. Because of this value, this reaction is effectively irreversible, and in fact serves to draw the whole process in the direction of formation of pyruvic acid in spite of the unfavorable equilibria involved in several of the intermediate steps.

Reversibility

Glycolysis from glucose 6-phosphate to phosphoenolpyruvate is a reversible process. Conversion of fructose 1,6-diphosphate to fructose 6-phosphate of course requires a separate enzyme from the kinase that catalyzes the forward reaction.

Under certain circumstances, it is possible to go from pyruvate to phosphoenolpyruvate, and thus overcome the very unfavorable equilibrium of the pyruvate kinase reaction. However, this requires a series of reactions that have no immediate bearing on glycolysis. We will save a consideration of this pathway until a later chapter.

8.5 Regeneration of NAD^+

We have mentioned above that reduced NAD is a product of the oxidation of glyceraldehyde 3-phosphate. Because the total amount of NAD in a cell is very small, this must somehow be re-oxidized or the whole process will rather rapidly grind to a halt. Also, the NAD, whether in oxidized or reduced form, is generally unable to move across cell membranes, so that even in multicellular organisms each cell must be able to carry out this re-oxidation process.

In the presence of an adequate supply of oxygen, the NADH may be re-oxidized by an elaborate process that results in the release of substantial amounts of energy in the form of ATP (about 3 ATPs for each NADH re-oxidized). We will examine this system in some detail in Chapter 10.

Anaerobic Re-oxidation of NADH

Many organisms spend their whole lives under conditions in which the supply of oxygen is nonexistent. In addition, many other cells face occasional conditions of limited oxygen supply and must be capable of sustaining life temporarily in its absence. The latter situation is commonly true of muscle cells under conditions of sudden strenuous activity.

While some special bacteria have developed their own methods of getting around this problem, there are two solutions common to a wide variety of cells.

Lactic Acid Fermentation

The first solution involves the simple reduction of the pyruvic acid to lactic acid, with the associated oxidation of NADH to NAD^+:

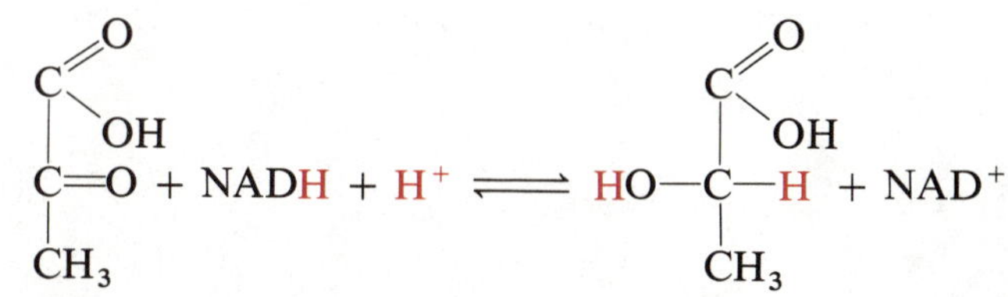

Pyruvic acid Lactic acid

The muscle cells of animals use this method, as do the lactic acid bacteria (used in making yogurt) and some other bacteria. One problem, of course, is the production of an end-product that is quite acidic. The animal cells solve this problem by sending the lactic acid into the bloodstream and then to the liver, where the oxygen supply is better and the lactic acid can be oxidized back to pyruvate and then on to CO_2 and water. The lactic acid bacteria simply perish when their own waste products reach too high a level, although they are much more resistant to acid conditions than most other microorganisms.

Alcoholic Fermentation

Most yeasts and many other microorganisms have solved the problem of NADH re-oxidation another way. They start by removing the carboxyl group of the pyruvic acid as CO_2, a process we call by the general term of decarboxylation. The coenzyme necessary for this type of reaction

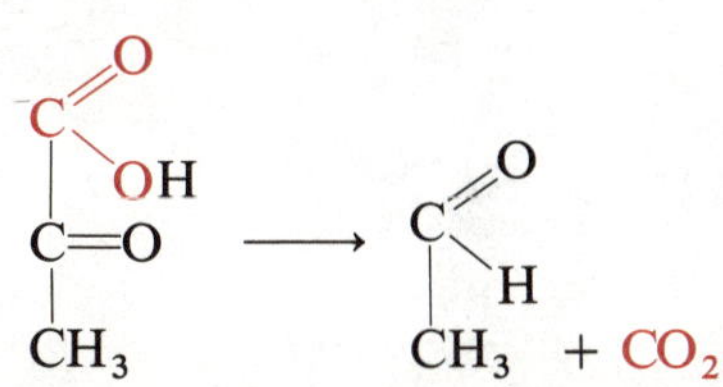

Pyruvic acid Acetaldehyde

is *thiamin pyrophosphate* (*TPP*), a compound produced from the vitamin thiamin. The major need for thiamin as a vitamin arises from the need to be able to make TPP as a coenzyme for a number of decarboxylation reactions involving α-keto acids.

Thiamin Pyrophosphate

Before its structure was clearly established, thiamine pyrophosphate (TPP) was called "codecarboxylase" in recognition of its involvement in every known case in which decarboxylation of an α-keto acid is catalyzed by an enzyme. A sequence of steps that may represent a typical mode of action of this coenzyme is shown below:

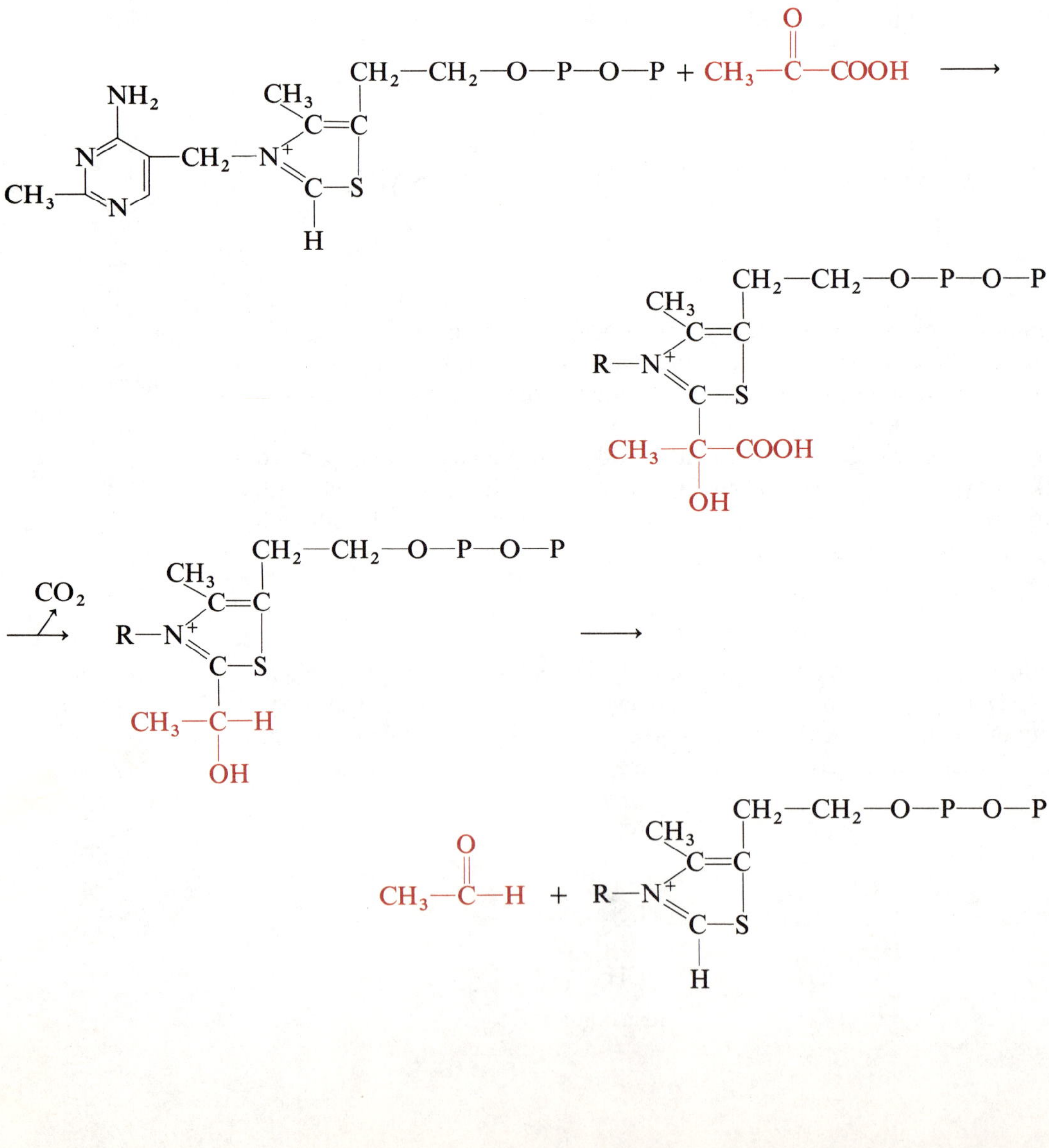

The addition of thiamin pyrophosphate to pyruvic acid is just a fancy example of addition of a polar compound to a carbonyl group. Compare the addition of TPP to pyruvate with the classical addition of HCN to acetaldehyde:

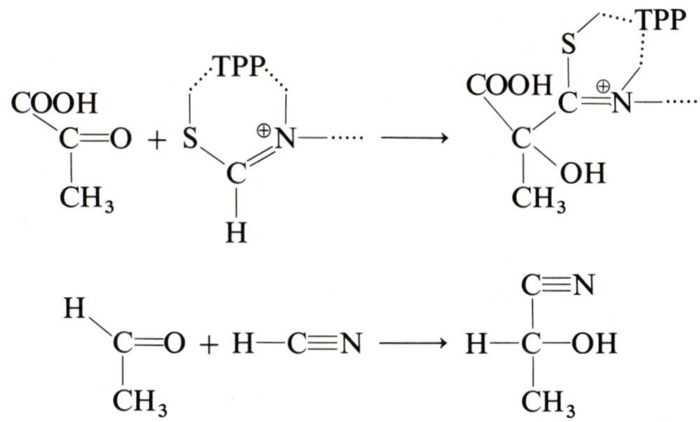

The second step in this approach is reduction of the acetaldehyde to ethanol with re-oxidation of NADH to NAD^+:

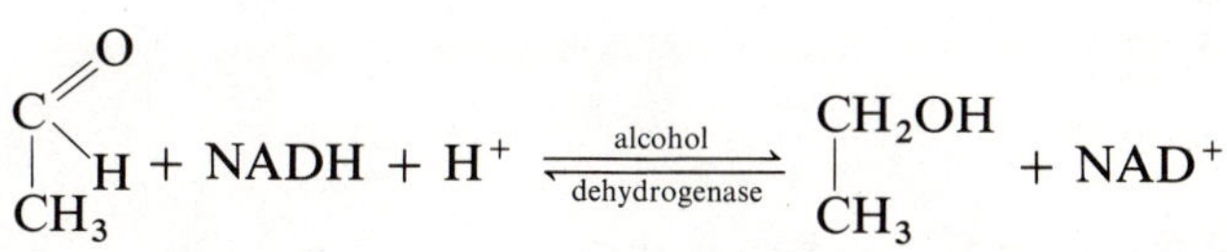

These two processes of anaerobic glycolysis (along with the other mechanisms used by microorganisms to exist in an anaerobic state) are often called **fermentation** processes. Reference to either lactic acid fermentation or alcoholic fermentation is common.

8.6 Energy Yield of Glycolysis

For the reasons mentioned above, we will postpone a consideration of the energy yield of aerobic glycolysis until later. However, we should stop and look at energy yield in the anaerobic process.

Figuring from a glucose newly entered into the cell, we would get

ATPs Used	Change in ATP
(a) glucose ⟶ glucose 6-phosphate	−1
(b) fructose 6-phosphate ⟶ fructose 1,6-diphosphate	−1

ATPs Formed	Change in ATP
(a) 2 glyceric acid 1,3-diphosphate ⟶ 2 glyceric acid 3-phosphate	+2
(b) 2 phosphoenolpyruvate ⟶ 2 pyruvate	+2
Net gain in ATP per mole of glucose	+2

If we consider that complete oxidation of glucose (to CO_2 and H_2O) yields some 2870 kJ of available energy per mole, this gain of 2 ATPs (about 61 kJ) seems a rather small percentage. It is, especially when compared to the yields under aerobic conditions that we will see in the next chapter. But it is enough to meet the energy needs of many cells for maintenance and slow growth.

Additional Readings

S. Dagley and D. E. Nicholson, *An Introduction to Metabolic Pathways* (Wiley, New York, 1970).

Any of the major texts listed at the end of Chapter 1.

B. Axelrod, "Glycolysis," in *Metabolic Pathways*, edited by D. M. Greenberg, 3rd edition, Vol. 1 (Academic Press, New York, 1967).

"Carbohydrate Metabolism," Vol. 17 of *Comprehensive Biochemistry*, edited by M. Florkin and E. Stotz (Elsevier Scientific Publishing Co., 1967).

Study Questions

1. Write out in proper sequence all of the reactions leading from glucose 6-phosphate to pyruvate.

2. Given that R = 8.31 J deg^{-1} mol^{-1} and T = 398°K, what would you expect to be the value of the equilibrium constant (K_{eq}) for each of the following reactions:
   (i) glucose + ATP ⟶ glucose 6-phosphate + ADP
   (ii) fructose 6-phosphate + ATP ⟶ fructose 1-6-diphosphate + ADP
   (iii) phosphoenolpyruvate + ADP ⟶ pyruvate + ATP

3. If glucose labeled with ^{14}C in position 1 is incubated with a cell-free homogenate of *Lactobaccillus casei* under anaerobic conditions, what carbon of the resulting lactic acid would be labeled with ^{14}C?

4. Anhydrides in general release more energy on hydrolysis than do esters. The degree to which a phosphate compound can be considered an

anhydride rather than an ester is a function of the pK_a of the attached group. From the available literature, determine the acid strength (pK_a) of each of the following groups:

(a) $-CH_2-OH$

(b) 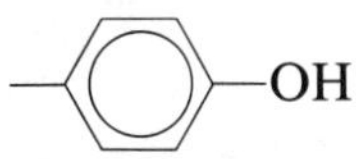

(c) $-CH=CH-OH$

(d) $-CH_2-SH$

(e) $CH_3-C(=O)-OH$

(f) 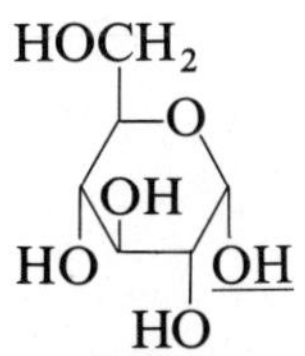

5. Assign each of the enzymes of the glycolytic pathway to one of the major classes of the systematic (IUB) nomenclature for enzymes.

6. When the glycolytic pathway operates anaerobically, several final products are possible.
(a) Name several possible products.
(b) In each case, show all of the reactions of NAD (or NADH) involved in arriving at those products.

7. Outline the individual steps by which glyceraldehyde-3-phosphate dehydrogenase brings about the following reaction:

$$\text{glyceraldehyde 3-phosphate} + NAD^+ + P_i \longrightarrow \text{1,3-diphosphoglycerate} + NADH + H^+$$

8. Given the values for $\Delta G^{\circ\prime}$ for each of the reactions of glycolysis in the text, what is one intermediate that you would expect to find present in very low concentration in cells under steady state conditions (for your purposes, assume an equilibrium situation).

9. Some microorganisms produce glycerol as one of their fermentation products when growing on glucose as sole carbon and energy source.

$$\text{(dihydroxyacetone-P} + \text{NADH} + \text{H}^+ \longrightarrow \text{glycerol-P} + \text{NAD}^+$$

$$\text{glycerol-P} + \text{H}_2\text{O} \longrightarrow \text{glycerol} + \text{P}_i\text{)}$$

Explain why glycerol is not likely to be the sole product of fermentation.

CHAPTER 9

Aerobic Oxidation: The Citric Acid Cycle

Having examined the processes of metabolism from glucose to pyruvic acid, we now wish to look at the processes by which the oxidation of pyruvate to CO_2 and water is accomplished. In eukaryotic cells, this occurs entirely within the subcellular organelles called **mitochondria**. We will have more to say about the structure and function of mitochondria after we have dealt with some of the reactions that occur within them.

In prokaryotic cells, which lack mitochondria, a similar sequence of reactions is catalyzed by enzymes which are firmly attached to the cellular membranes. In each case, the process involves a highly structured and highly organized system of enzymes carrying out a concerted sequence of reactions.

9.1 The Pyruvate Dehydrogenase Complex

The pyruvate dehydrogenase complex of enzymes is responsible for decarboxylation of pyruvate, for oxidation of the "acetaldehyde" portion to the level of acetic acid, and for the formation of a thioester between this "acetic acid" unit and a compound called Coenzyme A. The overall reaction is as follows:

$$\text{pyruvate} + \text{Coenzyme A (–SH)} + NAD^+ \longrightarrow \text{acetyl–S–Coenzyme A} + NADH + H^+ + CO_2$$

This process requires three separate enzymes, but they are so closely associated with one another that they can be isolated together as a structural unit. In addition, as we will see below, the overall reaction either goes completely or not at all; there are no free intermediate compounds formed. The process also involves four different coenzymes derived from B-group vitamins (see p. 52), although Coenzyme A here is really a cosubstrate rather than a coenzyme for the reaction.

Because there are so many new reactions and coenzymes involved, we will look at the pyruvate dehydrogenase system in detail. To this purpose, the overall process can be subdivided into a series of steps that give a better feeling for the mechanisms involved. Since this is such a unified enzyme system, the names of the individual enzymes have little meaning at this point, and we can simply call them Enzyme 1 (E_1), Enzyme 2 (E_2), and Enzyme 3 (E_3).

Step 1. Condensation of Pyruvate with Enzyme 1 and Loss of CO_2

Enzyme 1 is very similar to the enzyme, mentioned briefly in our discussion of anaerobic fermentation, which serves to decarboxylate pyruvate to acetaldehyde. Closely associated with the polypeptide portion of the enzyme is a molecule of thiamin pyrophosphate (TPP):

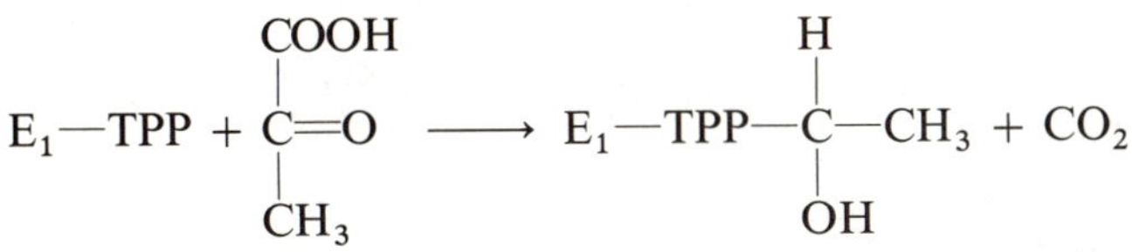

Step 2. Oxidation

One of the side-chain lysine groups of Enzyme 2 has attached to it by an amide bond a molecule of **lipoic acid**. This is an octanoic acid with two sulfur atoms in a disulfide linkage attached to carbons 6 and 8. Reaction of this group with the enzyme-bound product of Reaction 1 results in reduction of the disulfide to two sulfhydryl groups and oxidation of the "hydroxyethyl" group to the oxidation level of acetic acid. In the process, the acetate group becomes attached to one of the sulfhydryl groups as a thioester:

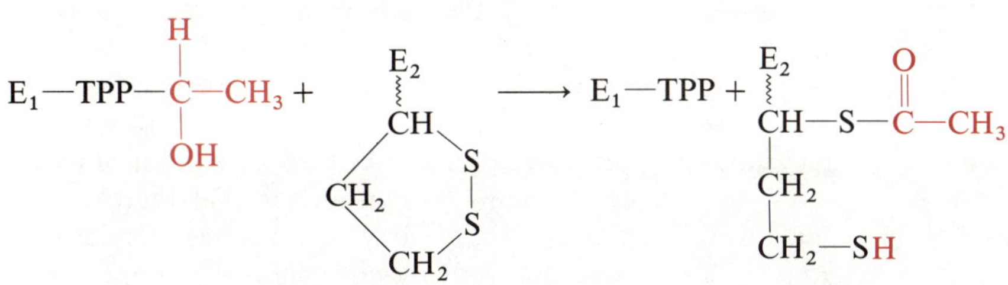

Lipoic Acid

This coenzyme is generally attached covalently to its enzymes through an amide linkage to the ε-amino group of one of the lysine residues in the polypeptide chain.

As with any disulfide compound, lipoic acid is capable of undergoing a reversible reduction to the dithiol compound, dihydrolipoic acid.

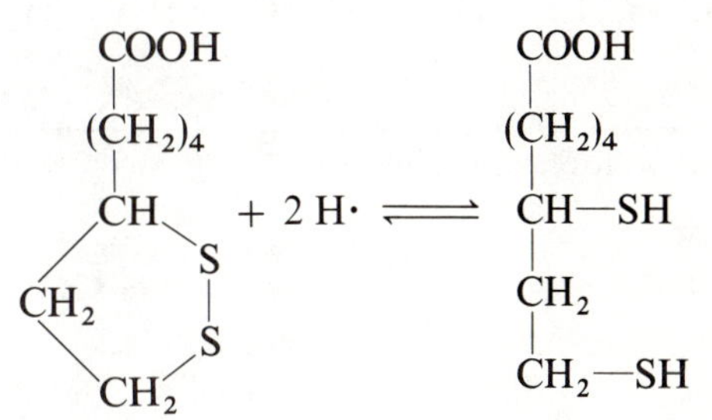

Lipoic acid — Dihydrolipoic acid

Step 3. Transfer of the Acetate Group to Coenzyme A

Step 2 represents only half of the reaction catalyzed by Enzyme 2. In addition to formation of the acetate thioester, this enzyme also catalyzes transfer of the acetate group from the sulfur of the lipoic acid molecule to form a similar thioester with the –SH group of Coenzyme A:

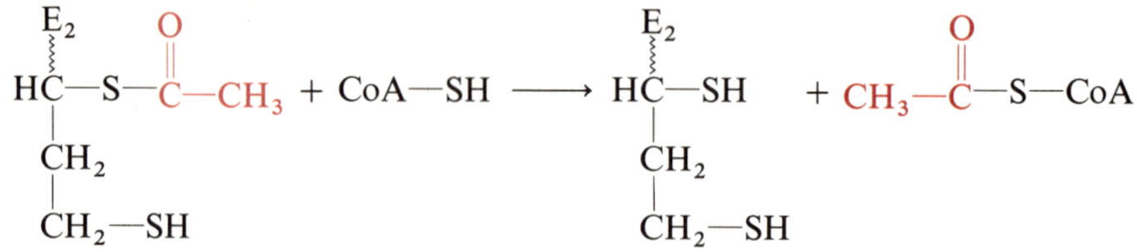

The only step necessary to complete the process and restore all of the enzymes to their original states is to re-oxidize the reduced lipoic acid (dihydrolipoic acid) of Enzyme 2 back to the disulfide form.

Coenzyme A

This is one more example of the problem of early names sticking to compounds. Coenzyme A received its name, before its structure or exact function was known, on the basis of the requirement for this compound in many of the reactions associated with the metabolism of acetate groups. The "A" comes from "Acetate," not an alphabetical listing of coenzymes.

Also, Coenzyme A rarely acts as a true coenzyme. It serves much more commonly as a cosubstrate in a reaction, carrying some substrate molecule through a series of enzymatic steps. This function as a cosubstrate is clear in the case of the pyruvate oxidase sequence of reactions, where the Coenzyme A comes out of the reactions in a new form, with the acetate group attached. However, perhaps we can rationalize the continued use of the name by considering that the Coenzyme A helps carry out a series of related reactions, rather than an individual step:

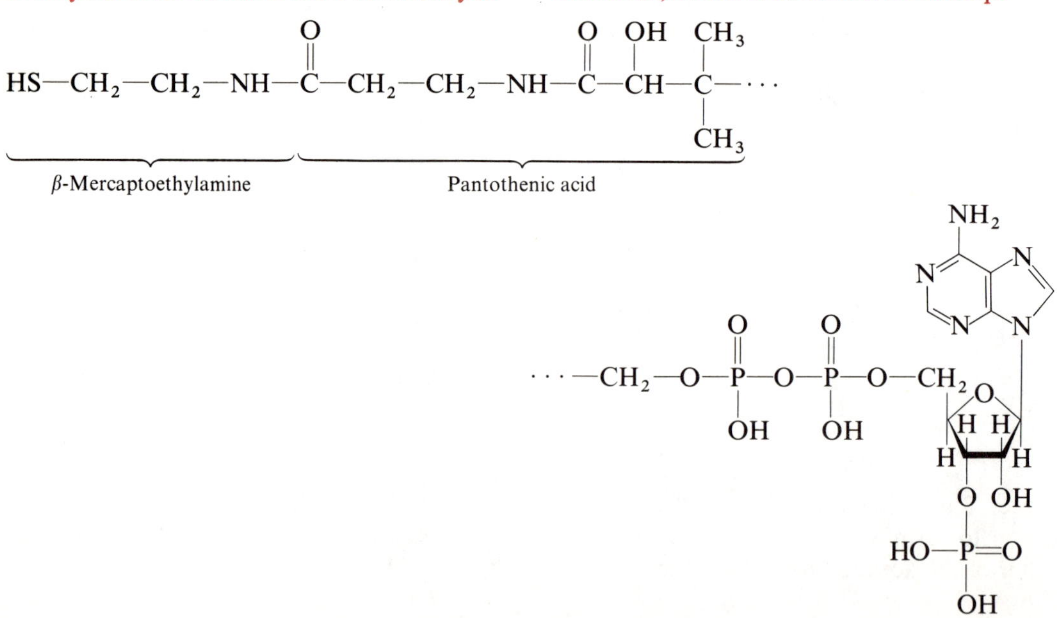

In writing reactions involving Coenzyme A we often show the free –SH group, since this is the part involved in reaction and it is convenient to be able to show that the product is a thioester or similar compound. You should bear in mind that the –SH group shown is part of the Coenzyme A structure.

Step 4. Re-oxidation of Reduced Lipoic Acid

The third enzyme catalyzes the following overall reaction:

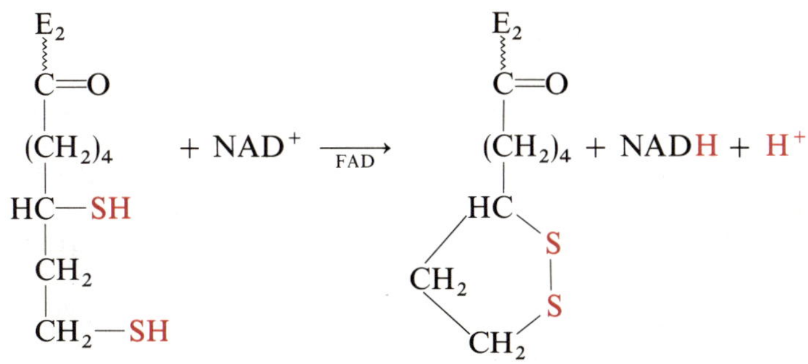

However, direct reaction of the NAD with the thiol compound is neither energetically nor mechanistically an effective process. Consequently, this enzyme employs another oxidation–reduction coenzyme as an intermediate. The compound involved in this case is flavin adenine dinucleotide (FAD) and the mechanism of the reaction involves first reduction of the enzyme-bound FAD by the dihydrolipoyl-enzyme-2. The reduced FAD is immediately re-oxidized by NAD^+, yielding the overall reaction shown above. Instant reaction of the $FADH_2$ with NAD^+ is required in this case since $FADH_2$ is not stable to molecular oxygen as is NADH.

Summary. Conversion of Pyruvate to Acetyl–Coenzyme A

We can then restate the effect of this organized multienzyme as follows:

$$\text{pyruvate} + \text{Coenzyme A} + NAD^+ \longrightarrow \text{acetyl–Coenzyme A} + CO_2 + NADH + H^+$$

$$\Delta G^{\circ\prime} = -33.5 \text{ kJ/mole}$$

Flavin Coenzymes

There are two coenzyme forms derived from the vitamin **riboflavin** which take part in oxidation–reduction reactions. These are often referred to as the flavin coenzymes; enzymes having these groups as coenzymes are referred to as flavoprotein enzymes:

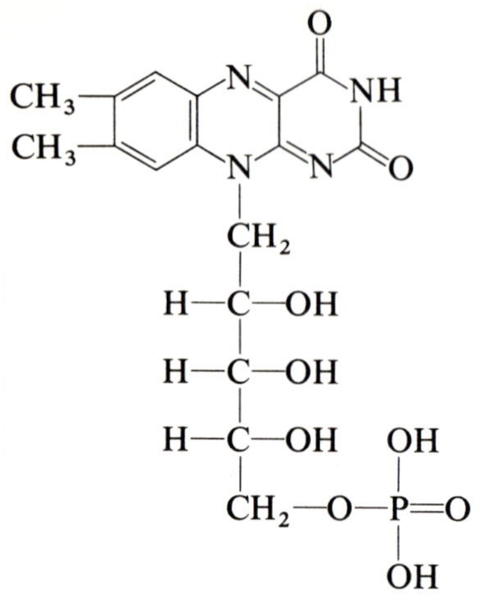

Flavin mononucleotide
(FMN)

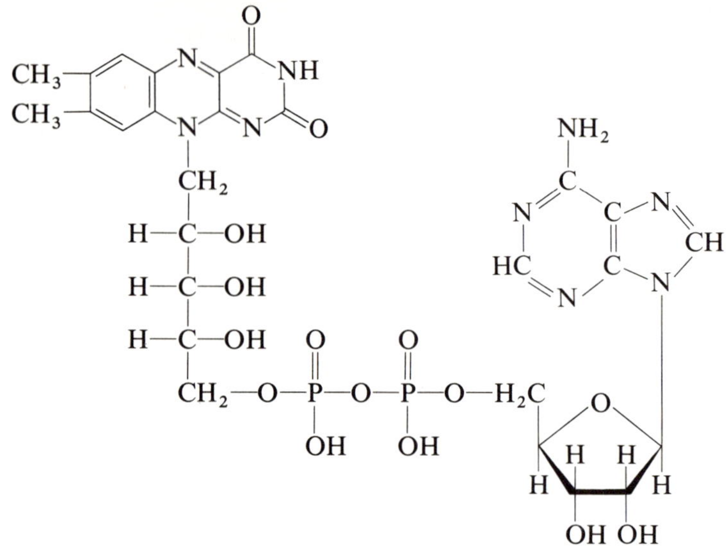

Flavin adenine dinucleotide
(FAD)

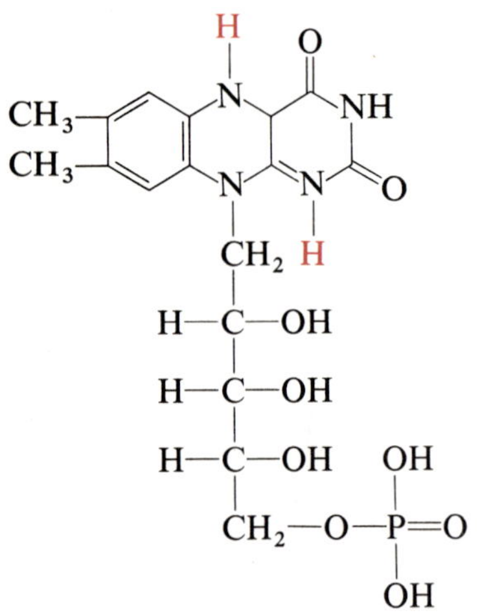

Reaction of either of these involves reduction of the isoalloxazine ring of the riboflavin portion. This is shown to the left for FMN.

The reduced forms, $FMNH_2$ or $FADH_2$, are immediately re-oxidized in the presence of molecular oxygen (unlike NADH, which is stable to oxygen). As a result, we find these coenzymes involved only in reactions where the reduced product can immediately take part in another closely related reaction such as we see in the case of the pyruvate dehydrogenase complex.

9.2 The Many Reactions of Acetyl–Coenzyme A

Just as we took time out to look at the reactions that a single molecule of glucose 6-phosphate might undergo in the cell, we should look at the similar role of acetyl–Coenzyme A as a branching point between pathways of metabolism.

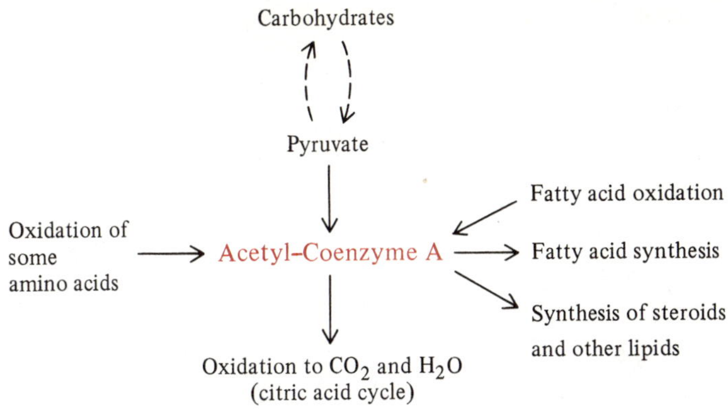

Figure 9-1
A simplified representation of some of the metabolic pathways in which acetyl–Coenzyme A is involved.

The various involvements of acetyl–Coenzyme A are outlined schematically in Figure 9-1. Note that the conversion of pyruvate to acetyl–Coenzyme A, like all other cases of "oxidative decarboxylation," is effectively an irreversible process.

It is important to note at this point that acetyl–Coenzyme A is the major intermediate linking together cabohydrate and lipid metabolism. As noted in the diagram, the energy yielding oxidation of fatty acids (the major constituents of dietary lipids) proceeds by conversion of the fatty acids to acetyl–Coenzyme A. From that point, the process of complete oxidation of the acetate unit to CO_2 and water is common to carbohydrate and lipid metabolism.

We will eventually look at all of the reactions of acetyl–Coenzyme A, but for the present we wish to concentrate on the energy-yielding process of complete oxidation to carbon dioxide and water.

9.3 The Citric Acid Cycle

The citric acid cycle is also known as the **tricarboxylic acid cycle** and, in older literature, the **Krebs cycle** after Sir H. A. Krebs, a British biochemist who contributed a great deal to our early understanding of metabolism.

This process is a true cycle in that we start with one reactant, take it through a series of reactions, and at the end regenerate the original starting material. The beginning compound in this case is oxaloacetic acid (which any student of organic chemistry recognizes should be 2-oxosuccinic acid, or at worst 2-ketosuccinic acid, but again we are done in by history and some inertia).

The reactions of the cycle involve basically the attachment of an acetate unit from acetyl–Coenzyme A onto the oxaloacetate, followed by oxidation and decarboxylation steps which result in loss of two carbons as CO_2. In the process the oxaloacetate is regenerated.

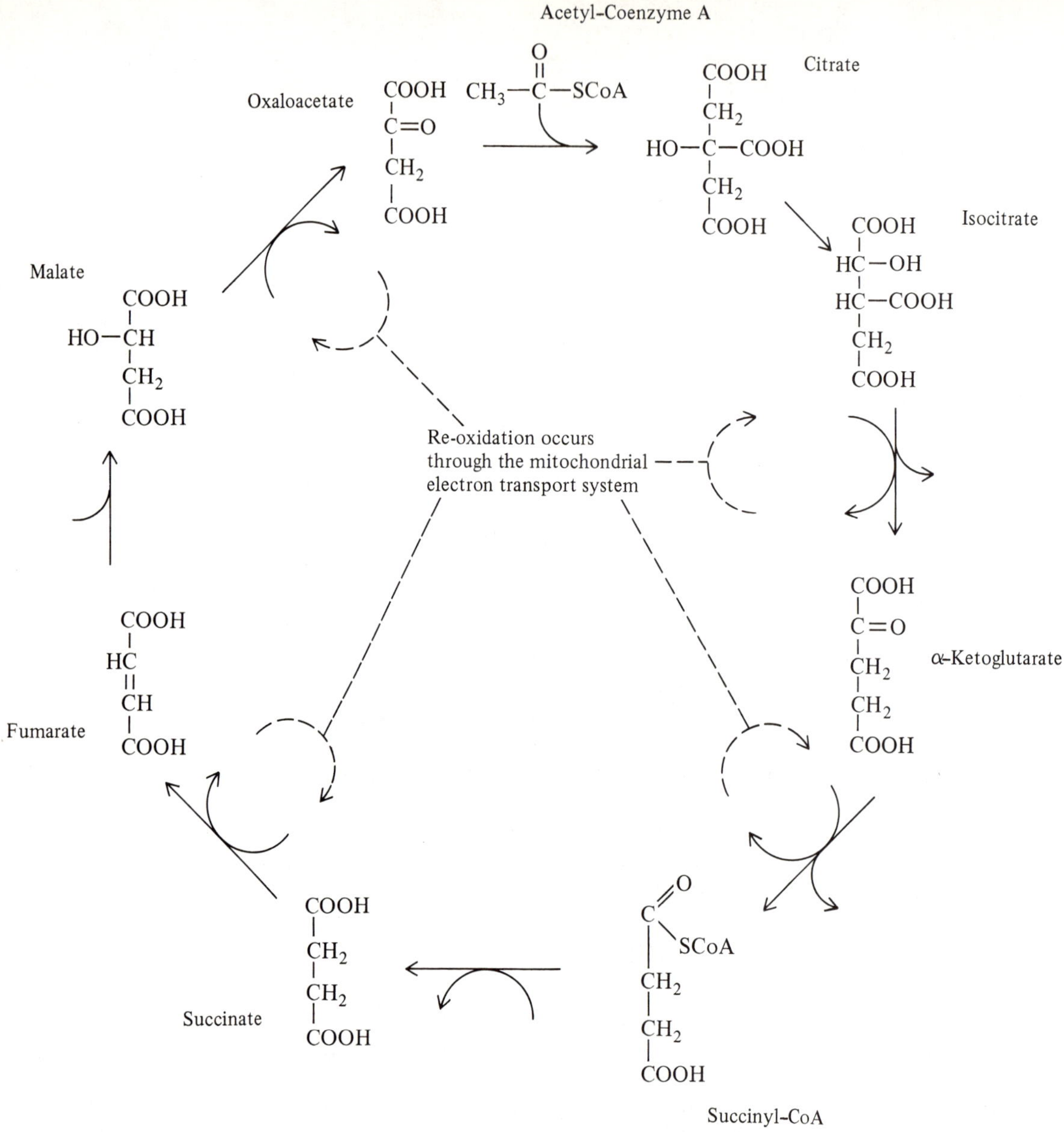

Enzymes

1. Citrate synthetase
2. Aconitase
3. Isocitrate dehydrogenase
4. α-Ketoglutarate dehydrogenase complex
5. Succinyl-CoA synthetase
6. Succinate dehydrogenase
7. Fumarate hydratase (fumarase)
8. Malate dehydrogenase

Figure 9-2
The reactions of the citric acid cycle.

The reactions of the cycle are shown in a very simplified form in Figure 9-2 in order to give you some feeling for the flow of carbon atoms involved.

Condensation to Form Citric Acid

Condensation of oxaloacetate and acetyl–Coenzyme A to form citrate is just another example of addition across a carbonyl double bond. In this case the group adding is the methyl of the acetate group, polarized by the existence of the activating thioester. The Coenzyme A molecule is lost by hydrolysis before the products are released from the enzyme surface:

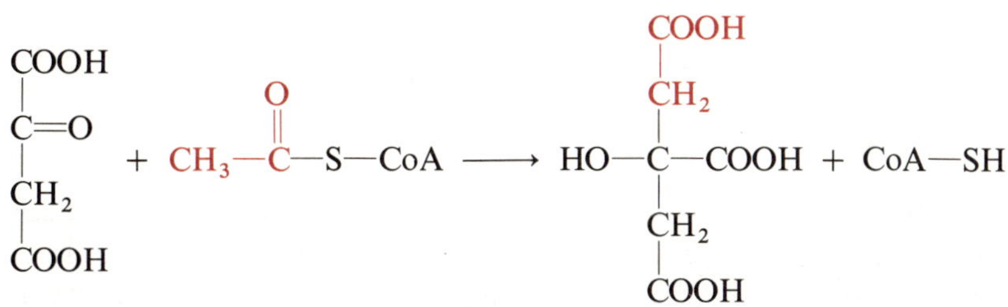

Oxaloacetic acid Acetyl–Coenzyme A Citric acid

$$\Delta G^{\circ\prime} = -32.2 \text{ kJ/mole}$$

Formation of Isocitrate

We commented in Chapter 4 on the ability of enzymes to distinguish between three-dimensional conformations of molecules, even when no steric or optical isomerism is apparent in the molecule. The first reaction of citric acid illustrates this very well. Although the citrate molecule appears to be symmetric, the nature of its attachment to the enzyme is not. Consequently, transfer of the hydroxyl group occurs only to the carbon that originated from oxaloacetate, rather than the carbon from acetate.

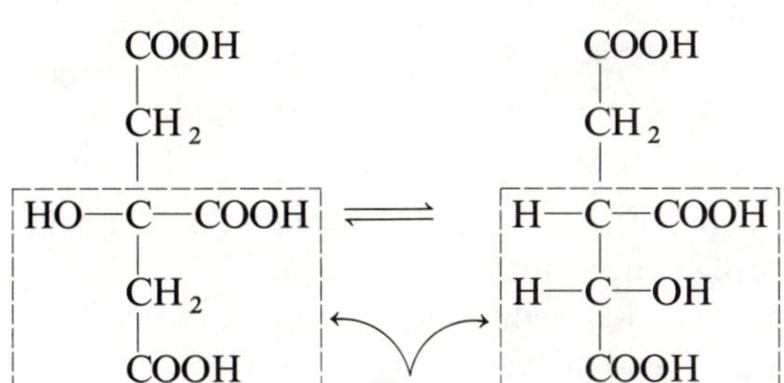

Citric acid From oxaloacetate Isocitric acid

$$\Delta G^{\circ\prime} = +6.3 \text{ kJ/mole}$$

The enzyme involved is called "aconitase" because some of the unsaturated acid, *cis*-aconitic acid, is formed during the reaction:

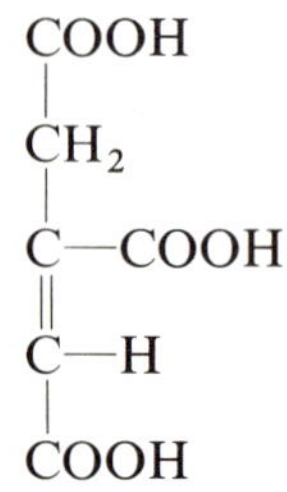

cis-Aconitic acid

Oxidation of Isocitrate and α-Ketoglutarate

Oxidation of the isocitrate is carried out using NAD as electron acceptor. The immediate product of this reaction is the appropriate keto form of the tricarboxylic acid. However, this is in part a β-keto acid and shares the tendency of all acids having a carbonyl or ester group in the β-position to undergo decarboxylation. This step requires the presence of manganous ion as a cofactor for the enzyme. Thus, we can write the reaction

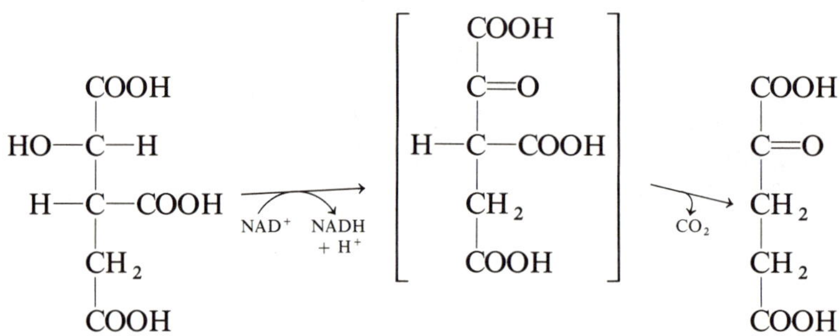

Isocitric acid   α-Ketoglutaric acid

$$\Delta G^{\circ\prime} = -7.1 \text{ kJ/mole}$$

with the square brackets enclosing a compound that under normal circumstances is only a transitory intermediate.

Viewed one way, α-ketoglutarate is simply a molecule of pyruvate with a molecule of acetic acid added. It might come as no surprise, then, to find that it undergoes a process exactly the same as that carried out by the pyruvate dehydrogenase complex, but in this case catalyzed by an "α-ketoglutarate dehydrogenase complex." Since we have already looked at the mechanisms involved, we can simply deal here with the overall results:

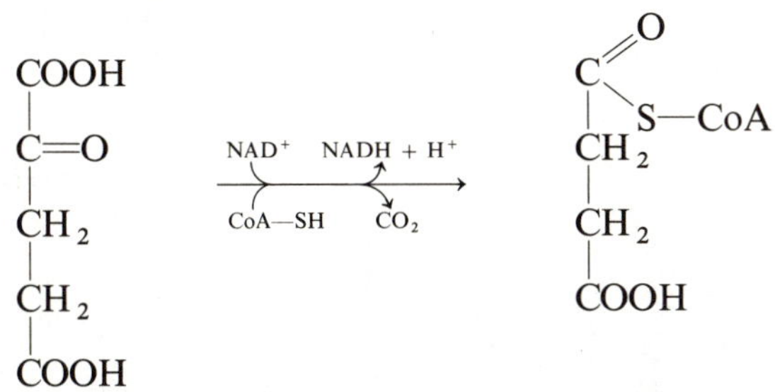

α-Ketoglutaric acid   Succinyl–Coenzyme A

$$\Delta G^{\circ\prime} = -33.5 \text{ kJ/mole}$$

Once again we end up with a product that is a thioester of a carboxylic acid. This time, part of the energy of this bond is used for the formation of either ATP or GTP (depending on the organism). The reaction is carried out by an enzyme which catalyzes the following process:

$$\text{succinyl–Coenzyme A} + \text{ADP} + \text{P}_i \longrightarrow \text{succinate} + \text{CoA} + \text{ATP}$$

$$\Delta G^{\circ\prime} = -3.3 \text{ kJ/mole}$$

The exact mechanism by which this process takes place is not important to our discussion at the present time.

The resulting molecule of succinic acid is truly symmetrical; it would be impossible to say at this point which of the carbons originated from oxaloacetate and which from acetyl–Coenzyme A.

Regeneration of Oxaloacetate

Once the cycle has reached the point of succinic acid, both of the carbons to be eliminated as CO_2 have already been lost. What remains is to regenerate the oxaloacetate by oxidation of succinate. We will see the same sequence of reactions occuring in other pathways of metabolism in which the β-carbon of the hydrocarbon chain of an acid is oxidized to form a ketone functional group.

Oxidation of succinic acid to the *trans* unsaturated compound, fumaric acid, is catalyzed by a rather unusual flavoprotein enzyme. Most flavoproteins have the flavin coenzyme simply associated with the polypeptide chain; it will be released on complete denaturation, for example. In the case of succinic dehydrogenase, however, the FAD molecule is covalently attached to a histidine residue of the polypeptide. The first step in the reaction involves succinate and the enzyme-bound FAD:

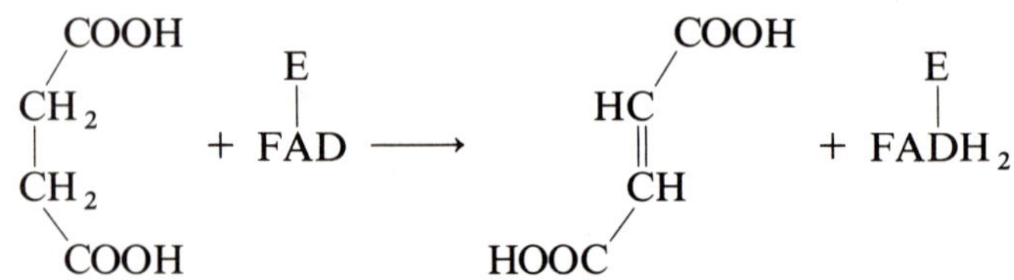

Succinic acid Fumaric acid

$$\Delta G^{\circ\prime} = 0 \text{ kJ/mole}$$

The reduced FAD then is immediately reacted with another oxidation–reduction coenzyme that will provide a reduced form relatively stable to molecular oxygen. This acceptor is called Coenzyme Q in the United States and ubiquinone in many other parts of the world. We will look at its role in the process of electron transfer and energy conservation shortly.

Coenzyme Q (Ubiquinone)

As the name suggests, Coenzyme Q is a quinone compound. In fact, it is simply a benzoquinone ring with a methyl group, two methyl ethers, and a specific hydrocarbon group attached. The length of the unsaturated branched side-chain varies somewhat from one organism to another. Thus, you may see the designation CoQ_6, CoQ_{10}, and so on, where the subscripts indicate the number of five-carbon isoprenoid units in the side-chain. In the other terminology, the equivalent names would be ubiquinone 30 or ubiquinone 50, the numbers here indicating the total number of carbons in the side-chain:

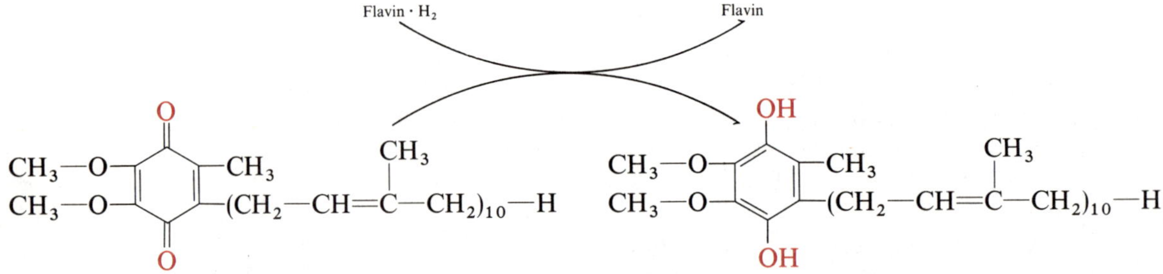

Coenzyme Q_{10} (quinone) Coenzyme Q_{10} (hydroquinone)

You will have learned in organic chemistry that direct addition of water across a carbon–carbon double bond is a very difficult and strenuous process. However, with the aid of the appropriate enzyme these reactions can be accomplished with facility and with a great deal of control over the direction of the addition and the optical asymmetry of the product. All of these reactions are readily reversible. In the present case, there is no question of the direction of addition, but the configuration of the product is clearly specified. Only L-malate is formed as a product:

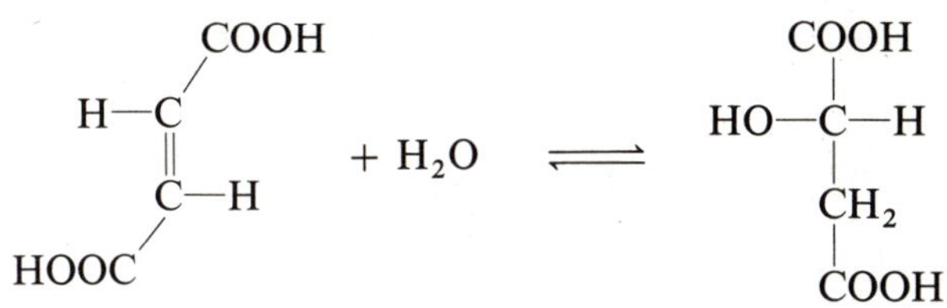

Fumaric acid L-malic acid

$$\Delta G^{\circ\prime} = -3.8 \text{ kJ/mole}$$

Oxidation of the L-malate is just one more example of an oxidation of an alcohol to a ketone, and as you would expect by now it requires NAD as a cosubstrate. The enzyme involved is called **malate dehydrogenase** in common terminology:

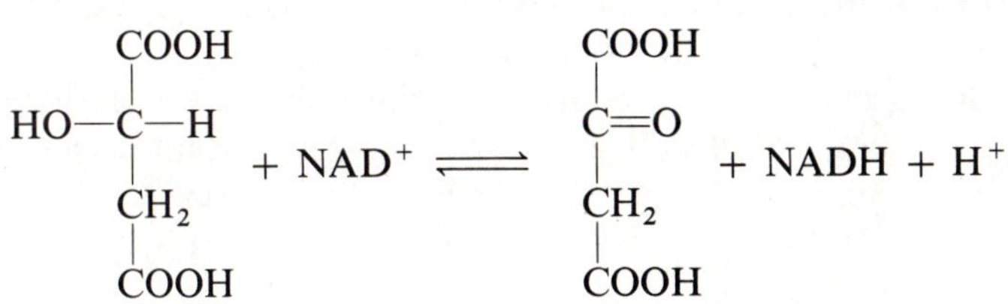

L-malic acid Oxaloacetic acid

$$\Delta G^{\circ\prime} = +29.3 \text{ kJ/mole}$$

Summary of the Cycle

In summary, we have seen in effect the oxidation of a molecule of acetate to two molecules of CO_2. The appropriate numbers of hydrogens will also be oxidized to H_2O eventually, but it is very hard to keep track of this process as it happens. At the same time, we have generated one ATP from ADP and P_i by the substrate-level phosphorylation occurring along with conversion of succinyl–CoA to succinate.

But we have also left over a number of reduced compounds. Starting with pyruvate, we have generated four molecules of NADH and one of $CoQH_2$. Just as we mentioned in the case of glycolysis, there must be some means of re-oxidizing these compounds or the whole system would rapidly grind to a halt. The organized system of enzymes and electron carriers that carries out the re-oxidation of these compounds at the expense of oxygen within the mitochondria is often called the **respiratory electron transport system.** We will examine this process in the next chapter.

9.4 Replenishment of Citric Acid Cycle Intermediates

Many of the intermediates that make up the citric acid cycle are also involved in other metabolic reactions. Consequently, there is a constant draining away of the intermediates; without some means for restoring the supply, the cycle would be unable to maintain an adequate level of activity.

First, we should note that in all animal cells the oxidation of the acetate units of acetyl–Coenzyme A is a one-way process. It is not possible to get any *net* synthesis of cycle intermediates from acetate units.

It *is* possible to obtain the two keto acids, oxaloacetate and α-keto-glutarate, by removal of the amino groups from the corresponding amino acids, aspartic and glutamic acids. Pyruvate is available in the same way from alanine. We will see these reactions in a later chapter.

Under most circumstances, however, the bulk of the cycle intermediates arise from products of glycolysis. The two most important reactions involve pyruvate and phosphoenolpyruvate as starting materials.

Pyruvate Carboxylase

We had occasion earlier to remember that carboxyl groups having a carbonyl, ester, or double-bond function *beta* to the carboxyl group are prone to nonoxidative decarboxylation. In some cases, this process can be reversed in the presence of the appropriate enzyme containing a coenzyme called **biotin**. In most cases, this coenzyme is covalently bonded to the polypeptide chain of the enzyme.

The energy necessary to carry out the addition is supplied by ATP. The overall reaction is

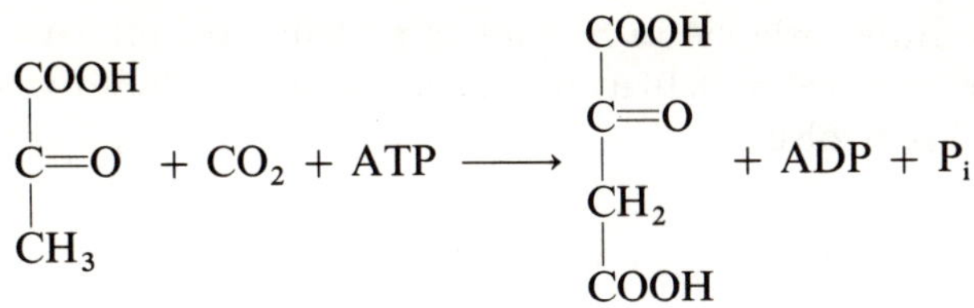

Please note that there are not very many of these biotin-mediated carboxylation reactions, and that they all require the energy of ATP or an equivalent. Most decarboxylation reactions are not reversible.

Biotin

This is another of the less well known B-group vitamins. It has the structure shown and is usually attached to the appropriate enzyme through an amide linkage to an ε-amino group of lysine.

Transfer of the CO_2 usually occurs first onto the biotin structure. It is then transferred to the acceptor molecule as shown below.

A number of the B-group vitamins are not well known simply because it is unlikely that the normal human diet will be deficient in them. Biotin very much fits in this category. However, it is unusual in that there is a specific protein in many avian eggs that binds biotin very tightly and renders it unavailable to the body. The protein is called "avidin" because of the tight binding involved. Persons with a diet containing raw egg at every meal have been known to develop biotin deficiency.

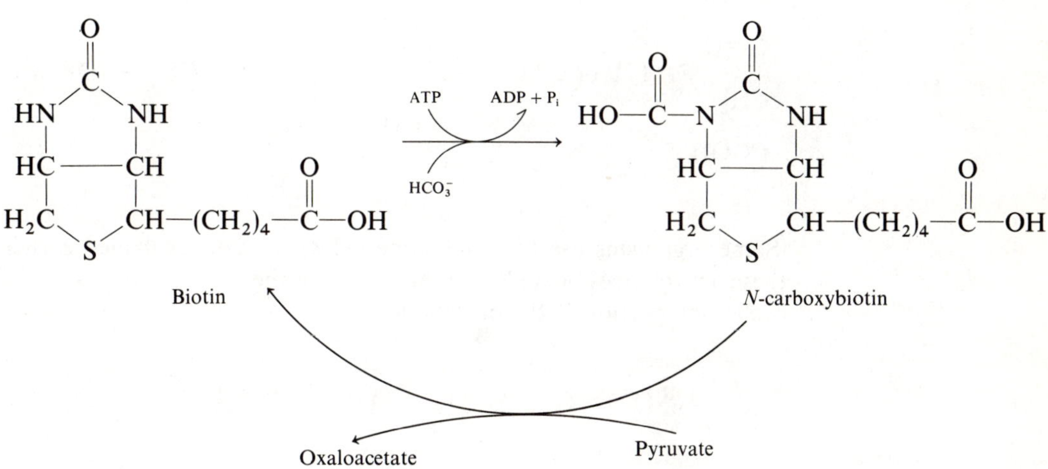

The pyruvate carboxylase reaction is probably the primary source of oxaloacetate in animal cells. In many plants and bacteria, however, phosphoenolpyruvate is involved.

Phosphoenolpyruvate Carboxylase

This reaction uses the large amount of energy released on hydrolysis of phosphoenolpyruvate to drive the addition of the bicarbonate ion:

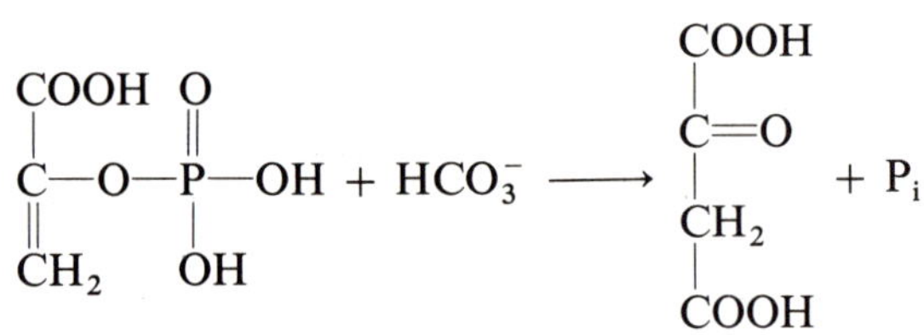

Phosphoenolpyruvate Carboxykinase

A somewhat similar enzyme does exist in animal tissues and some bacteria, but it is not effective in the synthesis of oxaloacetate because the equilibrium lies in the direction of *formation* of phospoenolpyruvate:

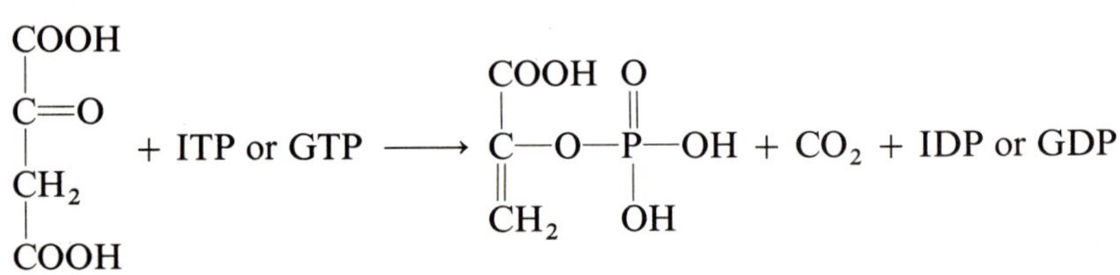

(Some organisms use ITP and some GTP; we have explained earlier that these are relatives of ATP that differ only in the substituents on the heterocyclic ring portion of the nucleotide).

9.5 Summary to This Point

Figure 9-3 outlines our progress to this point in looking at carbohydrate metabolism.

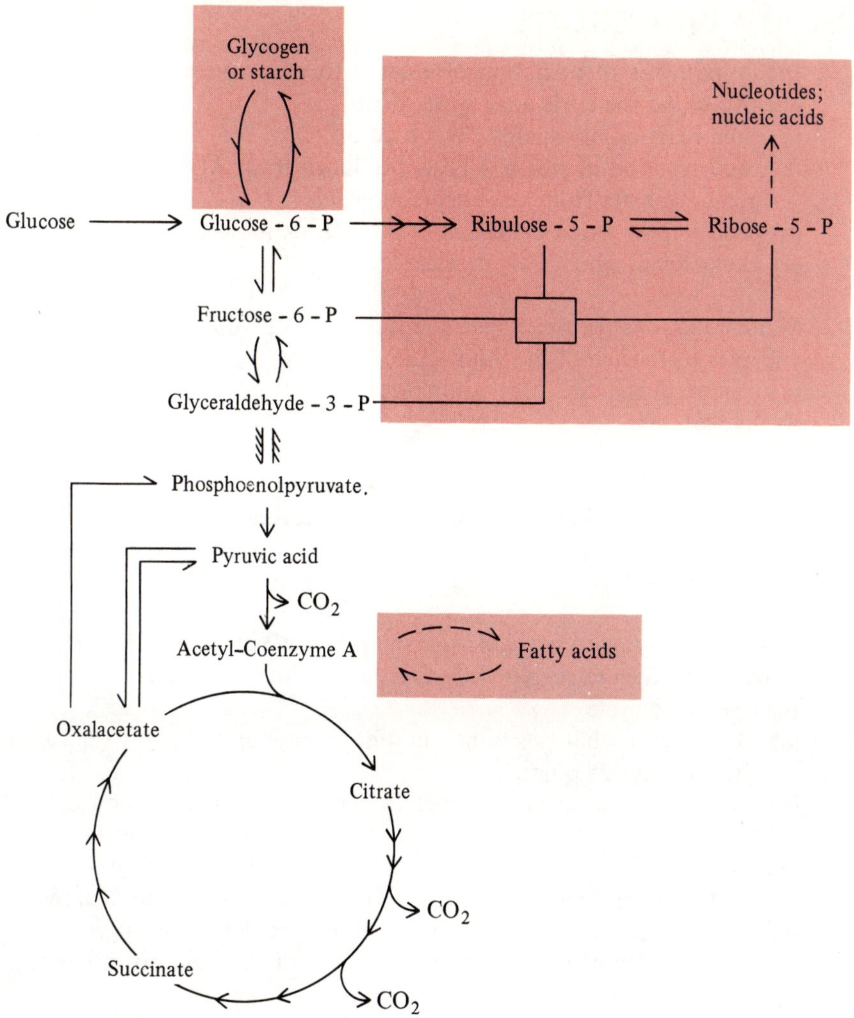

Figure 9-3
The pathways of carbohydrate metabolism that have been examined, and some (shaded) that remain to be explored.

Additional Readings

The standard texts mentioned at the end of Chapter 1, particularly with Lehninger for a functional view and Metzler for mechanism, stereochemistry, and energetics.

Current Topics in Cellular Regulation, Vol. 1 (Academic Press, New York, 1969). Contains a review by L. J. Reed describing the structure and function of the pyruvate dehydrogenase complex.

J. M. Lowenstein, "The Tricarboxylic Acid Cycle," in *Metabolic Pathways*, edited by D. M. Greenberg, 3rd edition, Vol. 1 (Academic Press, New York, 1967).

Many of the readings listed for Chapter 10 also have application to this chapter.

Study Questions

1. In the pathway of oxidation of glucose to CO_2 and water by way of glycolysis and the citric acid cycle, identify:
   (a) each reaction in which ATP is used up
   (b) each reaction in which ATP is produced from ADP by phosphate group transfer from a substrate molecule
   (c) each step in which oxidation of a substrate molecule occurs
   (d) each step in which CO_2 is released

2. Write equations for three specific reactions that would be catalyzed by enzymes from each of the following categories:
   (a) oxidoreductases (c) isomerases
   (b) hydrolases (d) lyases

3. List five different coenzymes and give for each:
   (a) a typical single-enzyme reaction in which it is involved
   (b) the vitamin (if any) required by humans in order to synthesize the coenzyme

4. Assume that some of the lactic acid of Question 3 of Chapter 8 is oxidized to pyruvate, then to acetyl–CoA, and passed through exactly one turn of the citric acid cycle.
   (a) Describe in what position(s) in what compound(s) you would find the label at this point.
   (b) After a second turn of the cycle, where would you find the label?

5. We mentioned in Chapter 6 that malonate is an effective inhibitor of succinate dehydrogenase. In mitochondria inhibited with malonate, which compounds would you expect to accumulate when pyruvate is added as substrate? (Consider $\Delta G^{\circ\prime}$ for each of the reactions involved.)

CHAPTER 10

Electron Transport and Oxidative Phosphorylation

It was known as early as the late 1940s that intact mitochondria are capable of transferring electrons either from NADH or from a substrate such as succinate to oxygen. It was also known that the mitochondria could use the energy released in this electron transport process to bring about the formation of ATP from ADP and inorganic phosphate, a process called "oxidative phosphorylation" to distinguish it from the formation of ATP by substrate-level phosphorylations such as those of glycolysis. It was even clear that re-oxidation of NADH by oxygen yielded three molecules of ATP while oxidation of succinate or similar substrates yielded only two. The intervening years have added greatly to our understanding of these processes, and of the close relationship between mitochondrial structure and function. However, many of the basic mechanisms involved have eluded clear definition and are still very much the subject of conjecture and theory.

10.1 Structure of Mitochondria

The essential features of mitochondrial structure are shown in Figure 10-1. The **inner membrane** is highly folded and encloses an interior space called the **matrix** of the mitochondrion. The inner surface of the inner membrane is covered with large numbers of small spheres believed to contain many of the factors necessary for the process of oxidative phosphorylation.

Enzyme Localization

Each of the areas of the mitochondrion contains a rather specific set of enzyme activities. Thus, the matrix space contains most of the enzymes of the citric acid cycle and of fatty acid oxidation (Chapter 14), while the inner membrane contains all of the enzymes and associated cofactors of electron transport. Some specific examples of localization of enzymes are given in Table 10-1.

Membrane Permeability

The outer mitochondrial membrane is permeable to almost any compound up to a molecular weight of about 10,000. The inner membrane, however, allows only a very limited number of compounds to pass—and even those only under very closely controlled conditions. For example, a molecule of ATP can leave the matrix space only if a molecule of ADP moves in to replace it. Some selected examples of compounds that can and cannot penetrate into the matrix are shown in Table 10-2.

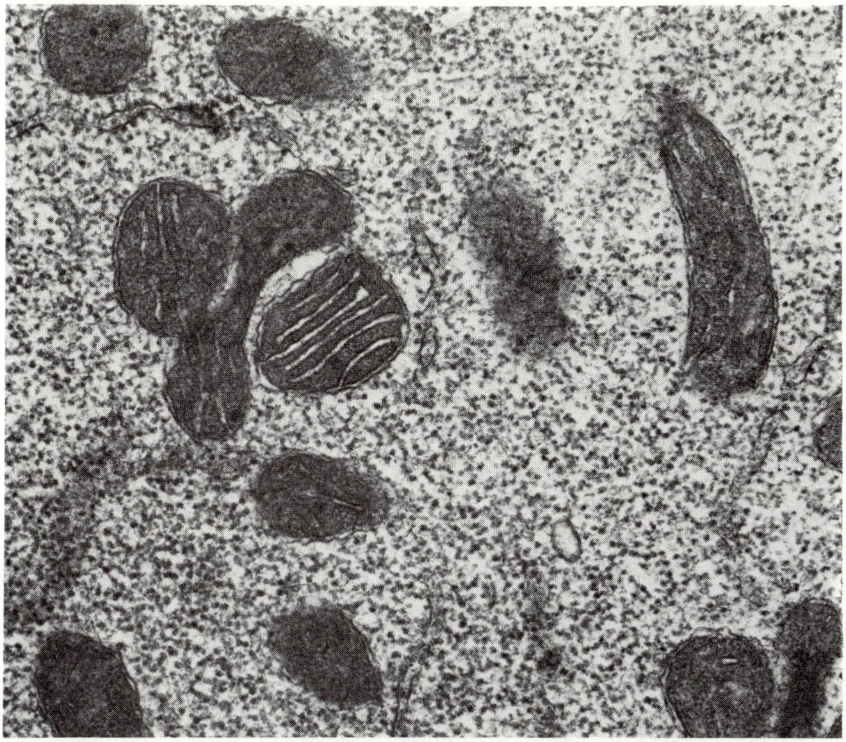

(A)

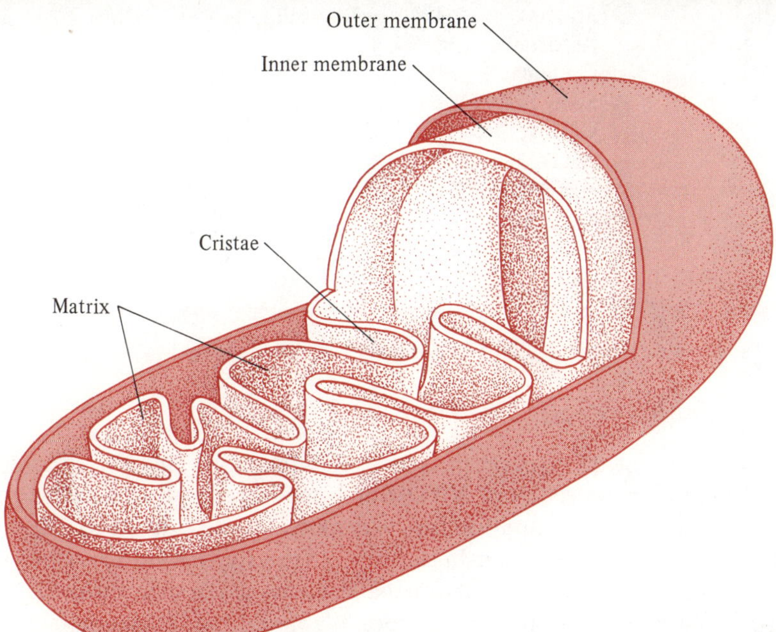

(B)

Figure 10-1
(A) Electron micrograph of a portion of a hemocyanin forming cell of the Horseshoe crab (*Limulus polyphemus*) showing a number of mitochondria with densely stained matrix spaces (× 100,000). (Courtesy of W. H. Farnbach, Oregon Regional Primate Research Center.)
(B) Conventional three-dimensional reconstruction of a mitochondrion as deduced from two-dimensional random thin sections of mitochondrial profiles in electron micrographs.

TABLE 10-1
Examples of localization of enzymes within mitochondria.

Outer Membrane
Fatty acyl–CoA synthetases
Nucleoside diphosphate kinase
Monoamine oxidase
Inner Membrane
All components of electron transport and oxidative phosphorylation
Succinate dehydrogenase
Carnitine acyltransferase[a]
Matrix
Most of the enzymes of the citric acid cycle except succinate dehydrogenase
Pyruvate dehydrogenase complex
Enzymes of β-oxidation of fatty acyl–CoA derivatives[a]
Glutamate dehydrogenase[b]
Aspartate transaminase[b]

a. See Chapter 14.
b. See Chapter 16.

TABLE 10-2
Selected examples of compounds that can and cannot cross the inner membrane of a typical liver mitochondrion.

A. Some that *can* cross the membrane:

ATP, ADP	Inorganic phosphate
Pyruvate	α-Ketoglutarate
Citrate	Glycerol phosphate
Succinate	Dihydroxyacetone phosphate
Malate	Fatty acyl–carnitines
Aspartate	
Glutamate	

B. Some that *cannot* cross the membrane:

Glucose
NAD, NADH
AMP, GTP, CTP
Coenzyme A

10.2 The Mitochondrial Electron Transport System

Operation of the pyruvate dehydrogenase complex and the citric acid cycle leaves four molecules of NADH and one of reduced Coenzyme Q in the matrix of the mitochondrion for every molecule of pyruvate oxidized. It is the role of the mitochondrial electron transport system to carry out the re-oxidation of these reduced intermediates by oxygen in such a way that the energy released by this process will be available for the formation of ATP.

We will examine first the process of electron transport itself, and then deal with oxidative phosphorylation.

As a starting point, we can make use of the ability to isolate subunits from intact mitochondria by means of treatment with appropriate detergents and salt solutions. One such subunit contains all of the materials necessary for re-oxidation of NADH or $CoQH_2$ by oxygen. It has been called the "electron transport particle." This particle can carry out the following processes:

$$\text{NADH} + \text{H}^+ + \tfrac{1}{2}\,\text{O}_2 \longrightarrow \text{NAD}^+ + \text{H}_2\text{O}$$

and

$$\text{succinate} + \tfrac{1}{2}\,\text{O}_2 \longrightarrow \text{fumarate} + \text{H}_2\text{O}$$

By very careful treatment, the particle can be divided into a series of subparticles, each carrying out a portion of the overall process. If we consider each of these subparticles to represent the activity of a single enzyme, we can construct a model of the electron transport particle as shown in Figure 10-2. Each of the boxes shown represents a specific subunit, which catalyzes a specific portion of the electron transport chain, and contains one or more coenzymes involved in the electron transport process. The four reactions

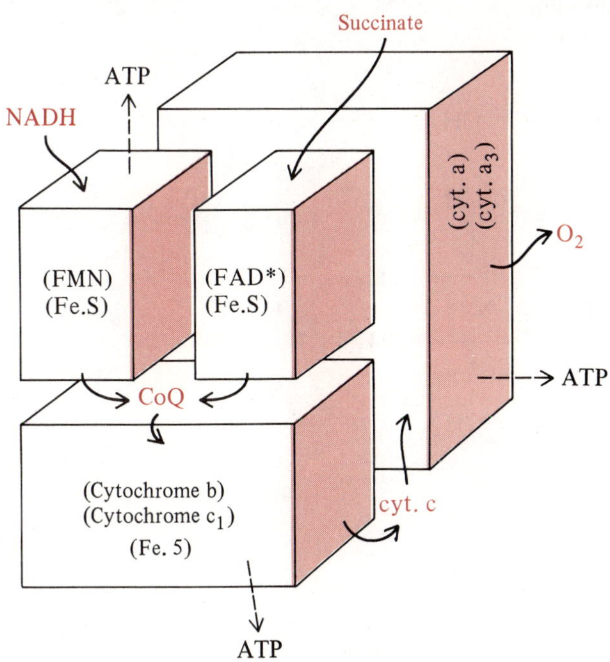

Figure 10-2
A model indicating the organized subunits that can be isolated from the mitochondrial inner membrane and their apparent roles in electron transport.

catalyzed by the four subunits can be summarized as follows:

$$NADH + H^+ + CoQ \longrightarrow NAD^+ + CoQH_2 \quad (1)$$

$$CoQH_2 + 2cyt.\ c\ (Fe^{+++}) \longrightarrow CoQ + 2cyt.\ c\ (Fe^{++}) + 2\ H^+ \quad (2)$$

$$2\ cyt.\ c\ (Fe^{++}) + 2H^+ + \tfrac{1}{2}O_2 \longrightarrow 2\ cyt.\ c(Fe^{+++}) + H_2O \quad (3)$$

$$succinate + CoQ \longrightarrow fumarate + CoQH_2 \quad (4)$$

Cytochromes

In 1886, MacMunn viewed cell suspensions through a prism spectroscope and observed that certain strong absorption bands appeared as oxygen was used up and disappeared when oxygen was admitted to the system. The term "cytochrome" was later coined for these substances, although their structures were unknown.

We now know that there are a number of these compounds in cells, at least four of which take part in mitochondrial electron transport. Each is a protein containing as part of its structure an iron–porphyrin compound similar to the heme of hemoglobin. The individual cytochromes may vary in structure of the protein and of the porphyrin ring, giving rise to differences in light absorption bands and in reactivity toward redox reactions. In each case, however, the cytochrome undergoes reversible oxidation and reduction through interconversion of the central iron atom between the Fe^{+3} and the Fe^{+2} states. The oxidation or reduction of a cytochrome thus involves a single-electron transfer.

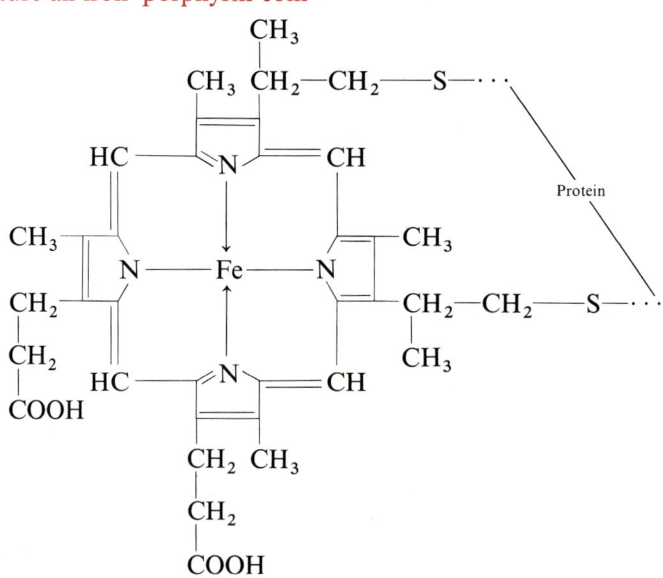

The iron–porphyrin portion of cytochrome c.

The involvement of succinate dehydrogenase as a member both of the citric acid cycle group of enzymes and as a member of the electron transport particle underscores the close structural relationships that prevail within the mitochondrion. It is, however, in part an accident that succinate dehydrogenase happens to bind strongly enough to the other components to be isolated as part of the electron transport particle. We will see shortly that there are several other enzymes of similar function that are lost during the isolation process but that play a very much related role in the functioning mitochondrion.

In the equations above, and in Figure 10-2, we have noted the coenzymes associated with each of the subparticles. Another way to look at the system is to follow the order in which the electrons are passed from one carrier to another. This is outlined in skeletal form in Figure 10-3, with the reservation that there is still a good deal of uncertainty about the details of the diagram.

Iron–Sulfur Centers

The symbol "Fe.S" represents the presence of a protein containing a complex of iron and sulfur atoms which decomposes with release of hydrogen sulfide when exposed to acid. The exact roles of these **iron–sulfur centers**, as well as their positions in the chain, are not yet clear. It is believed that they act by one-electron oxidations and reductions of the iron atom; that is, $Fe^{+++} + e^{-} \rightleftharpoons Fe^{++}$.

10.3 Oxidative Phosphorylation

In a freshly isolated and intact mitochondrion, provided with the appropriate substrates and cofactors, either oxidation of a substrate that is a source of NADH or oxidation of succinate is closely linked to conversion of ADP and P_i to ATP. The linkage is so close that in the absence of ADP to phosphorylate, oxidation will not occur. We say that the processes of oxidation and of phosphorylation are **coupled**, and refer to this phosphorylation occurring only when electrons are being transferred as being **oxidative phosphorylation**.

As isolated mitochondria are allowed to age, or if they are not treated gently enough, they will lose the ability to phosphorylate ADP but not the ability to carry out oxidation of the usual substrates. Under these conditions, the system is said to be **uncoupled**. A number of compounds can promote this uncoupling, including 2,4-dinitrophenol and arsenate ion.

Not all of the electron transfers occurring during operation of the electron transport system yield sufficient energy to drive the phosphorylation reaction. Figure 10-4 is an attempt to represent diagrammatically the relative energy differences between the oxidized and reduced forms of the various electron carriers. Note that it is only during the electron transfers involving substantial energy differentials that oxidative phosphorylation occurs.

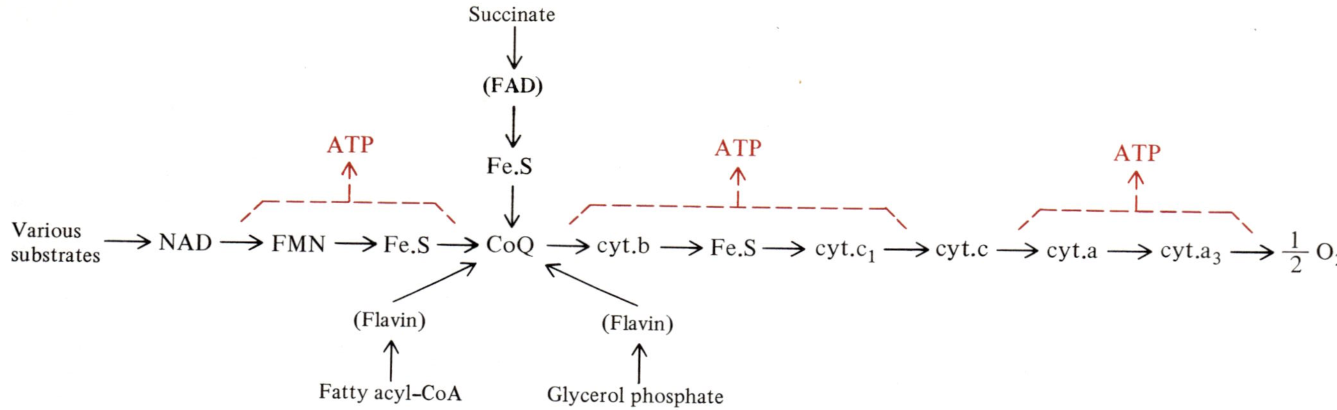

Figure 10-3
The apparent sequence in which electrons are passed from substrates to oxygen during mitochondrial electron transport. The approximate regions of the chain giving rise to formation of a molecule of ATP by oxidative phosphorylation are indicated.

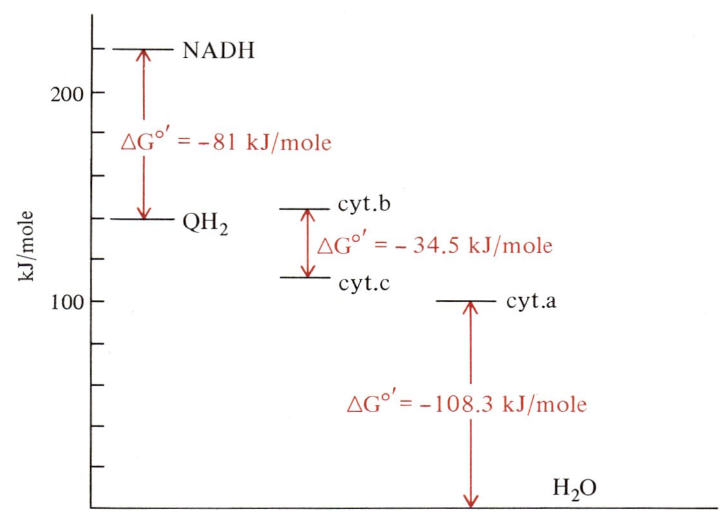

Figure 10-4
Energy differentials for various stages in the electron transport system. Values of $\Delta G^{\circ\prime}$ for those steps providing enough energy to permit formation of ATP from ADP and P_i are shown. Energy values shown for each reduced compound are the $\Delta G^{\circ\prime}$ values for oxidation of that compound by oxygen.

Theories of Oxidative Phosphorylation

There have been three basic theories proposed for a mechanism for oxidative phosphorylation. The first has been called the **chemical-coupling** hypothesis, and suggests that phosphate groups are first attached to an intermediate compound which is then activated and made capable of transferring the phosphate to ADP. Basically, the operation would be analogous to the glyceraldehyde 3-phosphate dehydrogenase reaction. The main problem for this theory is that in 30 years of research, no one has been able to find such an activated intermediate.

The second theory is the **conformational-coupling** hypothesis. In its simplest terms, this theory suggests that the energy of electron transport is stored in some fashion in a changed conformation of a molecule or group of molecules. This changed conformation must be such that it can promote the reaction of ADP and P_i, and in so doing be itself converted back to its original state. The theory suffers from a shortage of experimental evidence and of clearly defined mechanisms by which it might function.

The third, and currently the most popular, theory is the **chemiosmotic-coupling hypothesis** which owes most of its development to the English Nobel Laureate P. Mitchell. This concept proposes that the reactions of the electron transport system occur in such a way that protons are accepted from the matrix and expelled from the outer surface of the inner membrane. This requires that those steps in the transport process which use protons be located on the inner surface of the membrane, and those that release protons be located on the outer surface as indicated in Figure 10-5. The result was originally postulated to be a differential in proton concentration, that is in pH, across the membrane. Currently, it is suggested that there is created both a pH gradient and an electrical potential gradient across the membrane. These potentials then provide a condition in which reaction of ADP and P_i becomes energetically favored.

A great deal of experimental evidence supports the basic postulates of this hypothesis, but many of the precise mechanisms involved remain to be clarified.

Transfer of Reducing Equivalents between Mitochondria and Cytoplasm

Both NAD^+ and NADH are among the many compounds that are unable to travel through the mitochondrial membrane. However, there are several mechanisms by which the reducing power of an NADH may be transferred in either direction. Because these systems serve to carry electrons one way or the other on either side of the membrane, they are usually referred to as "shuttle" systems.

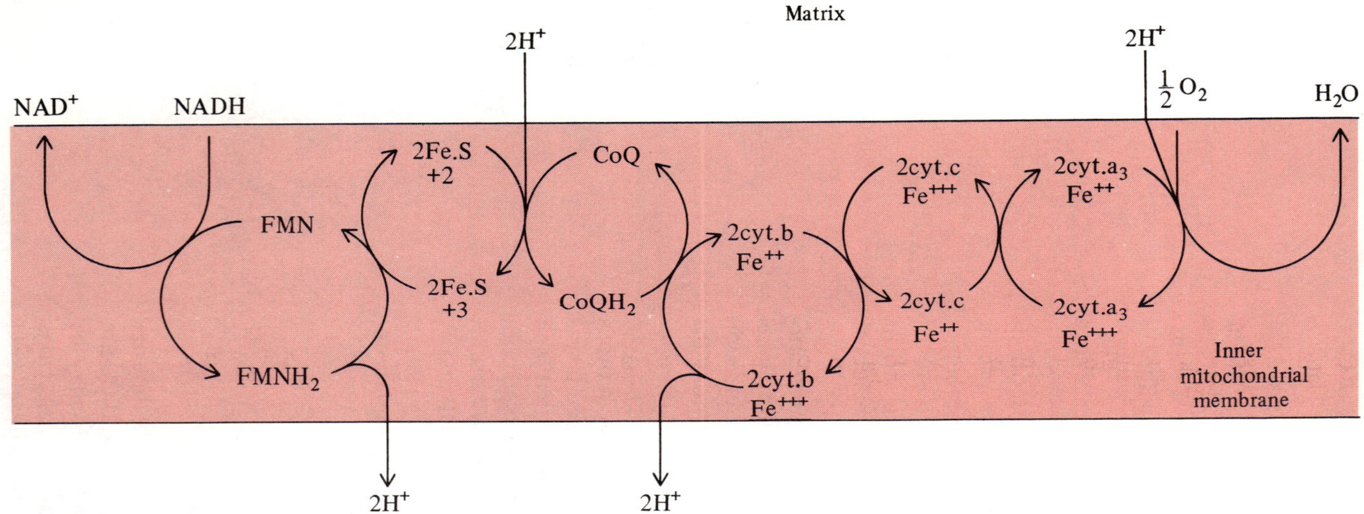

Figure 10-5
A simplified schematic drawing to show the basic concept of how protons may be effectively moved from the matrix to the intermembrane space during electron transport.

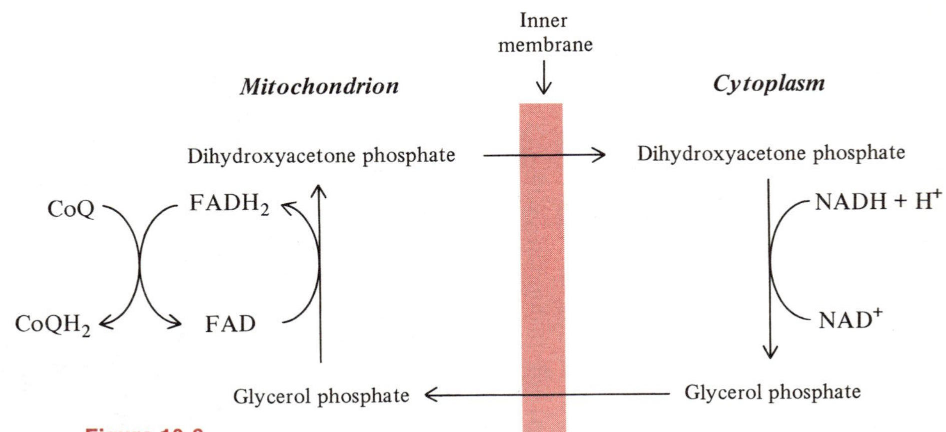

Figure 10-6
Operation of the glycerol phosphate shuttle as a means of moving the reducing power of an NADH molecule into the mitochondrial matrix. As the arrows indicate, this is a one-way process.

The Glycerol Phosphate Shuttle. This system works only in one direction—to get the reducing equivalent of NADH from the cytoplasm into the mitochondrion. The shuttle proceeds as shown in Figure 10-6 with a cytoplasmic enzyme called glycerol phosphate dehydrogenase catalyzing the reduction of dihydroxyacetone phosphate to glycerol phosphate. The glycerol phosphate is able to enter the mitochondrion, where it is subjected to oxidation, this time by a flavoprotein enzyme. The resulting dihydroxyacetone phosphate finds its way out to the cytoplasm, and the cycle continues.

Because of the characteristics we have already mentioned for flavoprotein dehydrogenases, the mitochondrial enzyme cannot be reversed; thus, the flow of reducing equivalents can occur only from cytoplasm to mitochondrion (and then through the rest of the electron transport chain to oxygen).

Malate Shuttle. The malate shuttle is very similar but somewhat more complex, involving aspartic and glutamic acids as well as α-ketoglutarate in the transport system. We will look at the reactions involving amino group transfers to keto acids in Chapter 16. In the meantime, you will have to accept the reactions on faith.

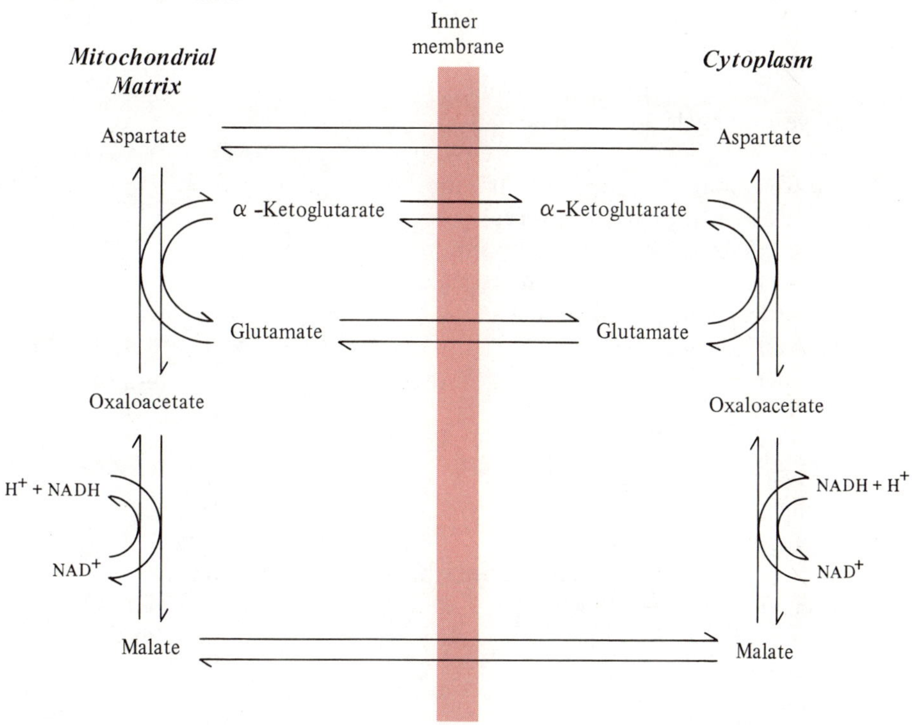

Figure 10-7
The malate shuttle. This process can move the reducing power of NADH in either direction.

In the case of this shuttle, both inner and outer dehydrogenases are reversible, and flow of reducing power can move in either direction depending on the energy state of the cell. The process is shown diagrammatically in Figure 10-7.

The relative activities of these two shuttle systems at any given time, and the direction of the malate shuttle, will depend on the metabolic state of the cell. If carbohydrate is available and glycolysis is proceeding rapidly, the NADH produced will be largely moved into the mitochondrion. However, if oxidative metabolism is slow compared to the demands for reducing power for synthesis, the inward flow will be slowed or reversed.

10.4 Energy Accounting

Starting with a molecule of glucose that has just entered a cell and ending with CO_2 and water by way of the citric acid cycle, we can set up the following energy balance sheet:

Cytoplasm:

glucose ⟶ glucose 6-phosphate		−1 ATP
fructose 6-phosphate ⟶ fructose-1,6-diphosphate		−1 ATP
2 glyceraldehyde 3-phosphate ⟶ 2 pyruvate	+2 NADH	+4 ATP
If we assume that the 2 NADHs are processed through the glycerol phosphate shuttle, each is worth 2 ATPs; if they go through the malate shuttle, each is worth 3 ATPs.		
2 NADH ⟶ 2 NAD^+	−2 NADH	+4 or 6 ATP
NET:		+6 or 8 ATP

Mitochondrion:

pyruvate ⟶ acetyl–Coenzyme A	+1 NADH	
isocitrate ⟶ α-ketoglutarate	+1 NADH	
α-ketoglutarate ⟶ succinyl–Coenzyme A	+1 NADH	
succinyl–CoA ⟶ succinate		+1 ATP (or GTP)
succinate ⟶ fumarate	+1 $CoQH_2$	
malate ⟶ oxaloacetate	+1 NADH	
4 NADH ⟶ 4 NAD^+	−4 NADH	+12 ATP
$CoQH_2$ ⟶ CoQ	−1 $CoQH_2$	+ 2 ATP
NET:		+15 ATP

However, remembering that each glucose gives two molecules of pyruvate, we get a figure for the process 2 pyruvate $\longrightarrow$ 4 CO_2 of +30 ATP. Thus, the total net ATP production for the complete oxidation of one molecule of glucose to CO_2 and water by way of glycolysis and the citric acid cycle is *36 to 38 ATPs*. Comparing this with the *2 ATPs* available from anaerobic glycolysis should give you some realization of why anaerobic organisms generally have much slower growth rates and activity levels than do aerobic ones.

Additional Readings

Scientific American articles

P. C. Hinkle and R. E. McCarthy, "How Cells Make ATP," 238(3), 104 (March 1978).

R. Margaria, "The Sources of Muscular Energy," 226(3), 84 (March 1972).

D. E. Green and H. Baum, *Energy and the Mitochondrion* (Academic Press, New York, 1970). A good short discussion of the nature and function of mitochondrial structures.

D. Keilin, *The History of Cell Respiration and Cytochromes* (Cambridge University Press, New York, 1966). A fascinating account of research on the cytochromes and related processes.

E. Racker, *A New Look at Mechanisms in Bioenergetics* (Academic Press, New York, 1976). A very enjoyable discussion of electron transport and oxidative phosphorylation written with wit and grace by one of the giants of the field.

D. E. Atkinson, *Cellular Energy Metabolism and Its Regulation* (Academic Press, New York, 1977). Primarily concerned with control aspects, but a very useful view of ATP production and utilization.

L. Peusner, *Concepts in Bioenergetics* (Prentice-Hall, Englewood Cliffs, N. J., 1974). A look at energy transfer reactions from a physical biochemistry approach.

W. Lovenberg, *Iron–Sulfur Proteins* (Academic Press, New York, 1973) (2 vols.). A set of thorough reviews on the nature and function of these proteins.

R. Lemberg and J. Barrett, *Cytochromes* (Academic Press, New York, 1973). An advanced discussion of the chemistry and biochemistry of the cytochromes.

Study Questions

1. Arrange the following compounds in the order in which electron flow takes place from substrate to oxygen in the mammalian respiratory electron transport system:

cytochrome c	coenzyme Q (ubiquinone)
cytochrome b	cytochrome oxidase
FMN	NAD

2. The reaction of glucose plus oxygen giving CO_2 and water has a $\Delta G^{\circ\prime} = -2871$ kJ/mole.
   (a) What is the percentage "efficiency" of energy utilization of a biological system using glycolysis and the citric acid cycle if we assume the $\Delta G^{\circ\prime}$ for hydrolysis of ATP to be -30.6 kJ/mole?

(b) What is the "efficiency" of a microorganism which converts glucose to lactate as its primary source of energy?

3. Certain microorganisms which normally require oxygen for growth can grow under anaerobic conditions in the presence of certain inorganic ions. Which of the following might theoretically be capable of serving such a function?

SO_4^{-2}	NH_4^+	F^-	NO_3^-
K^+	Fe^{+++}	Cu^+	ClO_3^-
Ca^{++}	VO_3^{-2}	AsO_4^{-3}	

4. Each of the following reactions (or reaction sequences) can be carried out in a rat liver cell. For each, as written, determine the maximum possible yield in useable energy in the form of ATP (or GTP).
(a) glucose + $O_2 \longrightarrow$ 2 pyruvate + 2 H_2O
(b) pyruvate + $\frac{1}{2}$ O_2 + CoA $\longrightarrow$ acetyl–CoA + CO_2 + H_2O
(c) lactate + 3 $O_2 \longrightarrow$ 3 CO_2 + 3 H_2O
(d) acetyl–CoA + 2 $O_2 \longrightarrow$ 2 CO_2 + CoA + H_2O

5. Show how Figure 10-2 might be modified to include the mitochondrial glycerol phosphate dehydrogenase.

6. Define or describe accurately each of the following terms:
(a) oxidation
(b) respiratory electron transport system
(c) substrate-level phosphorylation
(d) oxidative phosphorylation

7. A eukaryotic cell treated with an "uncoupling" agent, such as dinitrophenol, can carry out complete oxidation of glucose to CO_2 and H_2O, but without oxidative phosphorylation. What will be the yield in ATPs (and/or GTPs) per glucose in this situation? What will happen to the remainder of the energy released in the oxidative process?

CHAPTER 11

The Pentose Phosphate Pathways—and Their Function in Photosynthesis

We have chosen to use the plural in describing the pentose phosphate pathways, because there are really several metabolic systems, serving quite different cellular purposes, which share a group of reactions involving the pentose phosphates. It is useful first to look at the functions that are served by this set of pathways.

One sequence of reactions, known as the **phosphogluconate pathway**, can carry out an overall process similar to that of glycolysis followed by the citric acid cycle. That is, it can carry out the complete oxidation of glucose 6-phosphate to CO_2 and water with the release of a large amount of useable energy. However, this energy is obtained in the form of NADPH and is not readily convertible into ATP energy. Most of the NADPH formed is used in the reductive steps of biosynthesis, which very commonly require this specific nucleotide. Thus we may say that the primary function of the phosphogluconate pathway is the generation of NADPH for biosynthetic purposes.

Secondly, the pentose phosphate reactions allow for the conversion of glucose 6-phosphate into ribose 5-phosphate. Since ribose 5-phosphate is required for the synthesis of all nucleotides, including those that make up the nucleic acids that serve as the genetic informational material, this is a very important process for all cells. We will see that the ribose 5-phosphate can either be a product of the oxidative phosphogluconate pathway or can be formed without oxidative steps (an important advantage in the case of anaerobic organisms).

In green plants and other photosynthetic organisms, addition of only a very few unique reactions allows operation of a metabolic cycle by means of which the carbon of CO_2 can be built into molecules of glucose 6-phosphate. This is the third function of the pentose phosphate reactions.

11.1 The Phosphogluconate Pathway

Mercifully for the student, the number of reactions involved is not great. The oxidative portion of the pathway consists of only three reactions. First, glucose 6-phosphate is oxidized from the level of the aldehyde to the level of the acid. Since this occurs while the glucose is in the ring form, the actual product is a cyclic ester called 6-phosphogluconolactone. (You will remember that all cyclic esters are called lactones.) The lactone is hydrolyzed into the open chain form, 6-phosphogluconic acid, by another enzyme; then this product is also oxidized. Presumably, oxidation of the 6-phosphogluconate produces a keto group at the third carbon and the resulting β-keto acid rapidly decarboxylates.

This part of the pathway, from glucose 6-phosphate to ribulose 5-phosphate and CO_2, is shown in Figure 11-1. The hypothetical ketoacid intermediate is shown in square brackets to help in visualizing the nature of the reaction.

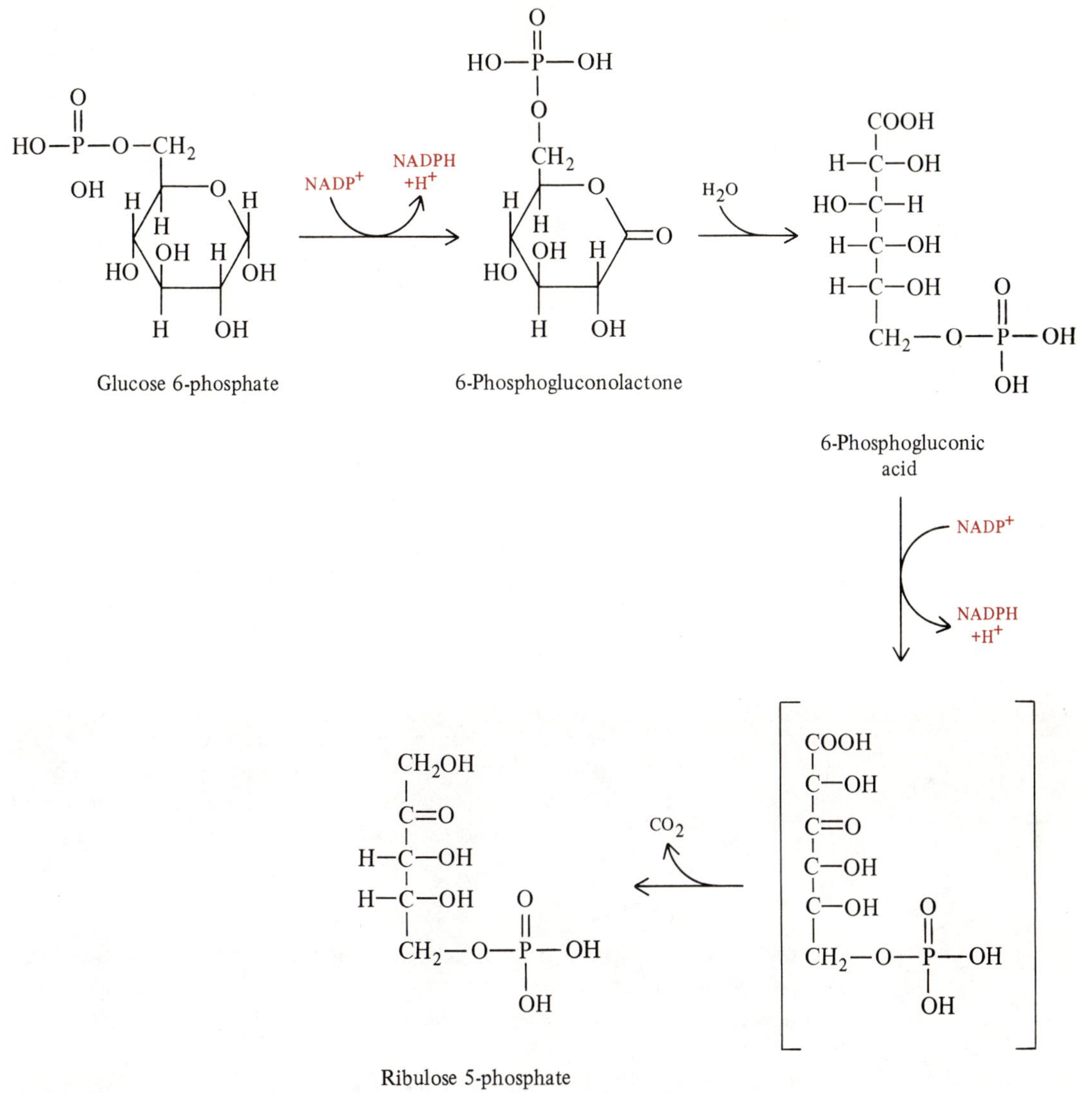

Figure 11-1
The phosphogluconate pathway from glucose 6-phosphate to ribulose 5-phosphate.

Ribulose 5-phosphate is of course simply the keto sugar corresponding to ribose 5-phosphate. As in other similar cases we have already seen, the interconversion of these two compounds occurs readily in the presence of an appropriate isomerase. The resulting ribose 5-phosphate can then be used in the formation of nucleotides, nucleic acids, and related compounds vital to the cell:

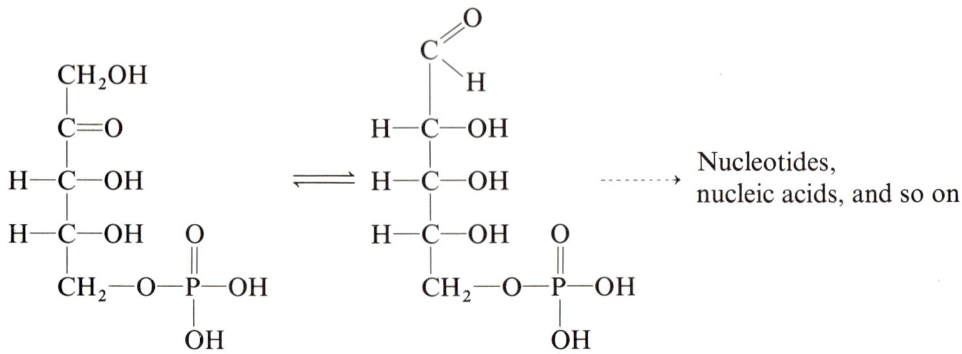

Ribulose 5-phosphate Ribose 5-phosphate

In order for these reactions to be capable of the complete oxidation of glucose 6-phosphate to CO_2 and water, with generation of NADPH, it is necessary that the molecules of ribulose 5-phosphate be capable of conversion back to glucose 6-phosphate. Then we have a truly cyclic process as demonstrated in Figure 11-2.

Provided that we can get the little box with the asterisk to work properly, the overall system shown takes one glucose 6-phosphate and converts it to six molecules of CO_2 with the production of 12 molecules of NADPH. You might note that if we were able to convert each of these NADPHs to 3 ATPs, the total yield for complete oxidation of the glucose 6-phosphate would be 36 ATP, almost exactly what we obtained by way of glycolysis and the citric acid cycle. In this case, however, the energy is used directly as NADPH and is not usually converted to ATP energy.

Pentose to Hexose—The Road Back

Operation of the box with the asterisk requires two new enzymes and the use of several of the reactions we have so far associated with glycolysis. The two new reactions are catalyzed by the enzymes **transketolase** and **transaldolase**.

Transketolase. The transketolase reaction can be represented as follows:

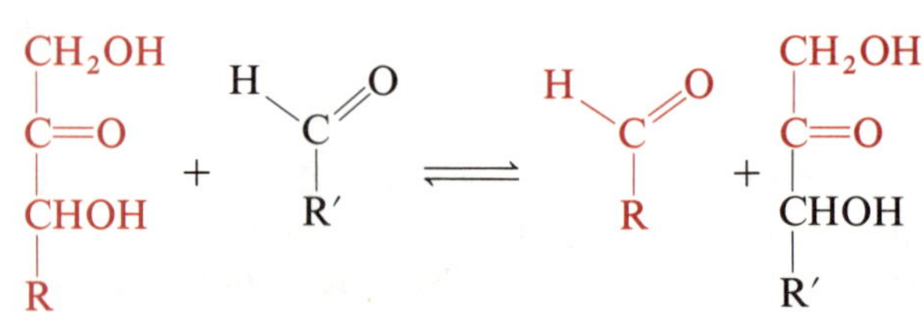

A particular example would be as follows (for clarity, the phosphate group is shown as **P**):

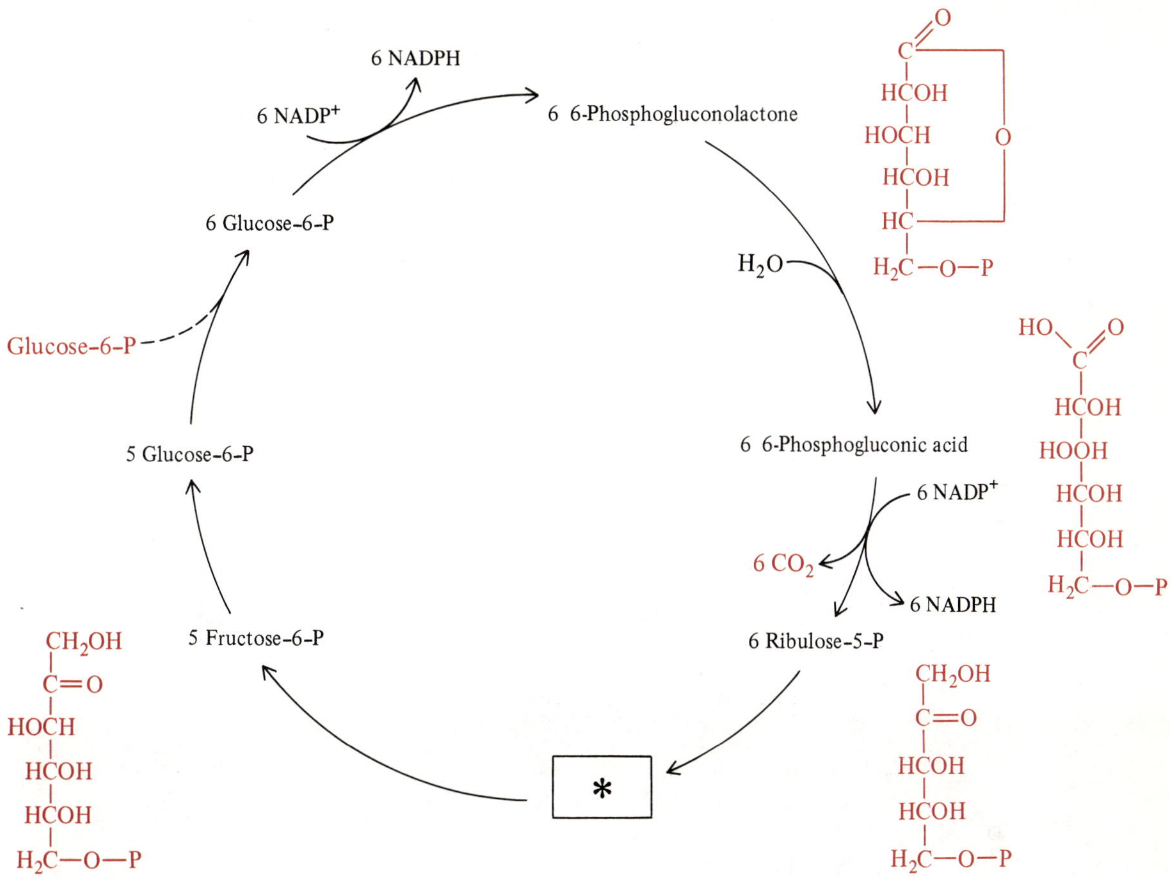

Figure 11-2
A simplified version of the phosphogluconate pathway showing its function as a means of carrying out total oxidation of glucose 6-phosphate and generating NAPDH. Figure 11-3 details the reactions represented here by an asterisk.

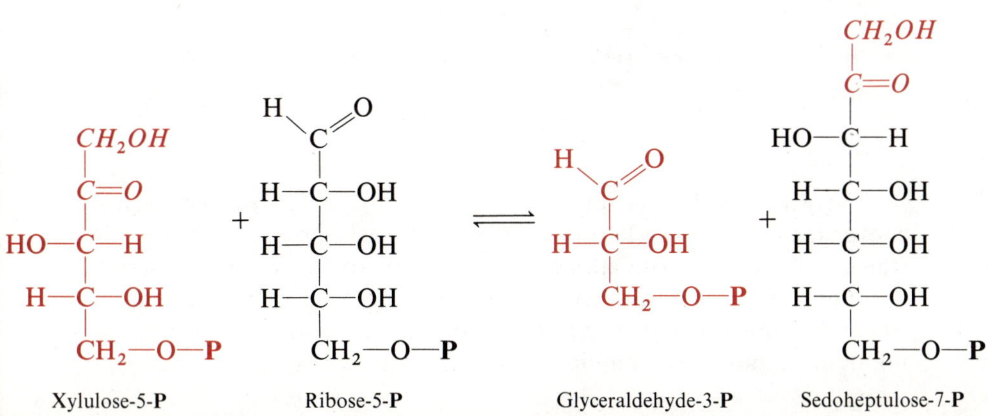

Usually the reactant and product sugars must have phosphate groups attached. In addition, one keto sugar and one aldose must appear on each side of the equation. Further, the keto sugar in every case must be one that has the hydroxyl group on the 3-position to the left in a Fischer projection. It is for this reason that ribulose 5-phosphate must be converted to xylulose 5-phosphate in order to take part in transketolase (or transaldolase) reactions.

Transaldolase. These reactions have very similar requirements for the reactant molecules, but the number of carbons being transferred in each case is three instead of two. The general reaction is shown below:

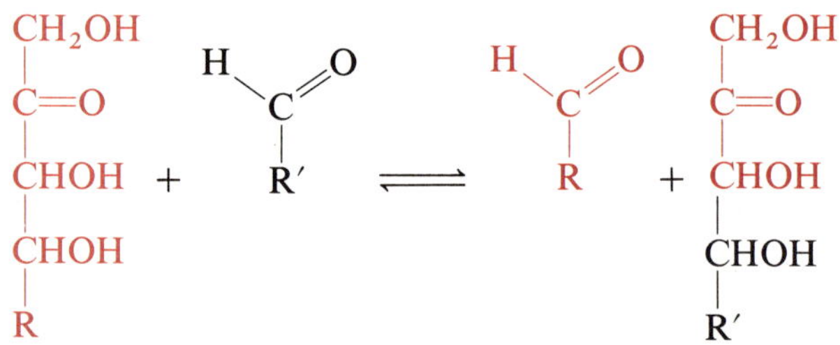

A corresponding specific example is

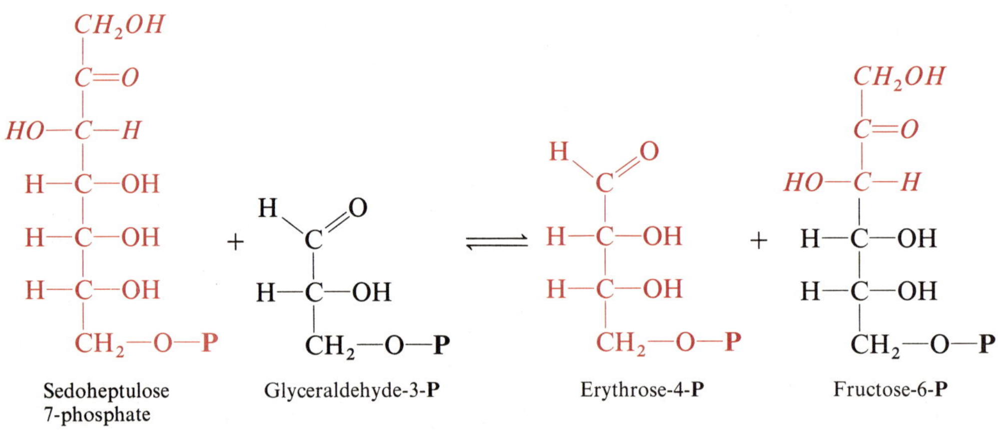

To get some feeling for how these reactions help to accomplish the conversion of pentose to hexose, a set of reactions is shown in Figure 11-3 by which 6 ribulose 5-phosphates can be converted to 5 glucose 6-phosphates.

Both the transketolase and transaldolase reactions have equilibrium constants close to 1. The road back to the point of glycolytic intermediates is not difficult, nor does it require an added energy source.

6 Ribulose-P (5C) —Isomerases→ 2 Ribose-P (5C) + 4 Xylulose-P (5C)

2 Ribose-P (5C) + 2 Xylulose-P (5C) —Transketolase→ 2 Glyceraldehyde-P (3C) + 2 Sedoheptulose-P (7C)

2 Sedoheptulose-P (7C) + 2 Glyceraldehyde-P (3C) —Transaldolase→ 2 Erythrose-P (4C) + 2 Fructose-P (6C)

2 Xylulose-P (5C) + 2 Erythrose-P (4C) —Transketolase→ 2 Glyceraldehyde-P (3C) + 2 Fructose-P (6C)

2 Glyceraldehyde-P (3C) —1. Aldolase, 2. Phosphatase→ Fructose-P (6C)

---

6 Ribulose-5-P → 5 Fructose-6-P

Figure 11-3
One set of reactions that will bring about the conversion of six molecules of ribulose 5-phosphate to five molecules of glucose 6-phosphate.

Anaerobic organisms must re-oxidize every reduced nucleotide they produce. In general, they can only allow the production of enough NADPH to meet the synthetic needs of the cell. Any excess production simply robs the cell of some of its ability to make ATP. These cells, however, must still be able to make ribose 5-phosphate. They do this by using the intermediates of glycolysis, reversing the transaldolase and transketolase reactions to get xylulose 5-phosphate, ribulose 5-phosphate, and ultimately ribose 5-phosphate by reactions we have already discussed. This occurs quite readily and requires no additional input of energy. One way in which the process can occur is shown in Figure 11-4.

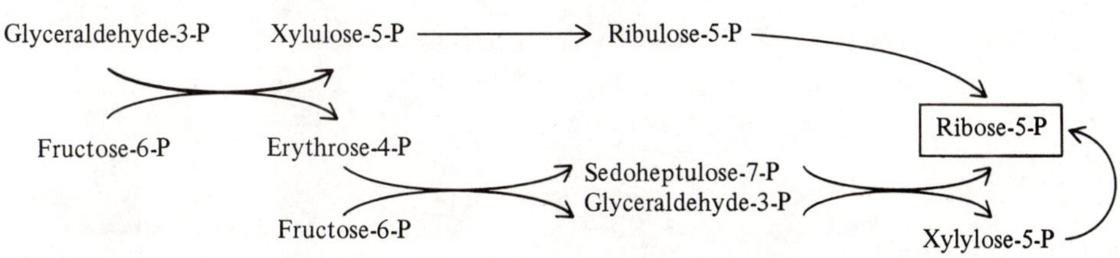

Figure 11-4
A schematic drawing of a set of reactions that permit glycolytic intermediates (here represented by fructose-6-phosphate and glyceraldehyde 3-phosphate) to be converted to ribose 5-phosphate essentially isoenergetically and without the need for any oxidative reactions.

11.2 Photosynthesis

Green plants, many algae, and some few bacteria are capable of carrying out the process known as photosynthesis. In a general sense, this means the ability to use the energy of light to carry out the process of converting carbon dioxide and water to molecules of glucose and oxygen. This is just the reverse of the oxidative metabolism we have studied up to this point. The process may be summarized as

$$6\,CO_2 + 6\,H_2O + \text{light energy} \longrightarrow \text{glucose} + 6\,O_2$$

This process is basic to life as we know it, since all other cells use as their sources of carbon and of energy the compounds formed by photosynthetic organisms. In short, energy from the sun is trapped in the form of bonding energy in a variety of molecules, then released when these molecules are metabolized in cells that lack the photosynthetic apparatus. We should also be aware that a plant in the dark is just like any other organism, and must oxidize some of its stored materials to obtain the energy needed to make it through the dark period.

Photosynthesis occurs within specialized subcellular organelles called chloroplasts. These are highly structured with much folding of internal membranes and resemble in many ways both the structures and specialized functions of mitochondria. A diagrammatic drawing of a chloroplast is shown in Figure 11-5.

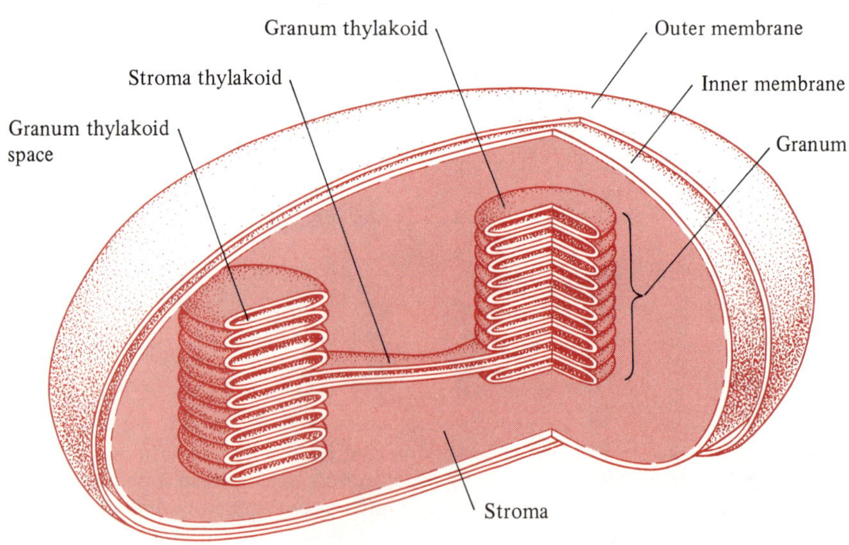

Figure 11-5
A diagram showing the essential structure of a chloroplast. The "light reactions" of photosynthesis are a function of the grana, with electron transport and photophosphorylation closely related to the thylakoid membranes

The process of photosynthesis involves one set of reactions that will occur only in the presence of light, and another set of reactions that can proceed in light or dark if the necessary levels of reactants are present. These two groups are generally referred to as the "light reactions" and the "dark reactions." Since the latter term is somewhat misleading, it is also common to speak of the "dark reactions" as representing the "path of carbon" in photosynthesis. This usage recognizes that the light reactions are responsible for generation of energy-rich compounds which are then used in the dark reactions to help bring about the fixation of carbon dioxide and its eventual conversion into sugar units.

The Light Reactions

The light reactions serve to enable the chloroplast to produce ATP and NADPH using available light as an energy source. In the process, the hydrogens of water molecules are used, with their electrons, to reduce the $NADP^+$. The remaining oxygen radicals form the oxygen gas which is released as a product of photosynthesis. The exact mechanisms by which these processes occur are complex and at the present time not well understood. The best we can do here is to provide a simplified general look at what is happening; a more detailed picture may be found in the references at the end of the chapter.

Chlorophyll

Chlorophyll is, of course, the pigment that lends its color to the green plants. As you will see from the structure in Figure 11-6, the basic molecule is made up of four pyrrole rings (five-membered rings containing a nitrogen) with specific side-chains attached. This type of structure is called a porphyrin ring and exists in a number of compounds, the most common of which are chlorophyll, the heme of hemoglobin and related compounds, and the various cytochromes. In each of these cases, there are substantial differences in the nature of the side-chain groups attached to the porphyrin ring. In addition, chlorophyll has a magnesium ion complexed in the center of the ring, while the heme and cytochrome compounds hold an atom of iron instead.

The long circular chain of conjugated double bonds provides a pi-orbital system that gives rise to the intense colors of the compounds. In addition, in the case of chlorophyll, this system provides a relatively stable home for electrons that have been activated by the absorption of the energy of a quantum of light. This activation energy can also be transferred from one chlorophyll to another through a whole stacked series until it reaches a center where the electron transport and energy-transfer process can begin. In this way, a large number of chlorophyll molecules may serve as part of an antenna system to absorb photons of light and transfer their energy to a central location for further use.

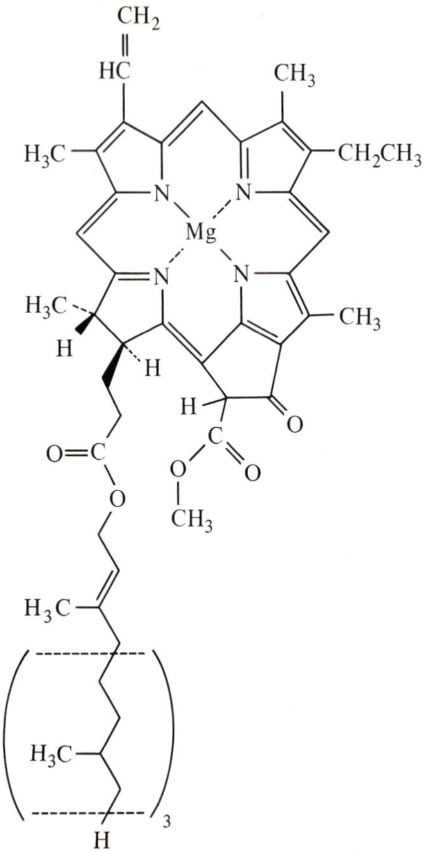

Figure 11-6
The structure of chlorophyll *a*.

The Electron Transport Systems

Because there is not yet a clear picture of the process, it is not possible to present a firm view of the electron-transport and energy-exchange reactions of photosynthesis. The drawing in Figure 11-7 represents one popular concept of how the system works. Two separate photosystems are believed to be involved. In each case, energy from light trapped by chlorophyll is used to promote some intermediate compound to a form in which it has a higher energy (that is, an "activated state").

The Path of Carbon in Photosynthesis

We started our discussion looking at the overall reaction of photosynthesis:

$$6\ CO_2 + 6\ H_2O + \text{light energy} \longrightarrow C_6H_{12}O_6 + 6\ O_2$$

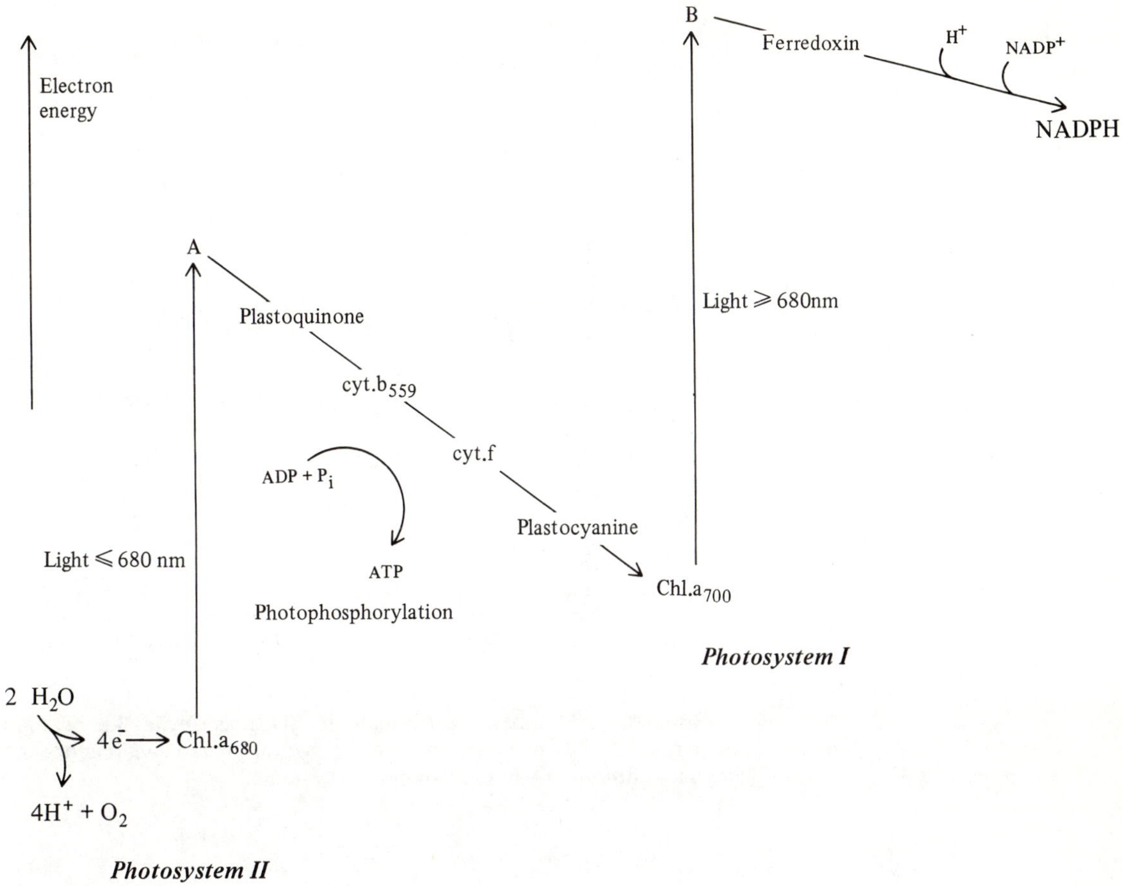

Figure 11-7
One concept of the nature of the electron transport system of photosynthesis. Note the two photosystems, each operating with light of different wavelength range. In this system, electron flow is from H_2O to $NADP^+$. Under some circumstances, a *cyclic* flow of electrons can occur from "B" to chl a_{700}; this produces ATP rather than NADPH. "A" and "B" are electron carriers of unknown identity. Chl a_{680} and chl a_{700} are chlorophyll *a* molecules in special environments.

So far, we have accomplished the splitting of water and release of oxygen, along with the production of ATP and NADPH. Now we wish to look at those reactions involved in making carbon dioxide into glucose. It is worth emphasizing once more that these reactions depend upon an adequate supply of ATP and NADPH, but are not directly involved with light. Thus the term "dark reactions" of photosynthesis.

This pathway is closely integrated with, and relies heavily upon, reactions of the phosphogluconate and glycolysis pathways. In order to

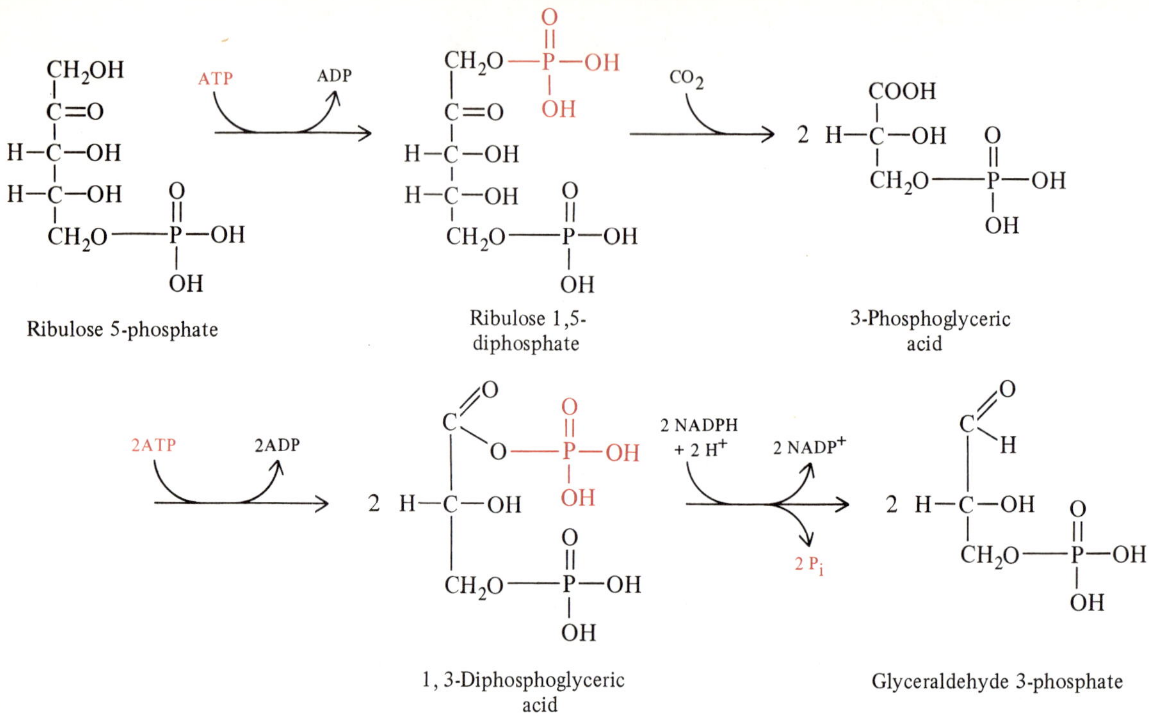

Figure 11-8
Reactions leading from ribulose 5-phosphate to glyceraldehyde 3-phosphate during photosynthetic carbon reduction. The first two of these reactions are the only new ones that are unique to photosynthetic systems.

operate as a true cycle and complete the reaction shown above, the process may best be viewed as starting with the product of phosphogluconate oxidation, ribulose 5-phosphate. The pathway from ribulose 5-phosphate and CO_2 to two molecules of glyceraldehyde 3-phosphate is shown in Figure 11-8.

The first two steps represent reactions that are new to us. First, ribulose 5-phosphate is converted to the 1,5-diphosphate by a specific kinase enzyme using ATP. The diphosphate then undergoes a rather unusual reaction in which CO_2 reacts with the second carbon. In the ensuing electronic rearrangement, the CO_2 becomes a carboxyl group and the molecule is split to yield two molecules of glyceric acid 3-phosphate (also called 3-phosphoglycerate).

From this point on, the reactions are simply those of glycolysis and its reverse. The use of NADPH instead of NADH is consistent with this in that plants generally use NADP for many reactions which in other organisms involve NAD.

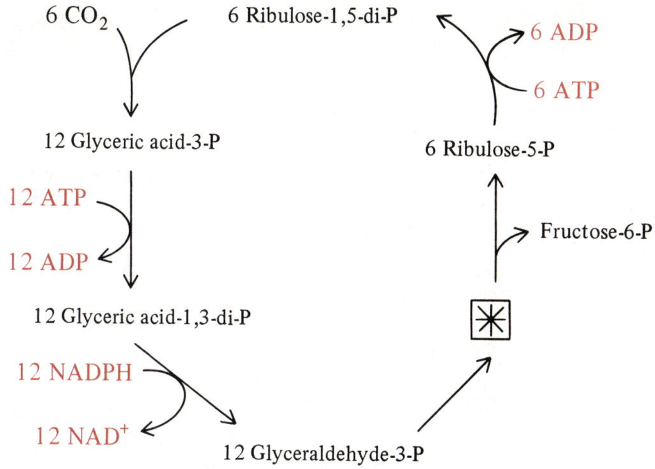

Figure 11-9
An abbreviated version of the photosynthetic carbon cycle (the Calvin Cycle). The transketolase, transaldolase, and related reactions have again been replaced with an asterisk for simplicity.

Once we reach glyceraldehyde 3-phosphate, we have completed all of the energy-requiring steps involved. The ATP generated in the light reactions has been used in the ribulose 5-phosphate kinase reaction and in forming 1,3-diphosphoglycerate from 3-phosphoglycerate. The NADPH has been used to reduce the diglycerate to the level of glyceraldehyde 3-phosphate. The remainder of the steps require no significant input of energy.

Of course, we could now simply take the two molecules of glyceraldehyde 3-phosphate and reverse glycolysis to produce a molecule of glucose 6-phosphate. But this would be cheating. We started with a molecule of ribulose 5-phosphate, and if the process is to continue indefinitely there must be some provision for regeneration of this starting material. The transketolase, transaldolase, and related reactions provide a means to this end. The integrated cycle (often called the Calvin cycle) is shown in Figure 11-9. For clarity, the various transaldolase and transketolase reactions have again been replaced by a boxed asterisk.

A balanced reaction for this cycle is interesting:

$$6\,CO_2 + 12\,NADPH + 12\,H^+ + 18\,ATP \longrightarrow \text{glucose-6-P} + 12\,NADP^+ + 18\,ADP + 17\,P_i$$

First we can look at the cost of joining these six CO_2 molecules into one glucose 6-phosphate. If we stay with the convention that one NADPH is worth three ATPs, then we could calculate a "cost" of 54 ATPs for this process. The reverse process, whether by way of the citric acid cycle or the phosphogluconate pathway, yields only about 36–38 ATPs. This reflects clearly the greater energy cost in forcing a synthetic pathway to go in the face of a very unfavorable equilibrium situation.

11.3 Summary to This Point

With the addition of the pentose phosphate pathways and photosynthesis, we have added substantially to the skeleton of carbohydrate metabolic pathways. The metabolic tree with these new branches added is shown in Figure 11-10.

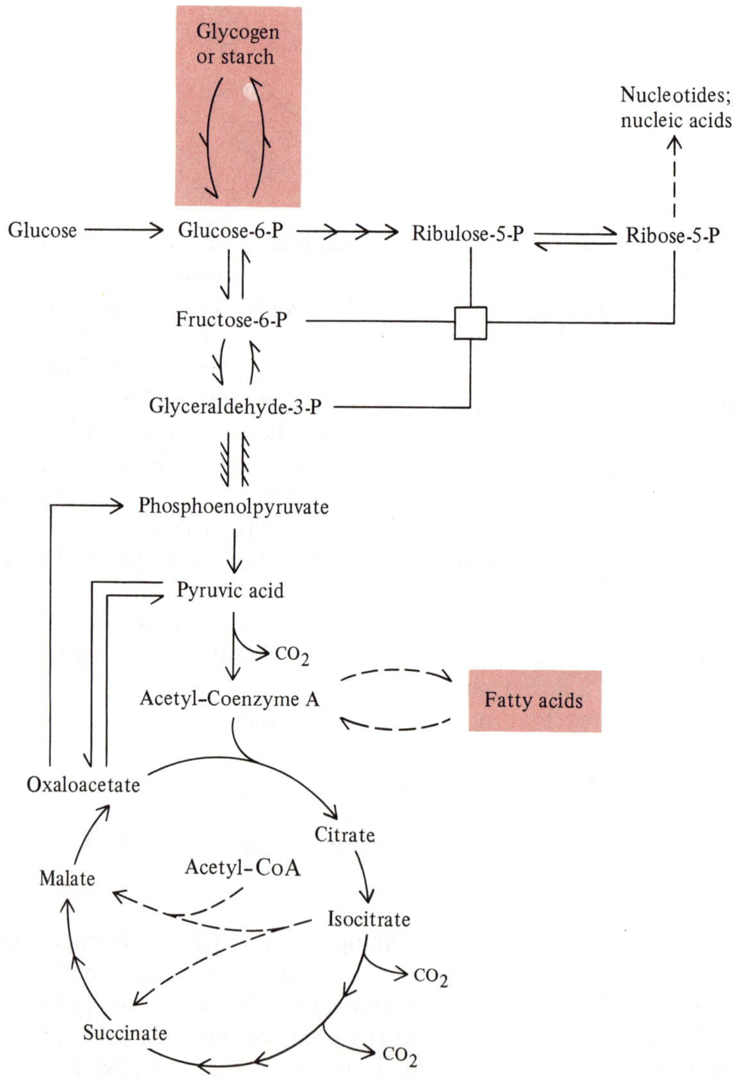

Figure 11-10
The growing "tree" of carbohydrate metabolism. The shaded areas remain to be discussed.

Additional Readings

Scientific American articles

J.A. Bassham, "The Path of Carbon in Photosynthesis," 206(6), 88 (June 1962).

Govindjee and R. Govindjee, "The Primary Events of Photosynthesis," 231, 68 (December 1974).

O. Björkman and J. Berry, "High-Efficiency Photosynthesis," 229(4), 80 (October 1973).

K. R. Miller, "The Photosynthetic Membrane," 241(4), 102 (October 1979).

The standard texts listed at the end of Chapter 1.

C. P. Whittingham, *The Mechanism of Photosynthesis* (Edward Arnold, London, 1974).

Govindjee (Ed.), *Bioenergetics of Photosynthesis*, (Academic Press, New York, 1975). A set of advanced review-type articles.

B. Axelrod, "Other Pathways of Carbohydrate Metabolism," in *Metabolic Pathways*, edited by D. M. Greenberg, 3rd edition, Vol. 1 (Academic Press, New York, 1967). A very thorough examination of the pentose phosphate pathways and related processes.

Study Questions

1. What are the three major functions of the pentose phosphate pathways?

2. Anaerobic organisms must have ribose 5-phosphate if they are to reproduce. Indicate by what sequence of reactions this is possible for such an organism growing on glucose as its sole source of carbon. (Note that you cannot leave any reduced nucleotides lying around.)

3. Write out a sequence of reactions that would constitute a pathway by which the carbons of succinic acid could be converted in large part to ribose 5-phosphate. Indicate which carbons of the succinate would appear in which positions of the ribose 5-phosphate.

4. In the *dark reactions* of photosynthesis, the ATP and NADPH produced in the *light reactions* is used to drive synthesis of carbohydrates. Write equations for
(1) the reaction(s) in which ATP is used
(2) the reaction(s) in which NADPH is used

5. Compare the yields in ATPs per mole of glucose for oxidation by either (a) glycolysis and the citric acid cycle, or (b) the phosphogluconate pathway operating purely as an oxidative mechanism. (Assume that one NADPH is worth three ATPs.)

6. Write out a set of reactions in a form similar to that of Figure 11-3 to replace the asterisk of Figure 11-9.

CHAPTER 12

Carbohydrate Storage and Some Related Reactions

Most cells store a certain amount of carbohydrate as a readily available reserve in the form of polysaccharides, which are usually made up of glucose units. The common forms in plants are amylose and amylopectin structures; in animals, the more soluble glycogen generally predominates. We have already reviewed the structures of these molecules in Chapter 7.

The polysaccharides provide a reserve supply of glucose units which can be used in times of sudden demand as a source of energy for the cell. As the cellular levels of ATP and related compounds drop, the cell must be able to increase the flow of carbohydrates through the oxidative pathways to regenerate the ATP. Generally, the stored carbohydrates are the most rapidly available sources of energy. The breakdown reactions feeding into the oxidative pathways must be capable of undergoing rapid activation and equally rapid damping when the energy state of the cell improves. There must also be a mechanism for the synthesis of the storage materials in the face of unfavorable equilibria. All of this leads naturally to separate pathways for synthesis and breakdown, each of them highly controlled by a number of the mechanisms which govern enzyme activity.

12.1 Glycogen Phosphorylase

It will probably be clearer in the long run if we begin our discussion with a look at the process by which the glucose units of a polysaccharide such as glycogen are fed into glycolysis and the oxidative cycles. The enzyme **glycogen phosphorylase** catalyzes the following reaction:

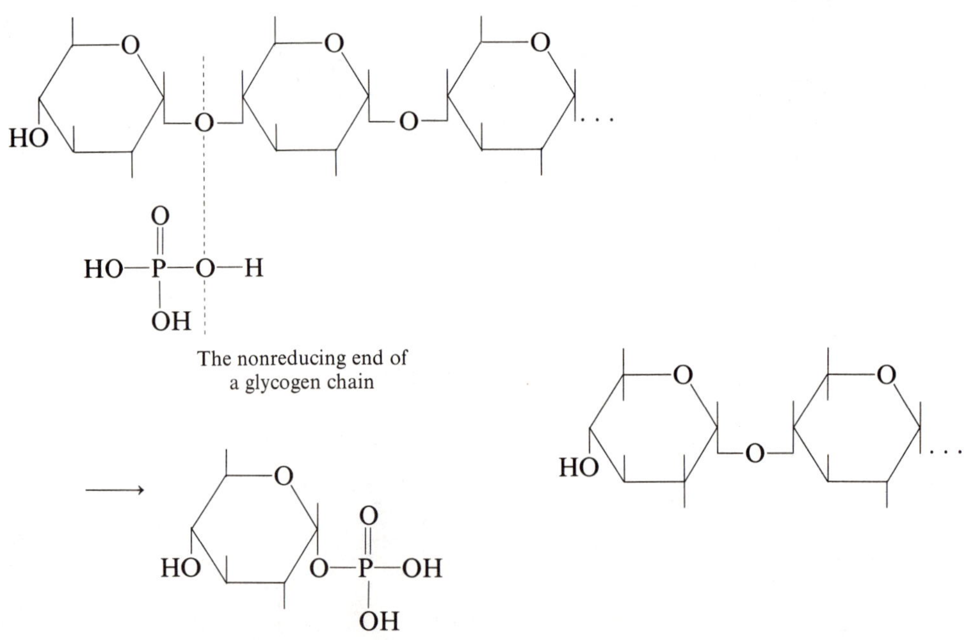

The nonreducing end of a glycogen chain

Glucose 1-phosphate

The reaction here is analogous to hydrolysis, except that phosphoric acid, not water, is added across the broken bond; hence, the term **phosphorolysis** for this type of reaction. Note that one glucose unit is removed from the nonreducing end of the polymer, and that the product of the reaction is glucose 1-phosphate.

The reaction as written is readily reversible, and it might on the face of it appear that it would provide a means both for glycogen breakdown and for its synthesis. However, a number of factors render its function very much a one-way street. First is the fact that the normal cellular level of glucose 1-phosphate is very low. It is rapidly converted to glucose 6-phosphate by a reaction we will look at shortly. In addition, the concentration of "ends" of glycogen chains to which a glucose can be attached is rather constant in the cell. Storing more glucose units involves lengthening the chains much more than it does increasing the number of them. Consequently, there is no effective way of forcing the reaction to go in the direction of polysaccharide synthesis.

On the other hand, phosphorolysis of the polymer chains will be encouraged in the presence of a high concentration of inorganic phosphate, as you might expect under conditions in which the cell would be low in ATP and high in ADP and P_i. Further, the phosphorylase molecule undergoes

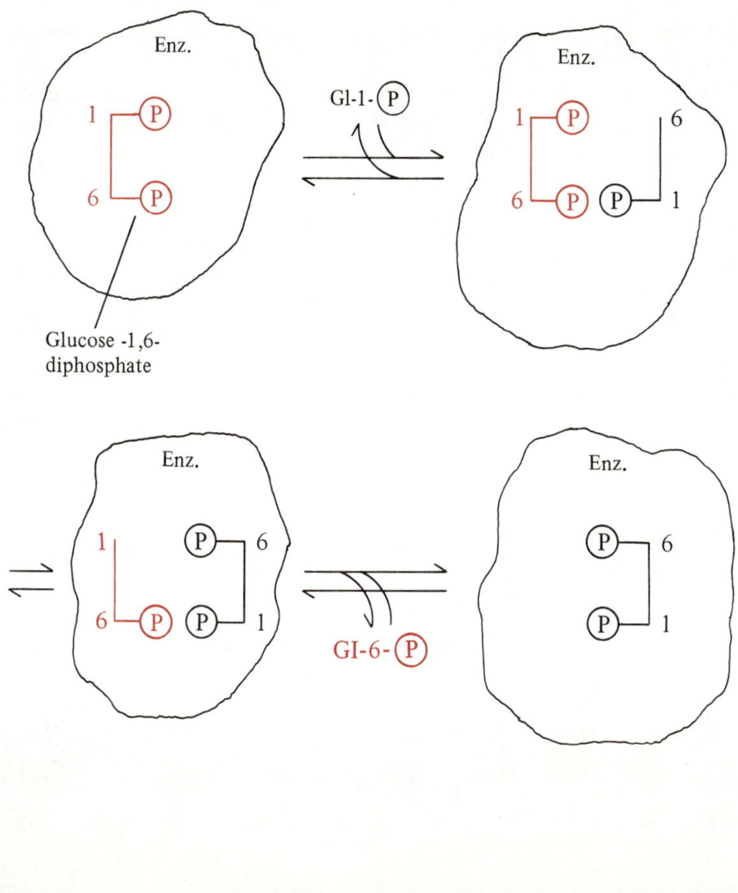

Figure 12-1
The role of the coenzyme, glucose 1,6-diphosphate, during the conversion of glucose 1-phosphate to glucose 6-phosphate by the enzyme phosphoglucomutase.

allosteric activation by ADP or AMP and in animal systems is affected by a number of hormones through a complex but extremely effective system of reactions leading to covalent modification of the structure of the enzyme. These and other aspects that make glycogen phosphorylase a fascinating example of metabolic control mechanisms will be discussed in Chapter 21.

Conversion of the glucose 1-phosphate to glucose 6-phosphate is catalyzed by an enzyme of the isomerase group with the common name of "phosphoglucomutase." The enzyme uses as a coenzyme a molecule of glucose 1,6-diphosphate which "trades" phosphate groups with the substrate molecules. The process is illustrated in Figure 12-1. The equilibrium for the reaction lies substantially in the direction of glucose 6-phosphate ($\Delta G^{\circ\prime} = -7.3$ kJ/mole).

12.2 Role of the Uridine Nucleotides in Glycogen Synthesis

The conversion of one of the glucose units of a polysaccharide such as glycogen to glucose 6-phosphate is clearly an energetically favored process. We should then expect that the reversal of the pathway will require some input of energy. In this case, the necessary energy comes not from ATP but from the closely related compound UTP (uridine triphosphate). UTP differs from ATP only in the structure of the nitrogen heterocycle involved, but its use permits a greater flexibility and a much more delicate control than would be the case were ATP used instead.

The first reaction is simply an anhydride exchange reaction, in which glucose 1-phosphate displaces the terminal pyrophosphate group from the UTP and is itself attached instead. The reaction may be written as follows:

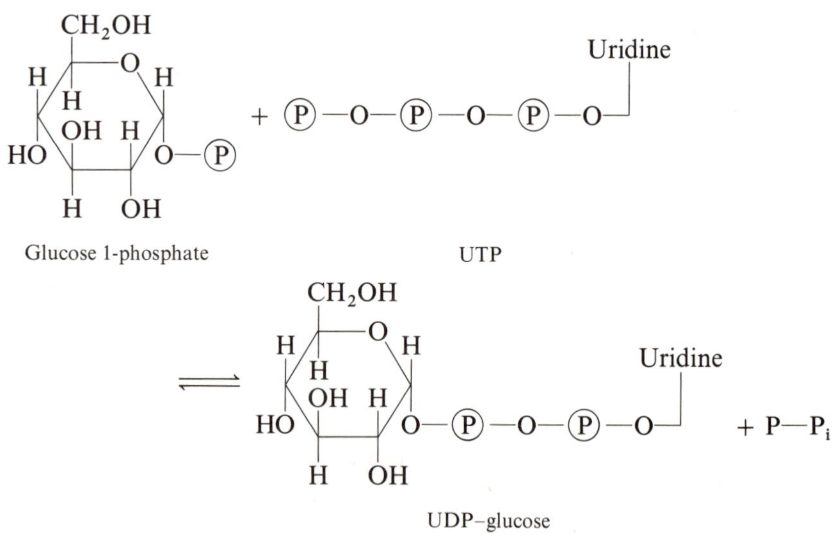

The second step is the transfer of the glucose unit from UDPG to the end of the growing amylose chain:

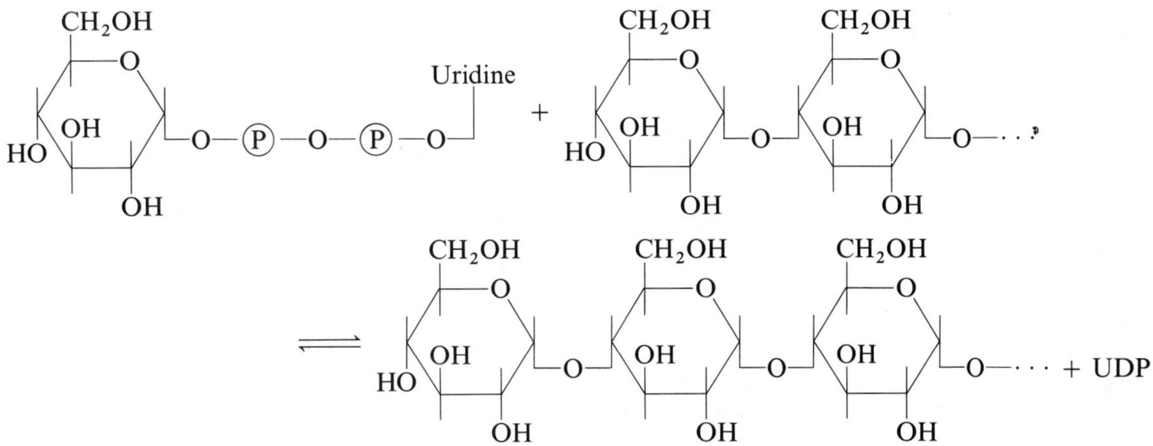

This leaves us with an overall process that can be summarized in the equation

$$\text{glucose 1-phosphate} + \underset{(\text{“}n\text{” glucose units})}{\text{glycogen}} + \text{UTP} \longrightarrow \underset{(\text{“}n+1\text{” glucose units})}{\text{glycogen}} + \text{UDP} + \text{P–P}_i$$

Since the UDP will have to be reconverted to UTP at the expense of "one ATPs worth" of energy, it is clear that this is the basic driving power for the reaction. However, it is useful to note another aspect of this process also. In the equation above, we have written the reaction of glucose 1-phosphate and UTP to give UDPG as fully reversible. As for most anhydride exchange reactions, this *is* almost an isoenergetic process as written, but in this case there is little likelihood of the reverse reaction in most cells. Enzymes called **inorganic pyrophosphatases** are widespread in tissues and highly active. Inorganic pyrophosphate has little opportunity to accumulate, being instead rapidly hydrolyzed to two molecules of inorganic phosphate. Consequently, formation of UDPG is effectively a one-way reaction.

The enzymes of this synthetic pathway are controlled by mechanisms very similar to those that we have mentioned in connection with the phosphorylase reaction. They serve largely to encourage storage when cellular energy reserves are high and to shut down the synthetic process when energy is needed.

The reactions involved in glycogen storage, and their relationship to the rest of carbohydrate metabolism, are indicated schematically in Figure 12-2.

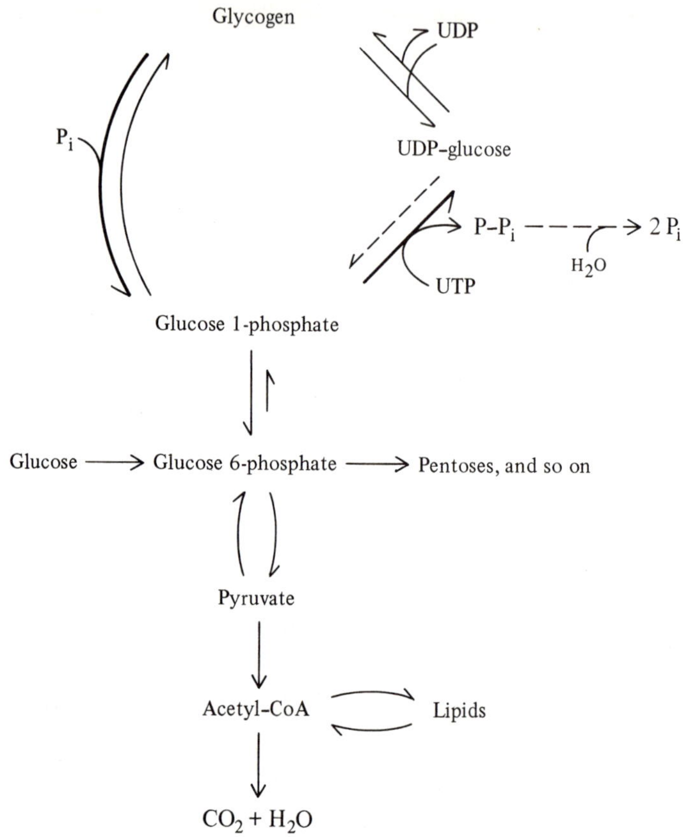

Figure 12-2
Glycogen storage and release and the relationship of these processes to the rest of carbohydrate metabolism and related reactions.

12.3 Sugars Other Than Glucose

In dealing with the pathways by which sugars other than glucose enter into carbohydrate metabolism, we can look first at the fate of various sugars in a typical animal. Then we can deal with some special reactions in plants and microorganisms.

The sugars most generally available in the diet of a human or similar animal are glucose, galactose, fructose, and mannose. Pentoses may be absorbed from the intestines, but are not utilized and are excreted unchanged in the urine.

The circulating blood supply normally contains very low levels of any sugar other than glucose. This results from the conversions that are carried out during passage through the intestinal wall cells and after absorption by the liver. The only sugar that the liver is capable of releasing into the blood is glucose.

Utilization of Mannose

If we look at what is likely to happen to hexoses other than glucose after entry into a liver cell, we find that mannose will fit into the glycolytic pathway rather easily. It is acted upon by hexokinase to form mannose 6-phosphate; reaction of this with an isomerase yields fructose 6-phosphate:

$$\text{mannose} + \text{ATP} \longrightarrow \text{mannose 6-phosphate} + \text{ADP}$$

$$\text{mannose 6-phosphate} \rightleftharpoons \text{fructose 6-phosphate}$$

Utilization of Galactose

Galactose encounters a special enzyme (galactokinase) and is converted directly to galactose 1-phosphate. This compound then undergoes an exchange reaction with UDP–glucose to give UDP–galactose and glucose 1-phosphate:

$$\text{galactose} \xrightarrow[\text{ADP}]{\text{ATP}} \text{galactose 1-phosphate}$$

$$\text{galactose 1-phosphate} + \text{UDP–glucose} \rightleftharpoons \text{UDP–galactose} + \text{glucose 1-phosphate}$$

$$\text{UDP–galactose} \rightleftharpoons \text{UDP–glucose}$$

A specific isomerase interconverts UDP–galactose and UDP–glucose and completes the process. The net result of all of this is the conversion of galactose 1-phosphate to glucose 1-phosphate.

Lack of the ability to synthesize the enzyme catalyzing the transfer reaction between galactose 1-phosphate and UDP–glucose gives rise to **hereditary galactosemia**, a severe problem in infants. This is not to be confused with **lactose intolerance**, a condition common among many adult populations in the world. This latter condition reflects an impaired ability to digest lactose and produces intestinal upsets when cow's milk or another lactose source is consumed.

Utilization of Fructose

The pathway of utilization of fructose is a little more complex. The enzyme fructokinase converts fructose to fructose 1-phosphate but no enzyme is known that will convert this compound to the 6-phosphate or to one of the other sugar-phosphate intermediates directly. There is, however, a special aldolase-type enzyme which will act on the fructose 1-phosphate to produce dihydroxyacetone phosphate and free glyceraldehyde:

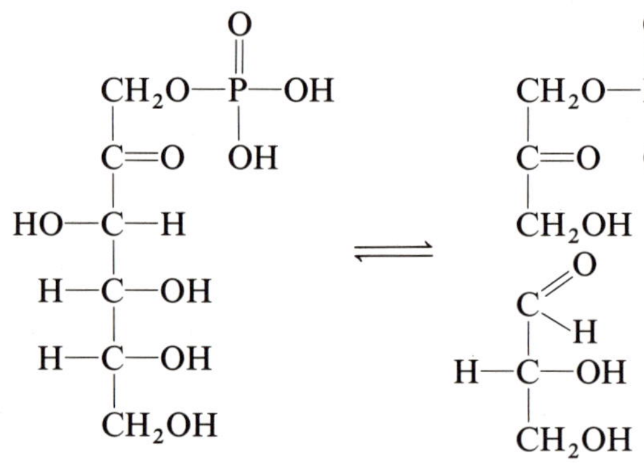

The free glyceraldehyde can be reduced to glycerol, phosphorylated to glycerol phosphate, and then re-oxidized to dihydroxyacetone phosphate in order to get into the mainstream of glycolysis.

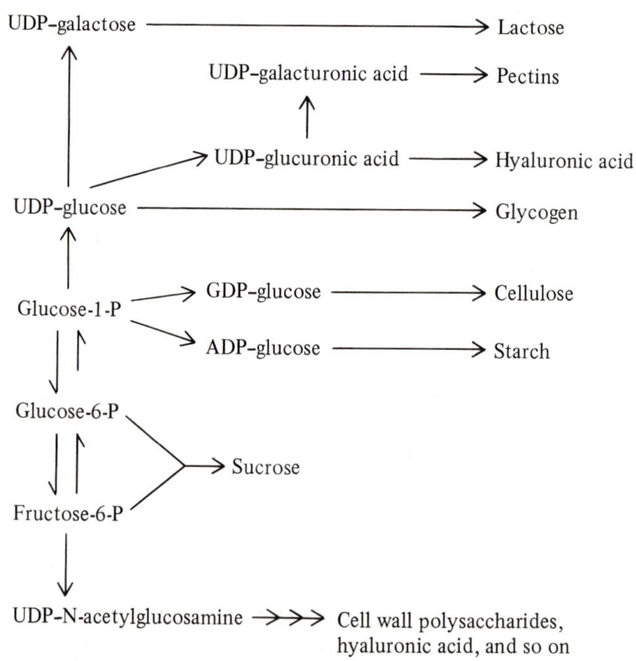

Figure 12-3
Examples of the involvement of nucleotide diphosphate derivatives in interconversions of sugars and in glycoside syntheses.

12.4 Synthesis of Other Sugar Derivatives

The pathways mentioned above will serve to indicate some of the mechanisms by which one sugar or sugar derivative can be converted into another. A wide variety of similar reactions exists for the production of the many special sugar derivatives used in cells for specialized purposes. Most of these reactions take place while the sugars involved are attached to one or another of the nucleotide diphosphates. Some examples of these conversions are indicated in Figure 12-3. You will note the involvement of specific nucleotides for different processes, even to the point of using UDP–glucose to make glycogen in animals but ADP–glucose to make starch and GDP–glucose to make cellulose in plants.

12.5 Reversal of Glycolysis—Gluconeogenesis

The ability to make glucose from other nutrients when total carbohydrate is in short supply is critical to an animal such as a human being that relies heavily on the circulating blood glucose as a source of energy for specialized tissues. It is equally important for many microorganisms existing solely on some di- or tricarboxylic acid as a source of carbon. We saw in looking at glycolysis that the major problem with reversal from pyruvate to glucose lay in the very unfavorable equilibrium of the pyruvate kinase reaction.

Then, in looking at the citric acid cycle, we found reactions by which to convert pyruvate to oxaloacetate and oxaloacetate to phosphoenolpyruvate, effectively bypassing this difficulty. Of course, any other compound that can be converted to oxaloacetate can be a source of glucose also. In the absence of incoming carbohydrate, the most usual source of glucose in animals is the breakdown of certain of the amino acids. Through reactions we will examine in Chapter 16, alanine, aspartic acid, and glutamic acid readily give rise to compounds that may be converted to oxaloacetate, and thus to glucose through the reversal of glycolysis. By more elaborate mechanisms, it is also possible for carbons from serine, isoleucine, and some other amino acids to get into the same pathway.

This process of forming glucose (and its relatives) from amino acids or other intermediates not usually part of carbohydrate metabolism is often called **gluconeogenesis** to indicate formation of glucose from a source that did not enter the cell (or the body) as carbohydrate. This process is indicated in simplified form in Figure 12-4.

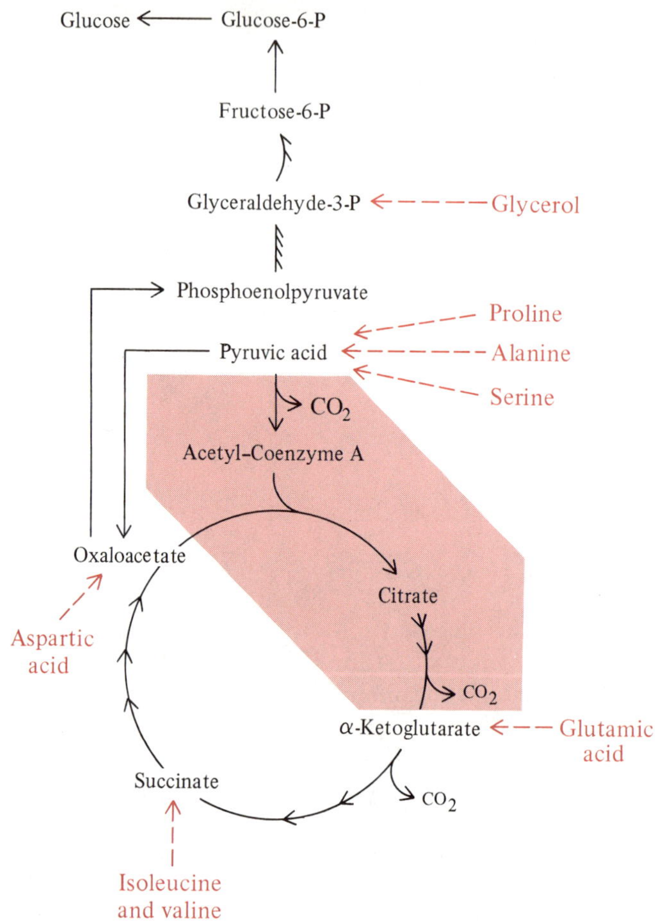

Figure 12-4
Gluconeogenesis. Some of the reactions by which amino acids and some other compounds can be converted into carbohydrates. The shaded reactions are not helpful in this process and are included only to maintain a complete frame of reference.

As long as we are on this topic of gluconeogenesis, it might be well to state again that the carbons of fatty acids, which are broken down to acetyl–Coenzyme A on the way to complete oxidation, cannot serve as a source of glucose in animals. This is one of the little difficulties that help explain why many of us put on fat a great deal more easily than we lose it.

12.6 The Glyoxylate Shunt Pathway

Many plants and some microorganisms *do* have a means by which they can accomplish the net synthesis of carbohydrates from acetyl–Coenzyme A. This has the virtue of allowing a very small seed to rely primarily on the very concentrated fats as carbon and energy sources, rather than having to pack in bulky carbohydrates. (In general, fats can generate 36 kJ/gram of available energy on oxidation, carbohydrates only 16 kJ/gram.) In any case, this pathway bypasses much of the citric acid cycle and short-circuits its operation; hence, the name *glyoxylate "shunt."* The pathway is shown in Figure 12-5, with the unused reactions of the citric acid cycle shown behind a color screen simply to provide a frame of reference.

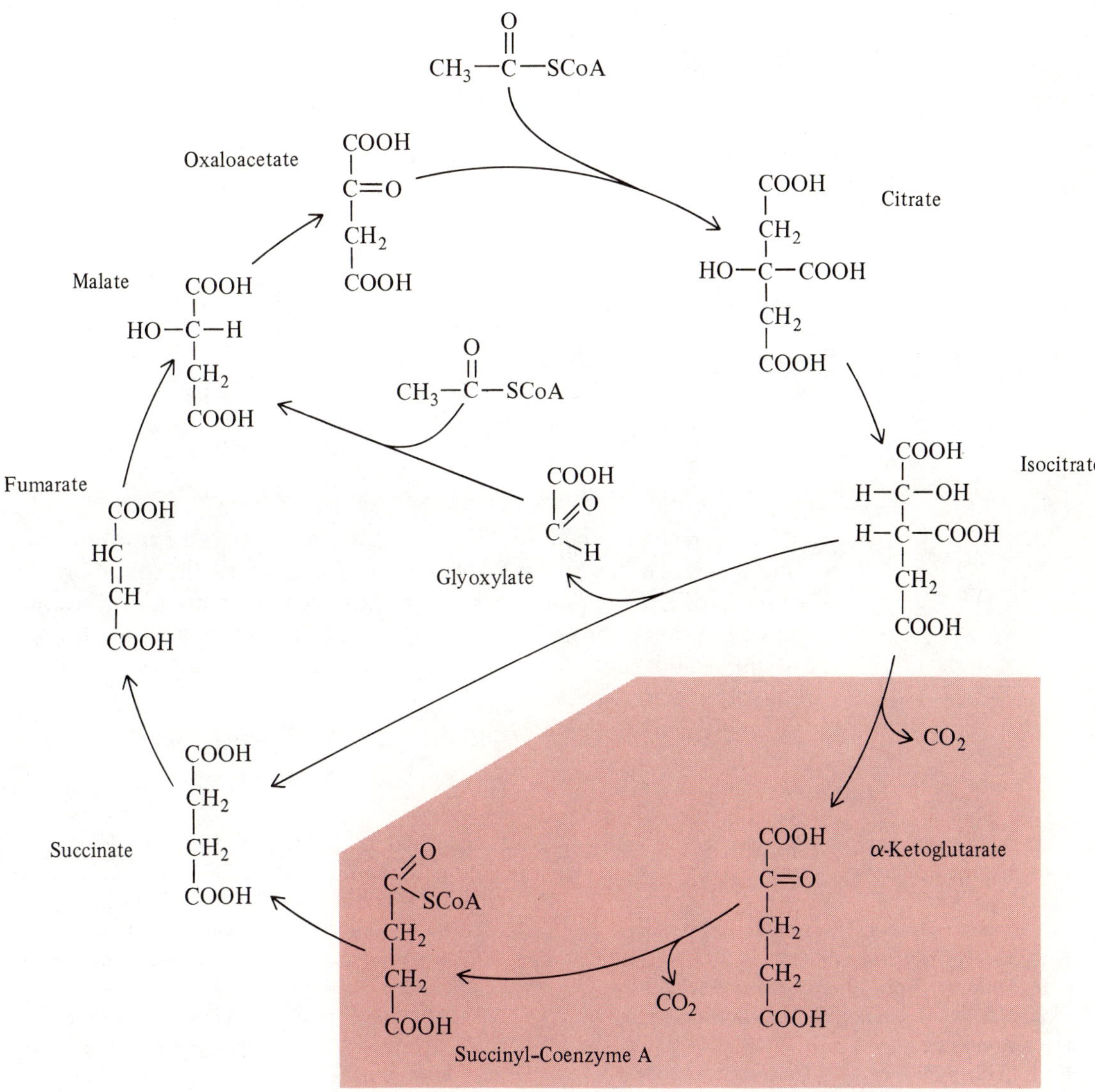

Figure 12-5
The glyoxalate shunt pathway. The shaded portion of the citric acid cycle is not involved in this scheme. This pathway is not known in animals.

Two new enzymes are involved, catalyzing the following reactions:

(a)

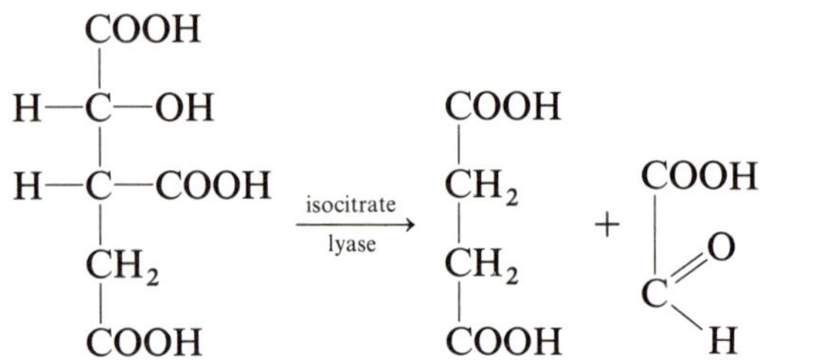

Isocitric acid Succinic acid Glyoxylic acid

(b)

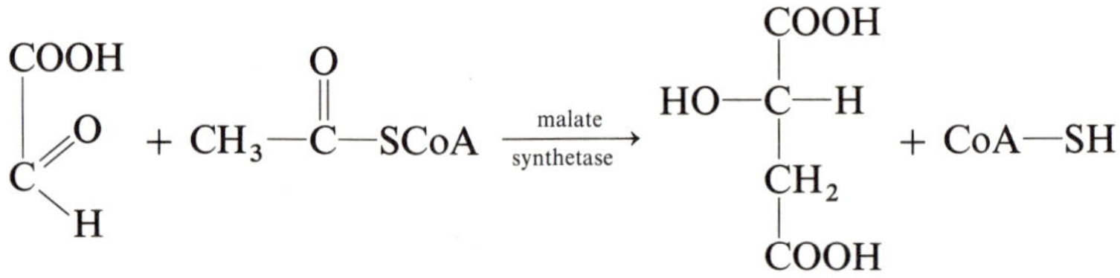

Glyoxylic acid Acetyl–Coenzyme A Malic acid

The net result of operation of this "shunt" pathway is the conversion of two molecules of acetate in the form of acetyl–Coenzyme A into one molecule of oxaloacetate. In a practical situation, the relative amounts of acetyl–Coenzyme A going through the regular citric acid cycle and going through the shunt would depend on the energy state of the cell and the need for carbohydrate compounds.

Additional Readings

W. Z. Hassid, "Biosynthesis of Oligosaccharides and Polysaccharides in Plants," *Science 165*:137 (1969).

E. H. Fischer, A. Pocker, and J. C. Saari, "The Structure, Function and Control of Glycogen Phosphorylase," in *Essays in Biochemistry*, edited by P. N. Campbell and F. Dickens, Vol. 10 (Academic Press, New York, 1974). An outstanding discussion of a most interesting enzyme.

B. Axelrod, "Other Pathways of Carbohydrate Metabolism," in *Metabolic Pathways*, Vol. 1, 3rd edition, edited by D. Greenberg (Academic, New York, 1967).

W. Pigman and D. Horton, *The Carbohydrates*, 2nd edition (Academic Press, New York, 1970–72) (4 vols.) An authoritative account of many aspects of carbohydrate chemistry and biochemistry, including a more extensive treatment of many of the topics of this chapter.

V. Ginsberg, "Sugar Nucleotides and the Synthesis of Carbohydrates," in *Advances in Enzymology*, Vol. 26, (1964), pp. 35–88.

Study Questions

1. Bacteria and molds grow in an amazing variety of media. In the laboratory, you may find microorganisms growing even in solutions of sulfuric, phosphoric, or acetic acids. Since every living cell must be able to synthesize ribose 5-phosphate in order to reproduce, show a sequence of reactions by means of which a microorganism growing in a solution of acetic acid (as sole carbon source) might make the required ribose 5-phosphate.

2. You know already that the overall energy state of a cell is directly related to the relative concentrations of ATP and of ADP and P_i. We have also seen that UMP or UDP can be rapidly converted to UTP at the expense of ATP. Draw diagrams showing the mass law effects of (a) relatively high levels of ATP, and (b) relatively low levels of ATP on the reactions of glycogen synthesis and utilization.

3. Write a complete sequence of reactions by which a molecule of fructose could be converted into a molecule of glucose 6-phosphate.

4. Write out a pathway of reactions for the synthesis of glucose 6-phosphate completely from pyruvate in a typical animal cell. Show each reaction clearly.

5. Similarly, show how a plant cell might utilize the acetyl–CoA molecules produced from fatty acid oxidation (which we will examine in Chapter 14) to carry out net synthesis of glucose 6-phosphate.

CHAPTER 13

Lipids: Structures, Nomenclature, and Hydrolysis

Attempts at a rigorous definition of the class of compounds to be called **lipids** usually become either hopelessly complicated or completely misleading. Generally, however, the lipids include those compounds which can be extracted from biological materials by organic solvents such as ether–alcohol mixtures, acetone, or various chlorinated hydrocarbons.

Fortunately, the **fatty acids** are much more readily defined. They are simply those monocarboxylic acids which are more soluble in organic solvents than they are in water; that is, essentially those with four or more carbon atoms.

The fatty acids occur in nature largely as components of more complex lipid materials. Before proceeding to look at various classes of lipids, we will therefore consider the various fatty acids which are of common occurrence.

13.1 Fatty Acids

The naturally occurring fatty acids include almost all of the saturated acids having an even number of carbon atoms in the range from 4 to 30 carbons. Other prominent fatty acid species are the 16-, 18-, and 20-carbon unsaturated fatty acids.

Saturated Fatty Acids

The general formula for these acids is $CH_3—(CH_2)_n—COOH$. Among those occurring most commonly are the following (the subscript numbers refer to the total number of carbon atoms):

Butyric acid	C_4			Palmitic acid	C_{16}
Lauric acid	C_{12}	Myristic acid	C_{14}	Stearic acid	C_{18}

Unsaturated Fatty Acids

The following are the most important of the unsaturated fatty acids:

Oleic acid (C_{18}, one double bond)

$$CH_3(CH_2)_7CH{=}CH(CH_2)_7COOH$$

Linoleic acid (C_{18}, two double bonds)

$$CH_3(CH_2)_4CH{=}CH—CH_2—CH{=}CH(CH_2)_7COOH$$

Linolenic acid (C_{18}, three double bonds)

$$CH_3CH_2—CH{=}CH—CH_2—CH{=}CH—CH_2—CH{=}CH(CH_2)_7COOH$$

Arachidonic acid (C_{20}, four double bonds)

$$\cdots—CH_2—CH{=}CH—CH_2—CH{=}CH(CH_2)_3COOH$$

$$CH_3(CH_2)_4CH{=}CH—CH_2—CH{=}CH—\cdots$$

Humans, along with many other animal species, are capable of inserting the one double bond to convert stearic acid to oleic acid, but are unable to convert oleic acid into either linoleic, linolenic, or arachidonic acid. The maintenance of normal structure and function, particularly of skin tissue, requires that cells have available at least one of these acids. Since they cannot be synthesized by the cells, they must be obtained from the dietary materials. However, most animal systems can interconvert linoleic, linolenic, and arachidonic acids. Therefore, the human diet need contain only one of these.

13.2 Glycerol Lipids

Acylglycerols

These are the simple neutral esters of glycerol and fatty acids, and include the mono-, di-, and triacylglycerols (mono-, di-, and triglycerides in older terminology). A large number of those compounds are possible, varying in the nature of the fatty acids involved and in the positions and order of their attachment to the glycerol molecule:

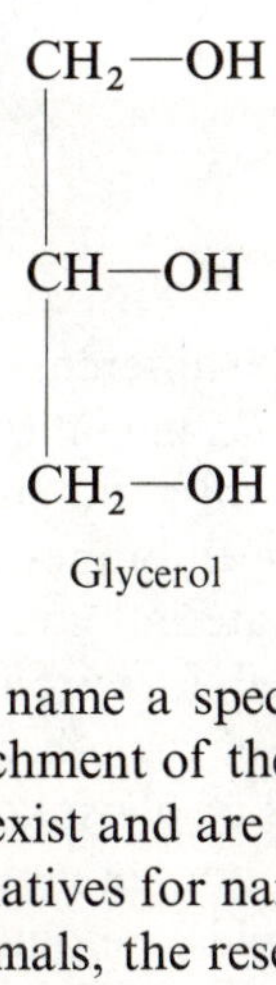

Glycerol

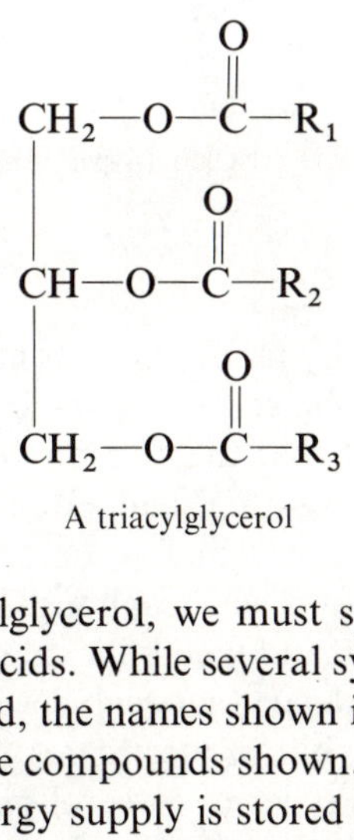

A triacylglycerol

In order to name a specific acylglycerol, we must specify clearly the positions of attachment of the fatty acids. While several systems of arriving at trivial names exist and are still used, the names shown in Figure 13-1 are the proper alternatives for naming the compounds shown.

In most animals, the reserve energy supply is stored predominantly in the form of triacylglycerols. This is relatively efficient, since lipids of this type yield, on oxidation, a great deal more energy for a given weight of stored material than either carbohydrates or proteins.

Glycerol Phospholipids

Any lipid containing phosphorus is included in the general class of **phospholipids**. Of these the glycerol phospholipids (sometimes referred to as

Figure 13-1
Examples of nomenclature for typical examples of triacylglycerols.

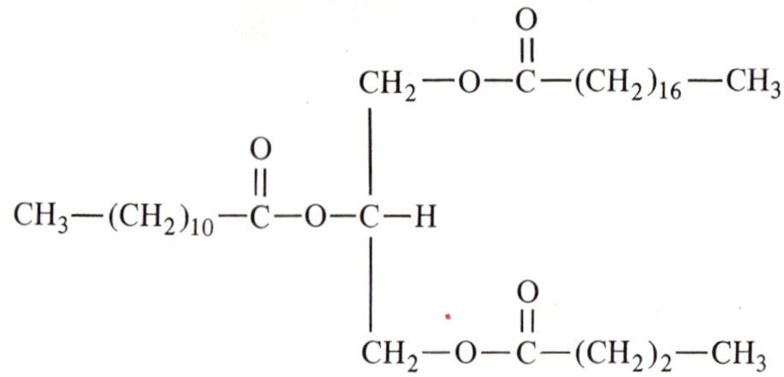

1-Stearoyl-2-lauroyl-3-butyroylglycerol

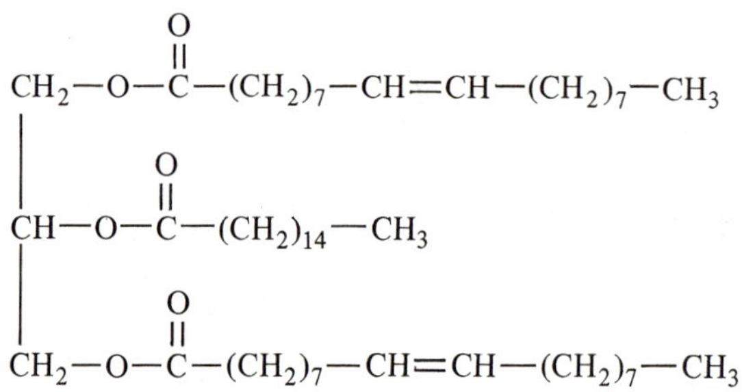

1-Oleoyl-2-palmitoyl-3-oleoylglycerol

phosphatides) are by far the most common. The different types of phospholipids appear to have certain biological functions in common. These can be related to the fact that the phospholipids are all good emulsifying agents. Large concentrations are usually found in cell membranes and in subcellular structures where lipids and water-soluble materials must be capable of intimate interaction. The phospholipids may therefore promote the transportation, absorption, and metabolic interactions of the other lipid substances. There is some evidence that each specific class of phospholipid has its own specific functions, but in most cases these are not yet clearly known to us.

The different types of glycerol phospholipids are outlined below. You will find that a number of them have the same essential structure, varying only in the nature of one of the groups linked to the phosphate portion of the molecule. Note also that most of the compounds are of phosphodiester nature and provide our first biological example of this type of bonding.

Phosphatidic Acids. These compounds are not found in tissue extracts in any significant concentration. However, they have been found to play important roles as intermediates in the biosynthesis of the other glycerol lipids:

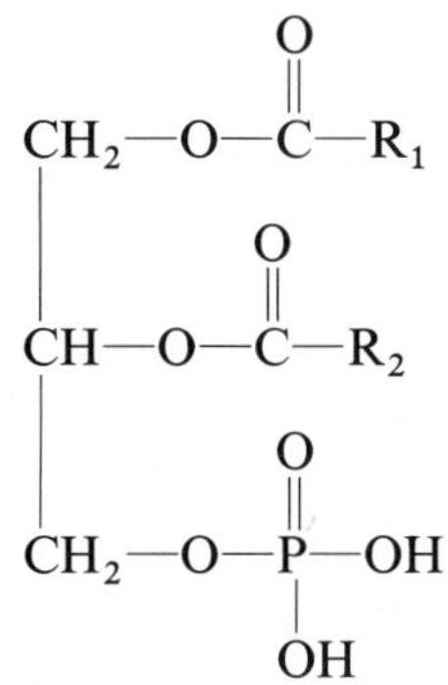

Phosphatidyl Cholines (Lecithins). This is probably the most common form of phospholipid, at least in animals. The lecithins are required for normal transport and utilization of other lipids, especially in the liver. Anything which interferes with the synthesis of choline will also block the synthesis of lecithins, and thus interrupt the normal transportation of lipids to and from the liver. This usually results in the accumulation of lipid material in the liver, giving rise to the condition called "fatty liver":

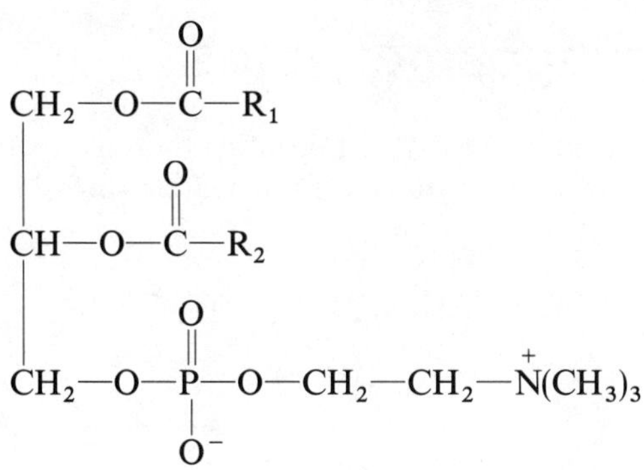

Free choline is a compound with an alcohol group. Its linkage to the phosphate portion of a lecithin, like that of the glycerol to the phosphate, is that of a phosphate ester:

$$HO{-}CH_2{-}CH_2{-}\overset{+}{N}(CH_3)_3$$

Choline

Phosphatidyl Ethanolamines, Phosphatidyl Serines, and Phosphatidyl Inositols. These phospholipids are similar in structure to the lecithins (above) but have the choline portion replaced by one of the compounds indicated:

$HO—CH_2—CH_2—NH_2$

Ethanolamine

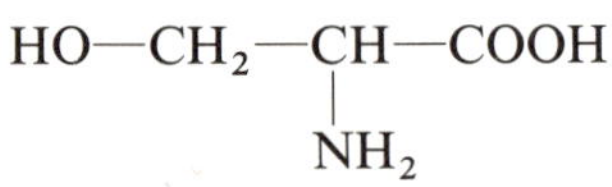

Serine

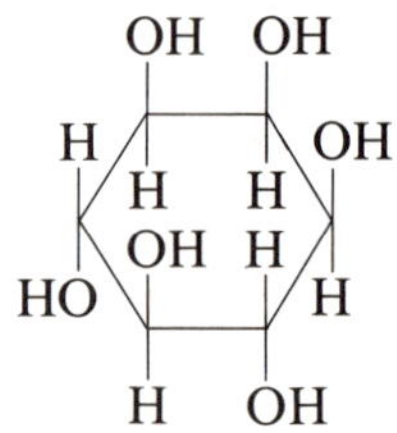

myo-Inositol

The ethanolamine and serine compounds are often referred to as the cephalins. They were originally distinguished from the lecithins on the basis of different solubility characteristics in mixed solvents.

In animals, the ethanolamine, serine, and inositol phospholipids are found in highest concentration in nerve tissues. Their exact functions there are not known.

13.3 Sphingosine Lipids

These contain as the alcohol portion either the long-chain amino alcohol **sphingosine** or the related saturated form **dihydrosphingosine**:

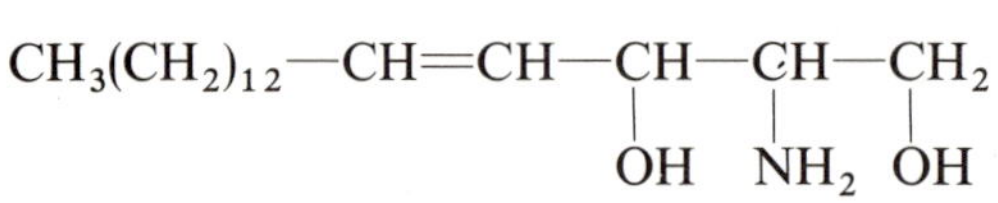

Sphingosine

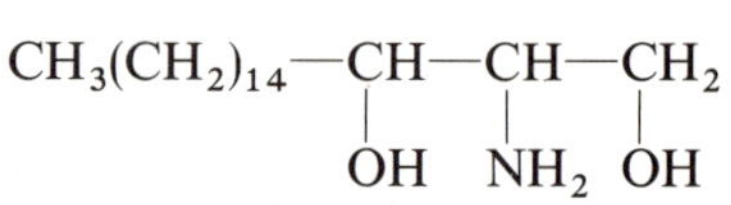

Dihydrosphingosine

The structures of many of these lipids have not as yet been well defined. Those listed below are examples of some of the types of sphingolipids that have been isolated and characterized.

Sphingomyelins. These lipids contain sphingosine plus a fatty acid attached in the form of an amide linkage and phosphorylcholine attached to the primary alcohol group:

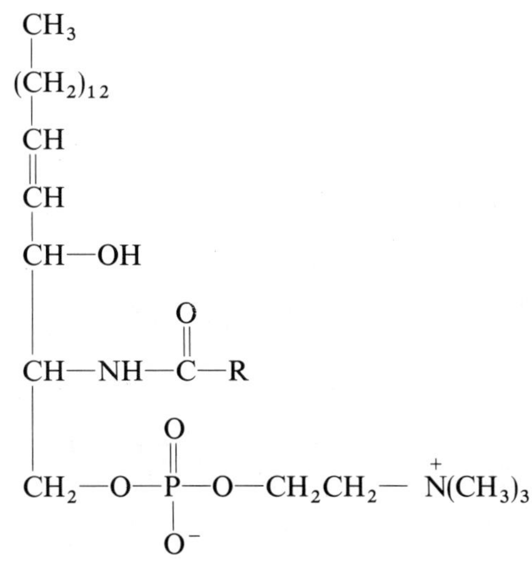

They are found commonly in nerve tissue, especially in the myelin sheath of the nerve.

Cerebrosides. These, too, are basically fatty acid amides of sphingosine, having instead of phosphorylcholine a galactose oligosaccharide joined to the primary alcohol group. This type of structure occurs widely but is most commonly isolated from brain tissue.

Gangliosides. This group is rather poorly defined. Attached to the fatty amide structure it is possible to find one or more molecules of glucose, galactose, *N*-acetylglucosamine, *N*-acetylneuraminic acid, and perhaps other sugar derivatives. The exact composition and structure of the molecules varies considerably with the source. The gangliosides are most commonly isolated from nerve or spleen tissues.

Phytoglycolipids. A wide variety of these are present in various plant sources. Generally, they appear to be based on the fatty acid amide of sphingosine or dihydrosphingosine, and contain in addition an assortment of sugars and often *myo*-inositol. Relatively few members of this group have been well characterized.

13.4 Steroids and Sterols

All steroid compounds have in common the basic ring structure which is known as a cyclopentanoperhydrophenanthrene nucleus:

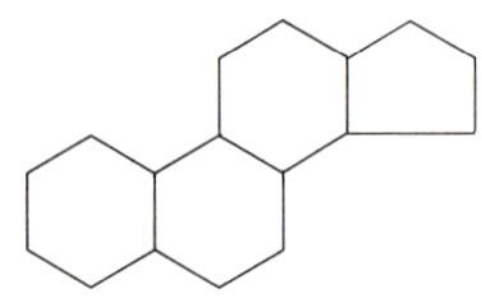

As the name implies, this is a fully saturated ring system. Various natural steroids, however, may have one or more double bonds as added embellishments. These compounds comprise one of the most ubiquitous and versatile families of compounds in nature. Small variations in structure or in the nature of attached groups are found to result in profound changes in biological activity. The structures of a number of different types of steroids having greatly varying biological activities are shown below:

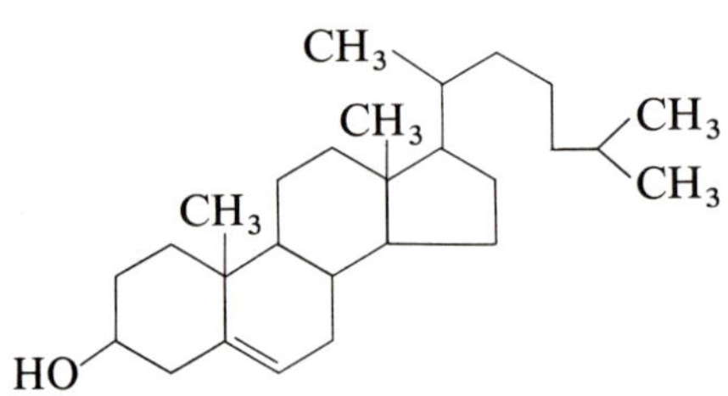

Cholesterol

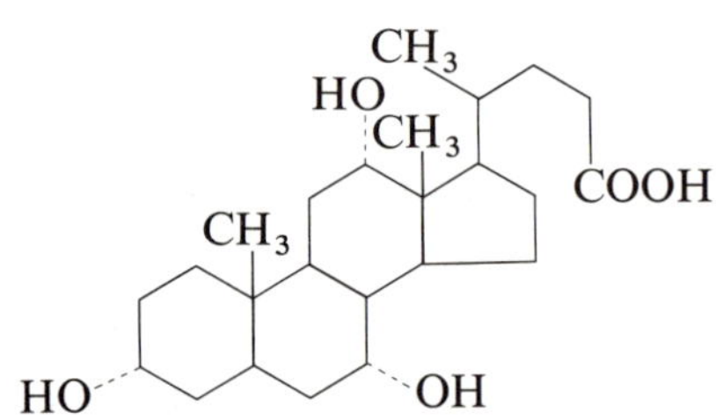

Cholic acid

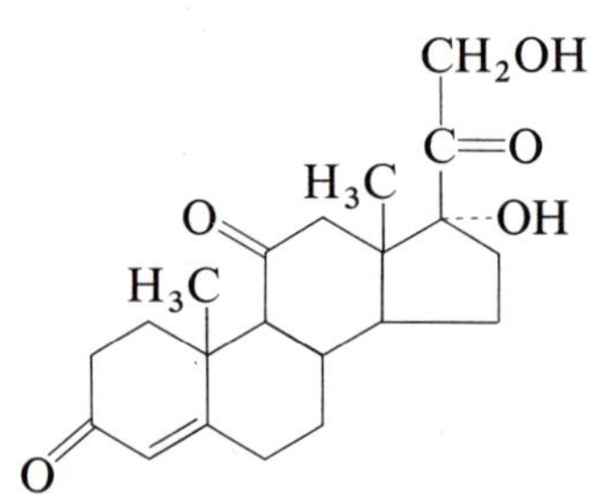

Cortisone

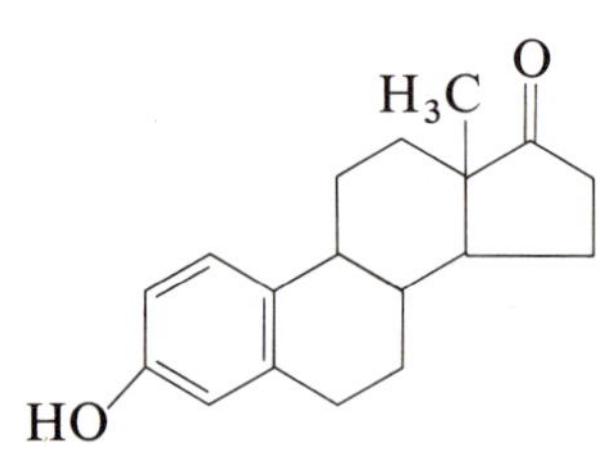

Estrone

While the total chemical synthesis of certain steroids has been accomplished, it is an extremely difficult process involving many reactions.

One of the problems involved is the necessity to obtain the correct optical isomer at each of at least five asymmetric centers in the molecule. Consequently, synthesis of new pharmacologically active steriods is normally accomplished by either chemical or microbiological modification of some suitable naturally occurring steroidal compound.

The bonds available for various substituents can be viewed as extending either above or below the overall plane of the molecule. In order to indicate this in the common formulas and names, those below the plane are denoted by a broken line and are called "α"-substituents. Solid lines are then used to denote bonds above the plane attaching "β"-substituents. These conventions have been used in the specific steroid formulas shown above.

Certain steroids, containing a hydroxyl group as their only functional one, are often referred to as **sterols**. The common examples of these are cholesterol, ergosterol, and stigmasterol; the last two occur in plants and the first in animals.

13.5 Fat Soluble Vitamins

Certain lipid compounds are vitamins; that is, they are required in small amounts in the diet for proper nutrition of some organisms (usually animals). Vitamins A, D, K, and E are required for humans and a variety of other animals.

In referring to Vitamins A, D, K, or E, we are not really referring to a single chemical compound but to a type of compound which will have a specified physiological effect. Thus, in each case there exist small families of closely related compounds, each of which will have, for example, "Vitamin A activity." In the brief discussion of the fat-soluble vitamins that follows, we will show structures only for the most common or most representative example in each case.

Vitamin A is involved in the processes of vision, maintenance of epithelial tissues, and growth in general. The chemical compounds having this activity are all closely related to the carotenoid plant pigments (which will be mentioned shortly) and, in fact, normally occur in the diet chiefly in the form of the various carotenes. Vitamin A is then formed from this material in the animal body:

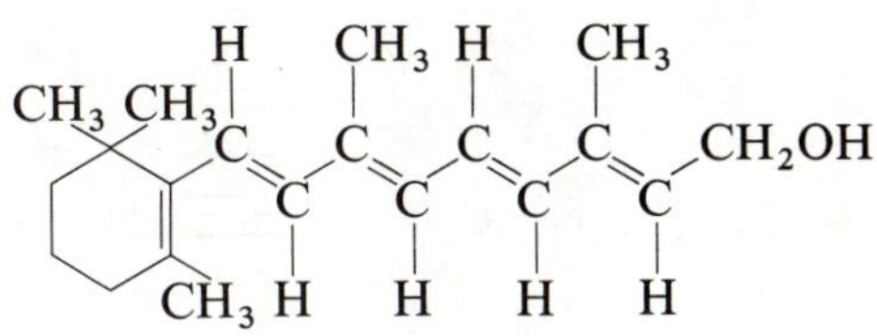

Vitamin A_1

Vitamin D is important for the control of absorption and the utilization of calcium. As a result, it is necessary for proper formation of bones and teeth:

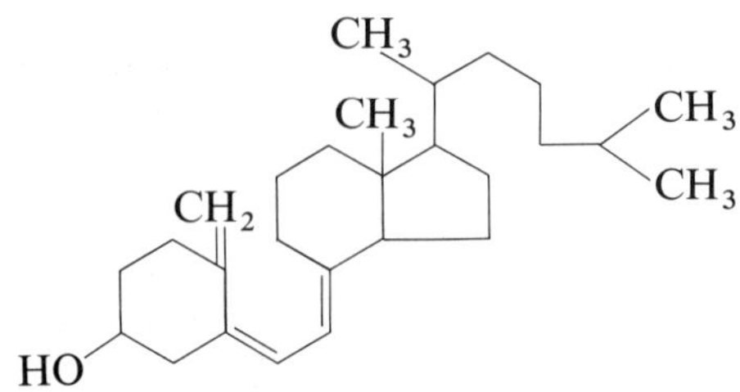

Vitamin D_3

Vitamin E is required for normal reproduction in many animal species, although this requirement has not been shown for humans. The chemical compounds having this activity are all very efficient antioxidants, particularly in preventing oxidation of unsaturated fatty acids. It is believed that a major portion of the function of these compounds within cells is the protection of cellular constituents from fortuitous oxidation:

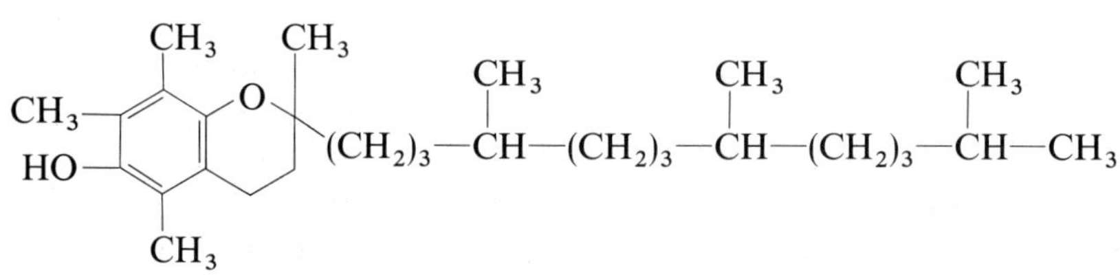

Vitamin E
(alpha-tocopherol)

Vitamin K is most notable in that it is required for synthesis of some of the compounds required for normal blood clotting. In addition, it is now believed that the Vitamin K compounds or similar structures may have other more general roles in the oxidative processes of cells:

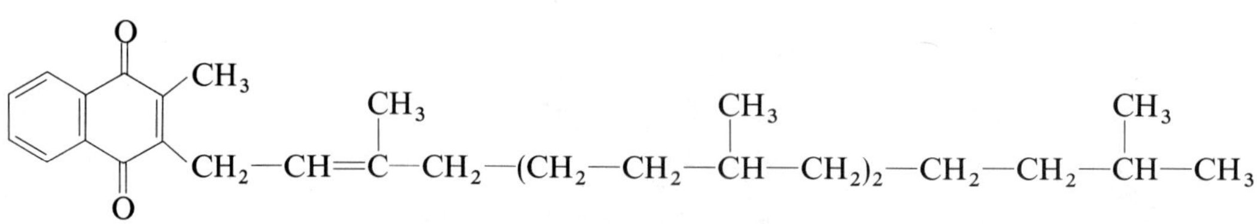

Vitamin K_1

13.6 Plant Pigments

This grouping includes a number of compounds that are alike only in that they are colored, occur in plants, and are extractable with organic solvents.

By far the most ubiquitous of the plant pigments are the **chlorophylls**. These contain a complex nitrogen-containing ring system called a **porphyrin**. The nitrogen atoms of the porphyrin are bonded to a magnesium ion. In addition, this whole complex is linked to a specific protein molecule. The similarity between the chlorophyll structure and the heme portion of the hemoglobins is striking and is illustrated in Figure 13-2. The chlorophylls play a key role in the process of photosynthesis, where radiant energy is trapped and utilized to provide energy for synthetic reactions.

The **carotenes** form another important group of pigments. Like some of the hydrocarbons mentioned above, these are essentially polymers of isoprene-type units; owing to the extended conjugated double-bond system, they are generally yellow in color. While their functions in plants are not clear, in animals the carotenes can be converted in whole or in part to

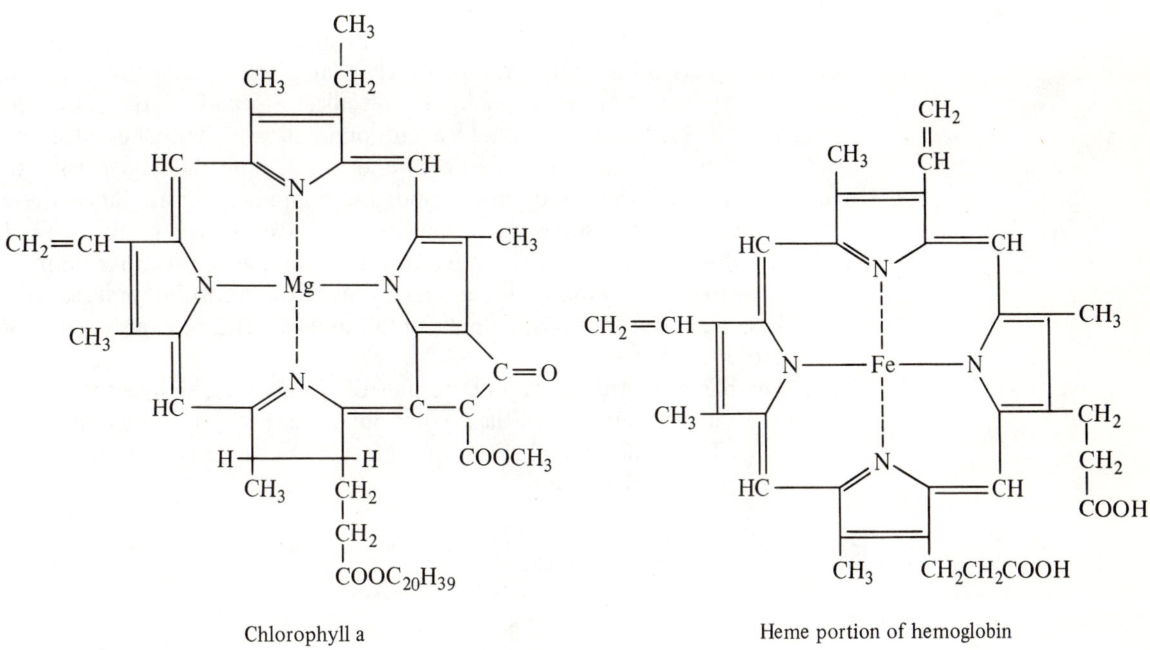

Figure 13-2
A comparison of the structures of chlorophyll *a* and the heme portion of hemoglobin.

Vitamin A structures:

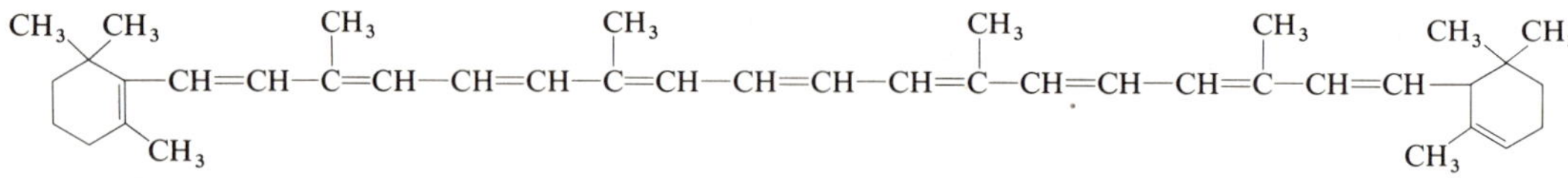

α-Carotene

The **xanthophylls** are oxygenated (hydroxy, keto, or epoxy) derivatives of the carotene-type structures.

13.7 Digestion and Absorption

We have commented earlier on the absolute necessity to have complex carbohydrates hydrolyzed completely to their constituent sugars, and polypeptides to the individual amino acids, before absorption into cells can take place. With the lipids, the need for hydrolytic breakdown is not so clearly defined. To a considerable extent, the rate of absorption of lipids appears to be a function of the degree of emulsification which has been achieved. When thoroughly emulsified with a very efficient synthetic detergent, most common lipid materials can be absorbed directly into intestinal mucosa cells.

The bile salts, which are the major emulsifying agents released into the digestive tracts of animals, are not nearly that efficient, and alone cannot ensure the absorption of intact lipid molecules, especially large saturated triglycerides. However, they usually act in conjunction with other emulsifying agents, such as dietary phospholipids and mono- and diacylglycerols, to facilitate the absorption of much undigested lipid material. Since these conditions will vary from time to time, about all we can say about the lipid mixture absorbed from the digestive tracts of humans and similar animals is that it will contain mono-, di-, and triacylglycerols, phospholipids, sterols, sterol esters, and fatty acids, all present in unknown and constantly varying proportions.

The bile salts are a group of steroids of closely related structures. The proportions in which each of these compounds are present in the bile varies somewhat from one animal species to another. In humans, the major bile salt is sodium glycocholate:

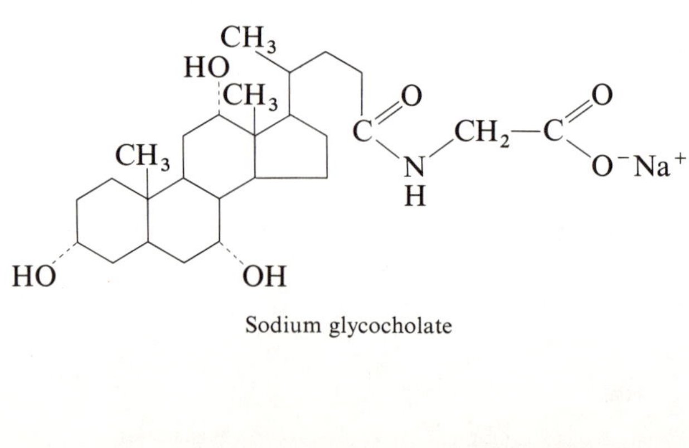

Sodium glycocholate

Hydrolytic Enzymes

A number of enzymes present in the digestive tracts of animals or in various plant sources catalyze hydrolysis reactions involving lipid materials. A few of the major ones are described below.

Lipases. These are enzymes that catalyze the hydrolysis of the fatty acid ester linkages at the ends of an appropriate tri-, di-, or monoacylglycerol. Thus, a triacylglycerol will first be converted to a 1,2-diacylglycerol and then to a 2-monoacylglycerol. These enzymes will catalyze the hydrolysis of the fatty acid ester at the 2-position, but at a very greatly reduced rate:

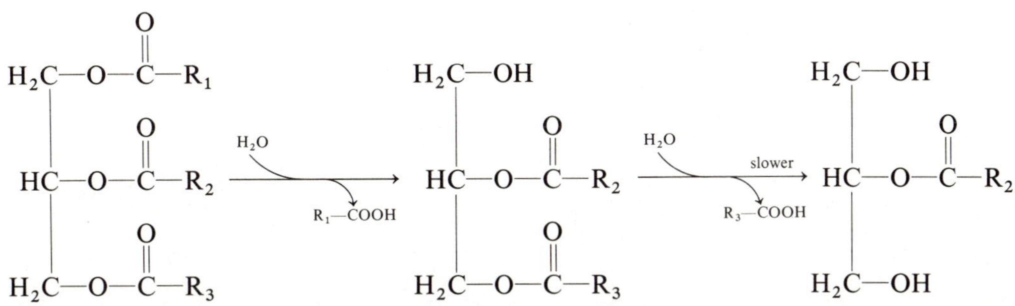

Phosphodiesterases. These enzymes catalyze the hydrolysis of a phosphate ester bond only when the phosphate group involved is doubly esterified; that is, it is a phosphodiester. They normally have no action on phosphomonoesters. In the case of a typical lecithin, a phosphodiesterase might catalyze one or the other of the reactions shown in Figure 13-3.

Phosphomonoesterases. These enzymes catalyze the hydrolysis of monoesters of phosphoric acid and are generally called simply "phosphatases." Some are not much affected by the structure of the substrate molecule, while others are highly specific for one compound or group of compounds.

The phosphomonoesterases found in animal digestive tracts are generally quite nonspecific in their action. Among other reactions, they will catalyze the removal of the phosphate groups from the products of phosphodiesterase action described above to yield (in the case shown) a diacylglycerol, choline, and inorganic phosphate. Of course, the diacylglycerol may be further acted upon by a lipase before it is absorbed.

Cholesterol Esterase. Much of the cholesterol in an animal's diet is present in the form of fatty acid esters. Cholesterol esterase is the specialized enzyme capable of catalyzing the hydrolysis of these esters to yield the free sterol and a fatty acid.

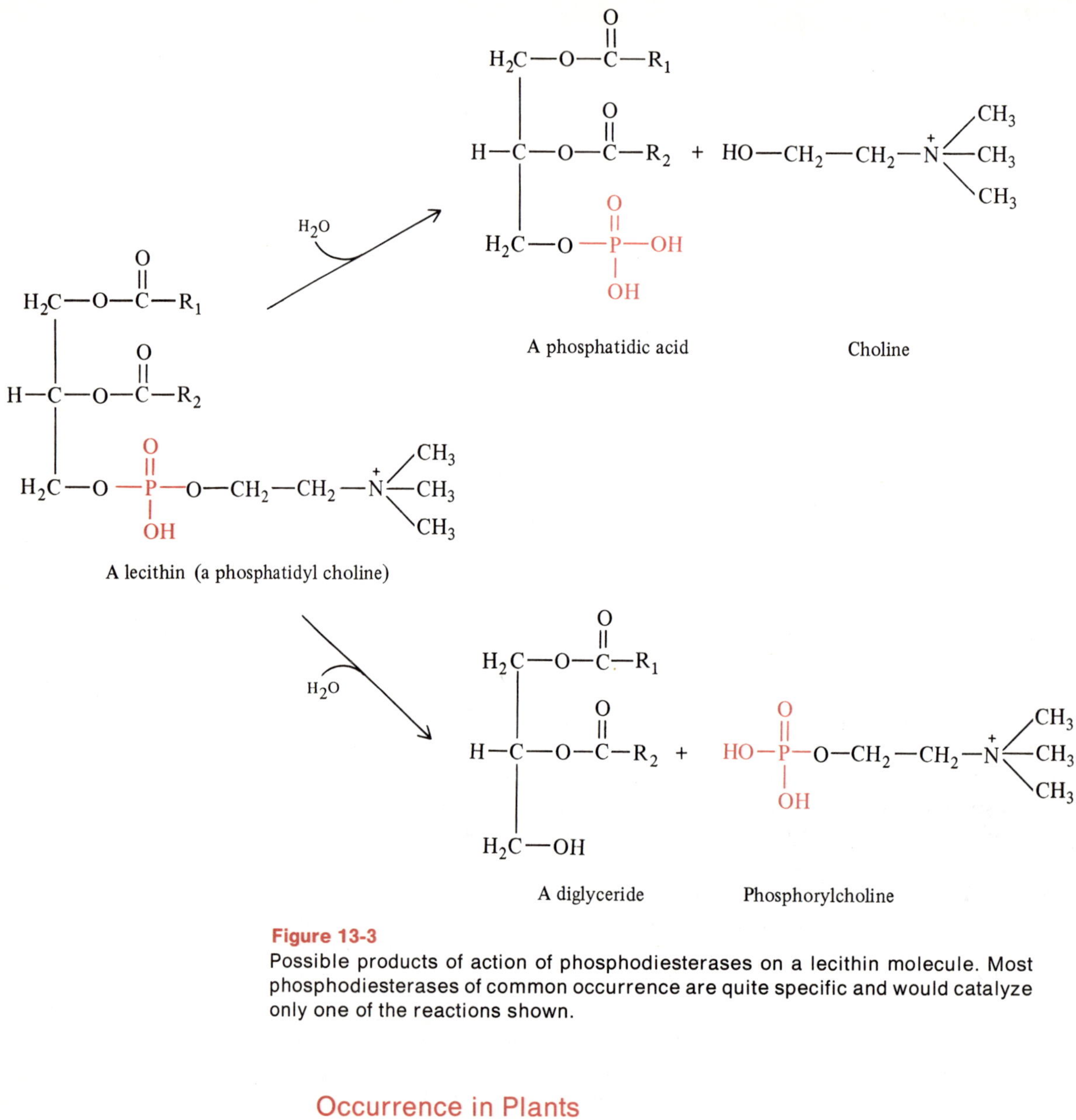

Figure 13-3
Possible products of action of phosphodiesterases on a lecithin molecule. Most phosphodiesterases of common occurrence are quite specific and would catalyze only one of the reactions shown.

Occurrence in Plants

Plant tissues normally contain at least the first three of these types of hydrolytic enzymes. They are present in especially high concentration in many seeds and appear to function in the utilization of stored lipids during germination and the early growth period.

Phospholipases

Enzymes capable of removing a fatty acid from an intact glycerol phospholipid are missing from most plant and animal tissues, perhaps a reflection of the cytotoxicity of the lysophosphatides that are the products of their action. Certain snake venoms, however, contain relatively high concentrations of a phospholipase which does remove one fatty acid from a phospholipid. When this venom reaches the bloodstream, the enzyme acts upon the phospholipids of the walls of the red blood cells, causing the release of lysophosphatides which are very potent surface-active agents. The result is extensive rupture of the red cells.

Additional Readings

Scientific American articles

L. F. Fieser, "Steroids," 92(1), 52 (January 1955).
N. Sharon, "Glycoproteins," 230(5) 78 (May 1974).
S. Clevenger, "Flower Pigments," 210(6), 84 (June 1964).

A. I. Gurr and A. T. James, *Lipid Biochemistry: An Introduction*, 2nd edition (Halsted Press, New York, 1976).

C. Hitchcock and B. W. Nichols, *Plant Lipid Biochemistry* (Academic Press, New York, 1971).

D. Chapman, *Introduction to Lipids* (McGraw-Hill, New York, 1969). Primarily deals with structure and chemistry at the introductory level.

E. Heftmann, *Steroid Biochemistry* (Academic Press, New York, 1970). A general survey of the steroids with attention to structure, metabolism, and physiological effects.

G. B. Ansell, J. N. Hawthorne, and R. M. C. Dawson, *Form and Function of Phospholipids*, 2nd edition (American Elsevier, New York, 1973). An advanced level discussion of the biochemistry of phospholipids.

M. Florkin and E. H. Stoltz (Eds.), *Comprehensive Biochemistry* (Elsevier Publishing Co., New York, 1965). Volume 6 is a compilation of review articles on lipid chemistry.

Study Questions

1. A whole mouse is homogenized and the resulting material is extracted exhaustively with chloroform. Name ten *specific compounds* that you might reasonably expect to find in the chloroform extract.

2. Draw a structure for one specific glycerol phospholipid containing at least one unsaturated fatty acid.

3. Name the compound you have drawn as accurately as you can.

4. List the compounds that would be produced on *complete* hydrolysis of your phospholipid.

5. "Soaps" are simply salts of fatty acids. They are typically produced by base hydrolysis of triacylglycerols (a process appropriately called "saponification"). Write a balanced equation for saponification of one specific triacylglycerol.

6. You have just isolated a pure lipid and find that acid hydrolysis produces only sphingosine, a fatty acid, and galactose. To what group of lipids does it probably belong?

7. Why are Vitamins A, D, E, and K still described by a letter when so many vitamins are now described by chemical structure instead? [For example, thiamin (B_1), riboflavin (B_2), pantothenic acid (B_3), biotin (H)]

CHAPTER 14

Fatty Acid Oxidation and Synthesis

For most organisms, a great deal of their energy reserves consists of lipid compounds, especially the triacylglycerols (triglycerides). The process which releases the stored energy is the oxidation of the fatty acids of these lipids to carbon dioxide and water. It is thus appropriate to start our consideration of lipid metabolism with a look at the process of fatty acid oxidation.

14.1 Fatty Acid Activation

Oxidative metabolism of fatty acids in eukaryotes is almost entirely a mitochondrial function. We have commented earlier on the fact that the outer membrane of the mitochondrion is permeable to anything with a molecular weight less than about 10,000. Thus fatty acids can find their way without difficulty into the space between the outer and inner membranes. It is at this point that they undergo conversion to thioesters of Coenzyme A, a process essential for any further metabolism of the fatty acids. The equation is shown in Figure 14-1.

The energy necessary for formation of the thioester is supplied by ATP. In order to carry this reaction substantially to completion, the same mechanism is employed as that used in Chapter 12 to drive the synthesis of glycogen. The ATP is similarly split at the internal anhydride linkage, releasing inorganic pyrophosphate. You will recall that the rapid hydrolysis of this compound to two molecules of inorganic phosphate is a facile function of most cells, rapidly removing pyrophosphate from the equilibrium and in this case rendering the reaction of ATP with the fatty acid effectively irreversible.

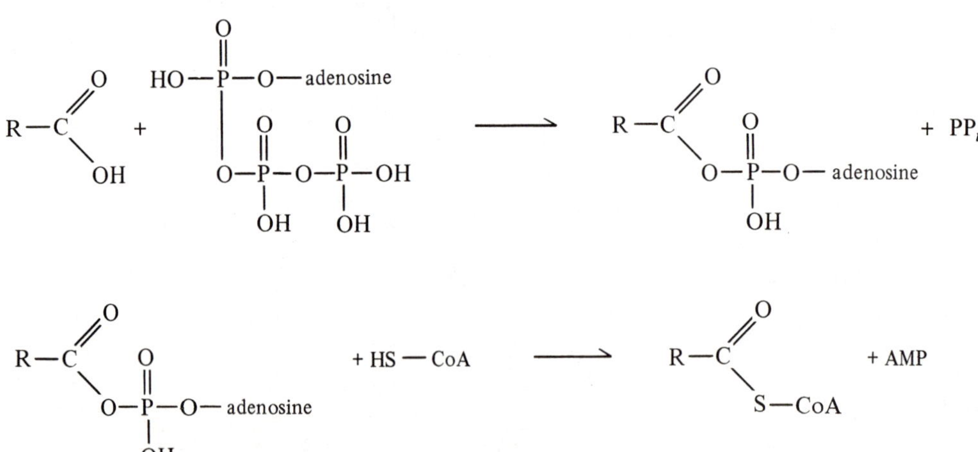

Figure 14-1
Activation of a fatty acid to a fatty acyl–Coenzyme A derivative.

Bear in mind that conversion of the by-products of the activation process, AMP and two molecules of inorganic phosphate, back to ATP will require 2 "ATP's worth" of energy. As a result, the net cost for the esterification must appear on our energy ledger as 2 ATPs rather than the one that appears in the equation.

Entry into the Mitochondrion

In the form of the fatty acyl–Coenzyme A ester, it is possible for the acid to move across the inner mitochondrial membrane. However, the presence of a little carnitine can be shown to speed the process greatly, and it appears that in many organisms the acid is transferred from Coenzyme A to carnitine for the trip across the membrane, and then back to Coenzyme A inside the mitochondrion. This process is shown in Figure 14-2.

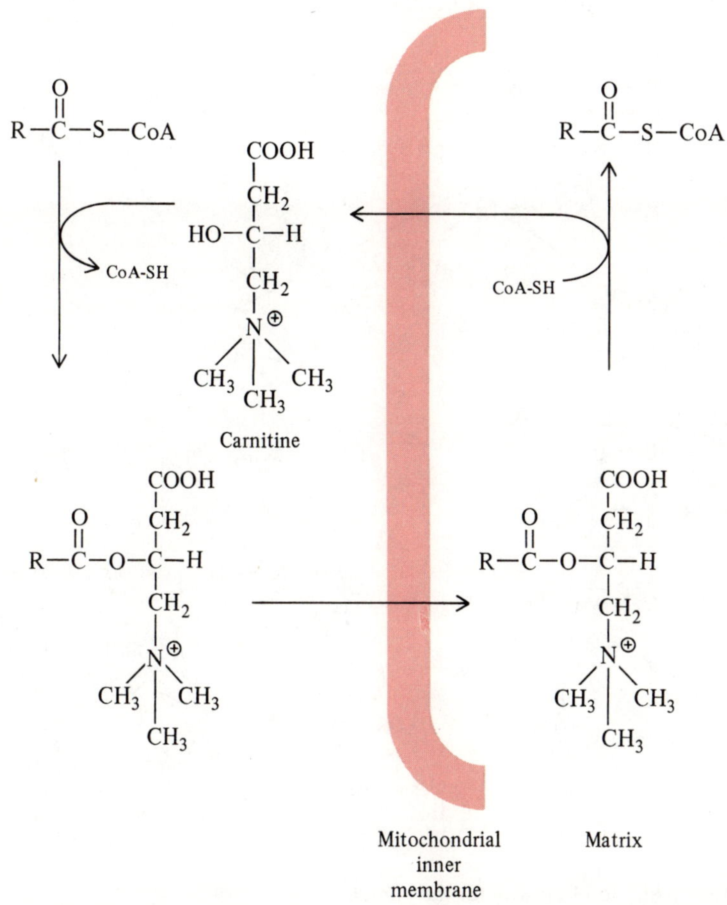

Figure 14-2
Role of carnitine as a transfer agent for entry of fatty acid esters into the mitochondrial matrix.

14.2 Oxidation to Acetyl–CoA

Once inside the inner membrane of the mitochondrion, the fatty acyl–CoA derivative is ready to begin the sequence of reactions usually referred to as the beta-oxidation pathway. The reactions are shown in Figure 14-3.

This pathway has the form of a spiral, one full turn accomplishing sufficient oxidation to permit the removal of two carbons in the form of a molecule of acetyl–CoA. What remains is a fatty acyl–CoA that differs from the starting material only in being two carbons shorter. It is ready for another turn of the spiral, and the process will continue in this fashion until all of the carbons of the fatty acid have been converted into acetyl–CoA.

You will have noted the similarity between this pathway and the sequence of reactions by which succinate is converted to oxaloacetate in the citric acid cycle. Again we have an FAD-requiring dehydrogenation to yield a double bond, directional addition of water, and then an NAD-linked oxidation of the alcohol to a ketone. One difference is that in a tissue such

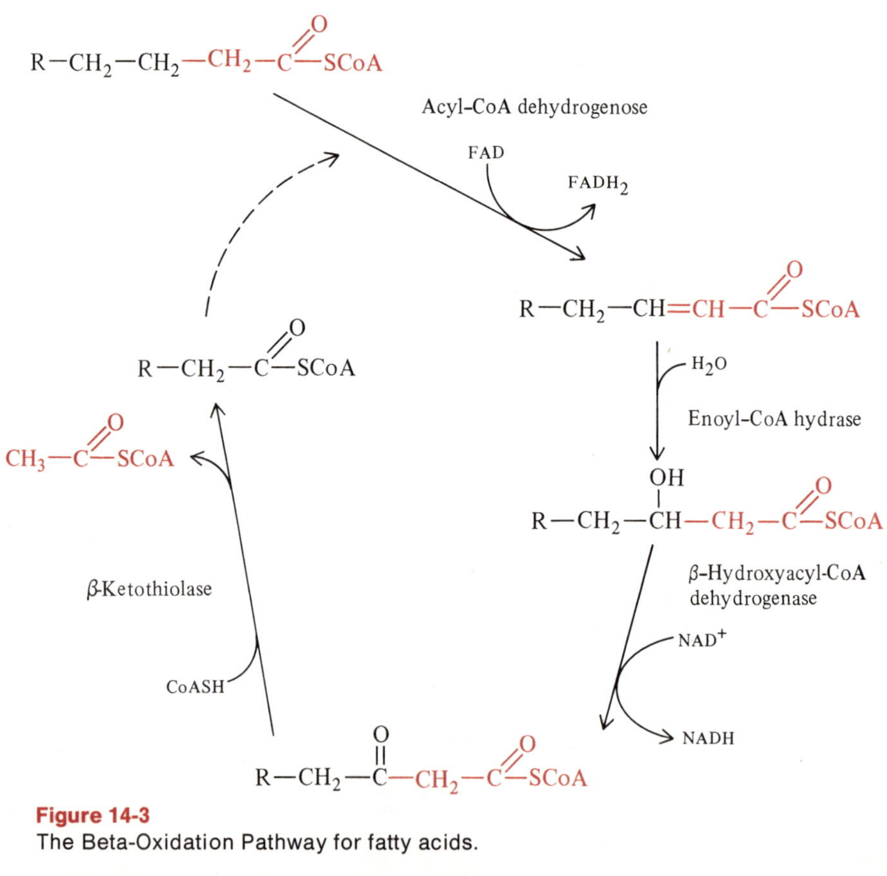

Figure 14-3
The Beta-Oxidation Pathway for fatty acids.

as animal liver, there are several "fatty acyl–CoA dehydrogenases," all flavoproteins but differing in their preference for certain lengths of carbon chains.

The remaining reaction, in which the β-keto compound is split to yield a molecule of acetyl–CoA, is often called an example of a "thiolysis" reaction (by comparison with "hydrolysis" and "phosphorolysis").

In terms of energy production, one turn of the beta-oxidation spiral produces one molecule of $FADH_2$ and one of NADH. Re-oxidation of these intermediates by the mitochondrial electron transport system permits the production of five molecules of ATP from ADP and P_i.

Oxidation of the Acetyl–CoA Units

Oxidation of the acetyl–CoA arising from the fatty acids proceeds through the citric acid cycle as we have already seen. Each acetyl–CoA will of course give rise to a total of 12 molecules of ATP during this process.

The Energy Bookkeeping

If we look at the energy yield from a six-carbon fatty acid, we find the following:

A. Activation to the CoA ester	−2 ATP
B. Two turns of the β-oxidation spiral	+10 ATP
C. Three acetyl–CoA molecules through the citric acid cycle	+36 ATP
Net	+44 ATP

This is reasonably comparable to the 38 or so ATP we were able to obtain from complete oxidation of a six-carbon sugar. Thus, the energy yield per atom of carbon is about the same. A mole of glucose, however, has a mass of 180 grams compared to 116 grams for the six-carbon fatty acid. On a *weight* basis, the fatty acid is almost twice as effective in supplying biologically usable energy.

Other Types of Fatty Acid Oxidations

Several other reactions are known by which fatty acids can be oxidized, partially or completely. We will not deal with these in any detail, but only note them here to show that β-oxidation is not the only oxidative pathway for a specific fatty acid or derivative.

α-Oxidation is a two-step process by which the α-carbon is first oxidized to the level of an alcohol using (in animals) molecular oxygen. The second step uses oxygen and NAD to carry out oxidation of the alcohol to the acid with decarboxylation of the original carboxyl group occurring along the way. This procedure is useful in dealing with certain branched-chain structures which are resistant to the action of the β-oxidation process:

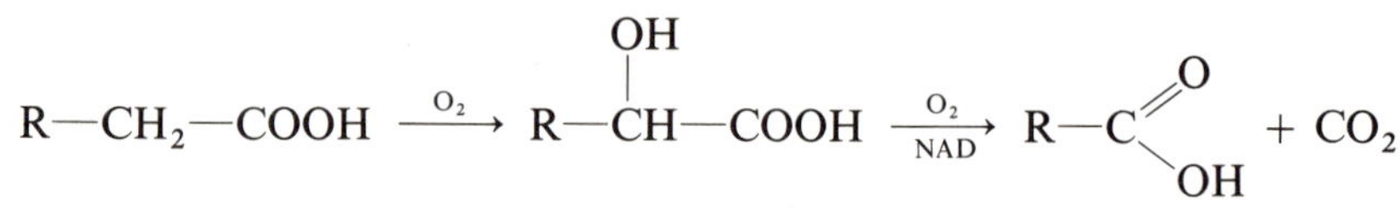

ω-Oxidation. A rather special group of enzymes and cofactors exists in special aerobic bacteria and in the microsomes of some animal tissues. These catalyze the insertion of an oxygen into the terminal methyl group of a hydrocarbon chain; further oxidation of the resulting alcohol yields a carboxylic acid:

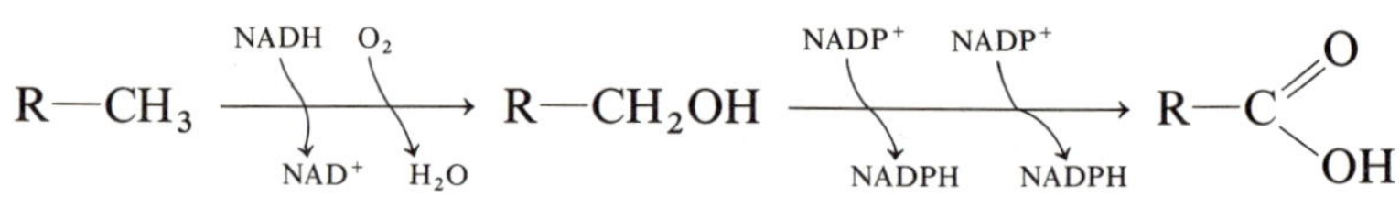

While these reactions may be of more current interest because the appropriate bacteria are able to oxidize spilled oil and similar hydrocarbons, they also represent a means by which fatty acids may be converted into dicarboxylic acids.

Oxidation of Unsaturated Fatty Acids. For a number of the polyunsaturated fatty acids, operation of the usual β-oxidation pathway results in intermediates that either have the double bond in the wrong place—or *cis* instead of *trans*—or have the hydroxyl group in the wrong optical configuration. In each case, isomerases exist which remedy the problem and permit normal β-oxidation to continue.

Some Interrelationships between Fatty Acid and Carbohydrate Oxidations

Under normal circumstances the bulk of the energy needed for formation of ATP from ADP is supplied by the citric acid cycle and reactions associated with it. This pathway is of course common to both carbohydrate oxidation and fatty acid oxidation through the common intermediate, acetyl–CoA. In addition, by reactions we have not yet studied, a number of the amino acids on oxidation yield citric acid cycle intermediates. We commented earlier on the ability of these intermediates to be involved in the process called gluconeogenesis, the formation of glucose molecules from noncarbohydrate sources.

When a cell is relatively well supplied with carbohydrate materials, both acetyl–CoA and ATP levels will be relatively high and as a result the rate of fatty acid oxidation will be minimal. Oxidation of amino acids will occur at a rate sufficient to deal with any excess beyond the needs for protein synthesis; however, there will be no need for any of this to be used for gluconeogenesis.

If the carbohydrate supply becomes deficient, the picture changes completely. This is most easily seen using a human liver cell as an example. This cell faces two major demands: sufficient energy to support its own activities, and a need to maintain the blood glucose at a level that will permit other cells, particularly those of the brain, to function. To meet these needs, the cell will of course mobilize its glycogen reserves and use this material for both purposes as long as it lasts. As that point nears, changes must take place.

First, in order to maintain the blood glucose level, pyruvate, oxaloacetate, α-ketoglutarate, and other products of amino acid oxidation will be used to generate glucose units. This process not only requires some energy in the form of ATP but also depletes the levels of all of the citric acid cycle intermediates. The result is that the citric acid cycle becomes relatively ineffective. The energy demands of the cell will then force an increase in the rate of oxidation of fatty acids to acetyl–CoA units. Since the citric acid cycle is impaired, these acetyl–CoAs will tend to pile up and force the formation of a substantial level of acetoacetyl–CoA (β-ketobutyryl–CoA):

$$CH_3-\overset{\overset{O}{\|}}{C}-SCoA + CH_3-\overset{\overset{O}{\|}}{C}-SCoA \rightleftharpoons CH_3-\overset{\overset{O}{\|}}{C}-CH_2-\overset{\overset{O}{\|}}{C}-SCoA$$

14.3 Ketone Bodies

There are always present in cells a certain number of enzymes capable of rather general hydrolysis of esters. A certain amount of this acetoacetyl–CoA is then hydrolyzed to the free acid, acetoacetic acid. This cannot readily be converted back into the CoA derivative in the liver, and is thus released into the bloodstream. Muscle cells can normally handle the reconversion to the CoA ester, but when large amounts are produced, the rate of transport into muscle cells is inadequate, and much of it must be excreted in the urine. Since this is a β-keto acid, some decarboxylation may also occur, with the formation of acetone. These two compounds, acetoacetic acid and acetone, make up the major "ketone bodies" that are excreted in the urine when carbohydrate metabolism is severely reduced. They can also be produced to some extent by a different mechanism when a diet high in protein is consumed.

14.4 Acidosis

If the production of acetoacetate is severe or prolonged, it can lead to major physiological problems. Excretion of the acid requires neutralization by some cation; excessive needs may deplete the bodily stores of cations and cause a shift in the pH of the blood. This results in a condition called "acidosis" with a number of effects, including impairment of the ability to transport oxygen.

The human body has a remarkable ability to adapt to changing circumstances. A person on a starvation diet will excrete "ketone bodies" but will not normally develop acidosis at a moderate activity level. The more severe reactions are, however, seen in conditions such as diabetes mellitus when there is a lack of ability of cells to transport glucose through the cell membrane.

14.5 Synthesis of Fatty Acids

Biosynthesis of fatty acids in an overall sense is the reversal of β-oxidation, in that acetyl–CoA units are joined together and then reduced in a spiral process eventually yielding the desired chain length. However, the process occurs at a different place within the cell, uses a special ATP-requiring reaction to overcome the unfavorable equilibrium of the β-ketothiolase reaction, and involves a special protein called acyl carrier protein (ACP) to hold the growing fatty acid chain while it goes through the reactions of the synthetic spiral.

Formation of Malonyl–CoA

The equilibrium of the β-ketothiolase reaction lies very much in the direction of formation of acetyl–CoA. In order to overcome this energetic barrier, acetyl–CoA is first carboxylated to malonyl–CoA in a biotin-requiring reaction and later decarboxylated after condensation with another acetate unit. The reaction forming the malonyl–CoA is as follows:

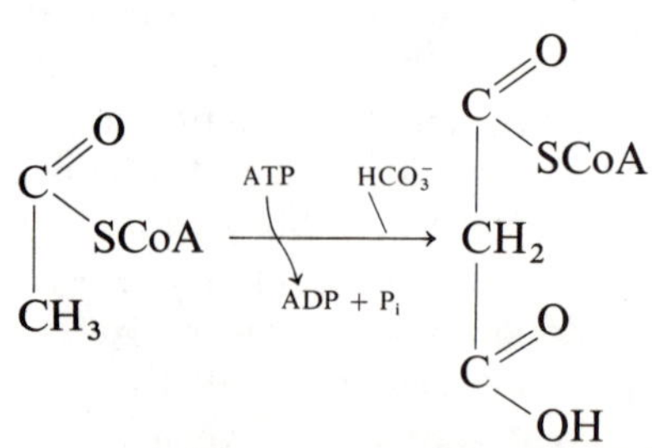

Malonyl–Coenzyme A

Presumably the use of malonyl–CoA to condense with the growing fatty acid chain serves at least two purposes: first, to allow the use of the energy of ATP to carry the reaction toward completion, and second, to create a methylene group more susceptible to condensation than would be the methyl group of acetyl–CoA.

The Fatty Acid Synthetase System

While the reactions involved are the same, two different types of synthetic systems seem to exist. In many microorganisms and plants, the enzymes involved are soluble and show little or no tendency to associate. In this case, the growing fatty acid chain is carried from one enzyme to another while attached to the acyl carrier protein (ACP).

Acyl Carrier Proteins

The acyl carrier proteins (ACPs) are all small, heat stable proteins with molecular weights in the range of 5000–10,000. Attached to the polypeptide chain is a molecule of 4′-phosphopantetheine. Note that this molecule is identical to a major portion of Coenzyme A (page 142): The phosphopantetheine is bound by an ester linkage between the phosphate group and the hydroxyl group of a serine residue. As with Coenzyme A, the sulfhydryl group is the active part of the molecule in the acceptance and transfer of acyl groups.

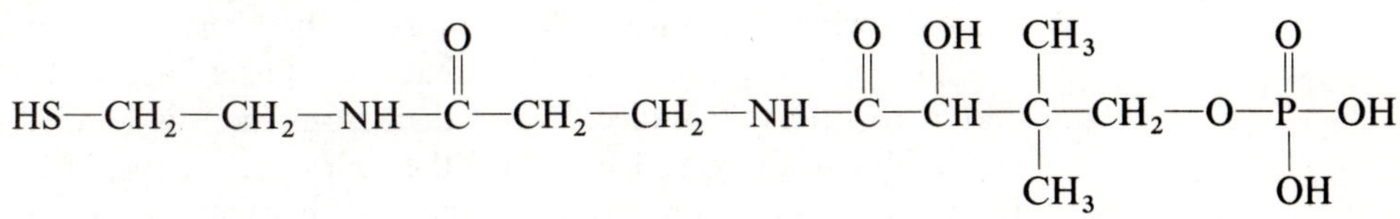

4′-Phosphopantetheine

In yeast and mammalian cells, the synthetase system occurs as a complex similar to the keto acid oxidase complexes involved in oxidative decarboxylation of pyruvate, α-ketoglutarate, and so on. All of the necessary enzymes occur associated together in a single functional complex having an aggregate molecular weight on the order of one million. In this case the acyl carrier protein molecule is built in as part of the complex and appears to serve the function of transporting the substrate unit from one enzyme to the next within the complex.

Acetyl–CoA + ACP-SH $\xrightarrow{①}$ Acetyl–S-ACP + CoA

Acetyl–S–ACP + Enz ③ $\longrightarrow$ Acetyl-S-Enz③+ ACP

Malonyl–CoA + ACP-SH $\xrightarrow{②}$ Malonyl–S-ACP + CoA

Acetyl–S-Enz③+ Malonyl-S-ACP $\longrightarrow$ Acetoacetyl–S-ACP + Enz③+ CO_2

Acetoacetyl-S-ACP + NADPH + H^+ $\xrightarrow{④}$ β-Hydroxybutyryl–S-ACP + $NADP^+$

β-Hydroxybutyryl-S-ACP $\xrightarrow{⑤}$ trans-Crotonyl-S-ACP + H_2O

trans-Crotonyl–S-ACP + NADPH + H^+ $\xrightarrow{⑥}$ Butyryl–S-ACP + $NADP^+$

Butyryl–S–ACP + Enz ③ $\longrightarrow$ Butyryl–S–Enz③ + ACP

Butyryl-S-Enz③ + Malonyl–S-ACP $\longrightarrow$ β-Ketohexanoyl–S-ACP + Enz③ + CO_2 and so on.

① Acetyl transacylase	④ β-Ketoacyl ACP reductase
② Malonyl transacylase	⑤ Enoyl ACP hydrase
③ β-Ketoacyl ACP synthetase	⑥ Enoyl ACP reductase

Figure 14-4
A listing of the sequence of reactions carried out by the enzymes of the fatty acid synthetase complex.

In either case, the reactions involved are listed in Figure 14-4. The schematic drawing in Figure 14-5 attempts to show the functioning of the mammalian synthetase complex. Note that the "starter" acetyl–CoA is first transferred to the ACP molecule and then from there is handed off to a cysteine residue of Enzyme 3. The ACP then goes back to pick up a malonyl group and the enzyme proceeds to condense this with the waiting acetyl group on the cysteine residue. Decarboxylation takes place as part of this reaction, and the product is acetoacetyl–ACP.

In mammalian systems (Figure 14-6), both reduction steps require specifically NADPH. This would seem to reflect two factors. First, in any oxidative cell, there would have to be some stable reducing compound to carry out the reduction of the double bond to the saturated product. Simple reversal of the flavoprotein oxidase of β-oxidation is not possible due to the rapid nonenzymatic oxidation of $FADH_2$ in the presence of oxygen. Second, the use of NADPH rather than NADH is our first example of the rather general phenomenon of cells using NAD for oxidative processes (energy production) and NADP for synthetic purposes. It has been shown that in a

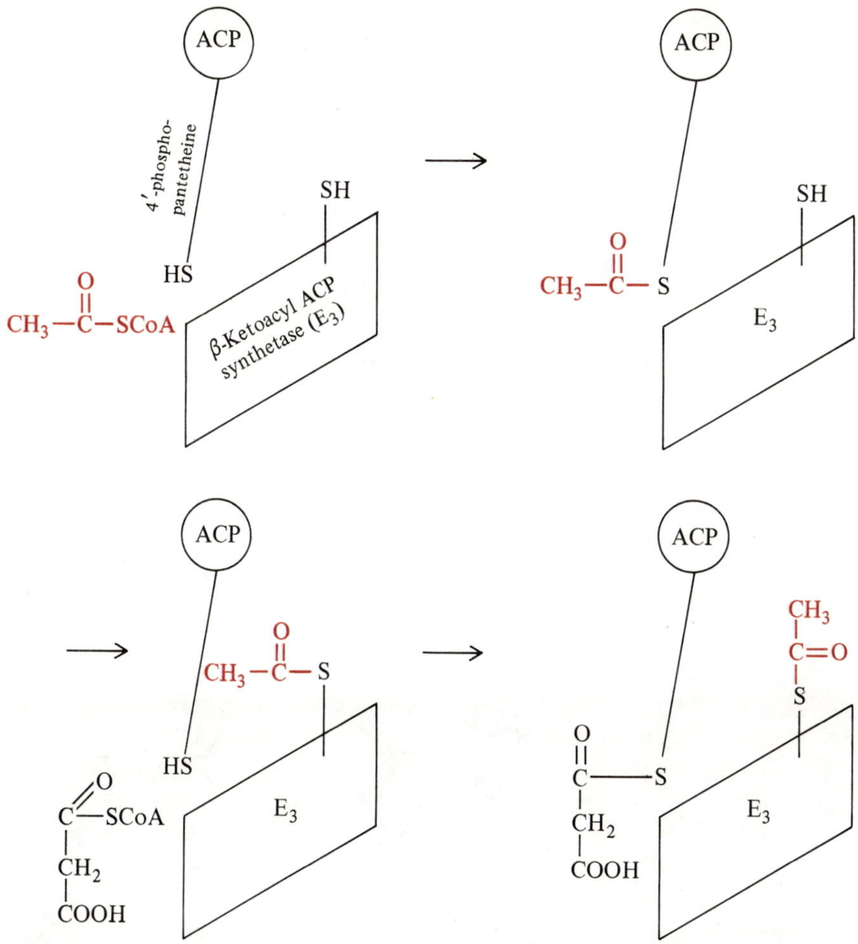

Figure 14-5
Action of the central ACP-type protein of a fatty acid synthetase complex in loading the first unit of acetyl–CoA onto the β-ketoacyl–ACP synthetase enzyme prior to accepting a molecule of malonyl–CoA

normal oxidative cell with optimal energy sources, the level of NAD^+ is high compared to NADH, while the exact opposite is true for $NADP^+$ and NADPH. The ability to maintain this high ratio of NADPH/$NADP^+$ allows the cell to drive its synthetic reductions toward completion without disturbing the normal production of energy through the respiratory electron transport system and similar mechanisms.

To start the second turn of the spiral, the butyric acid group must again be transferred from the ACP to Enzyme 3, where it can be held while the ACP picks up the second molecule of malonate from malonyl–CoA.

Associated with the system is an enzyme having the job of removing the chain from the ACP when it reaches an appropriate size. This enzyme will be

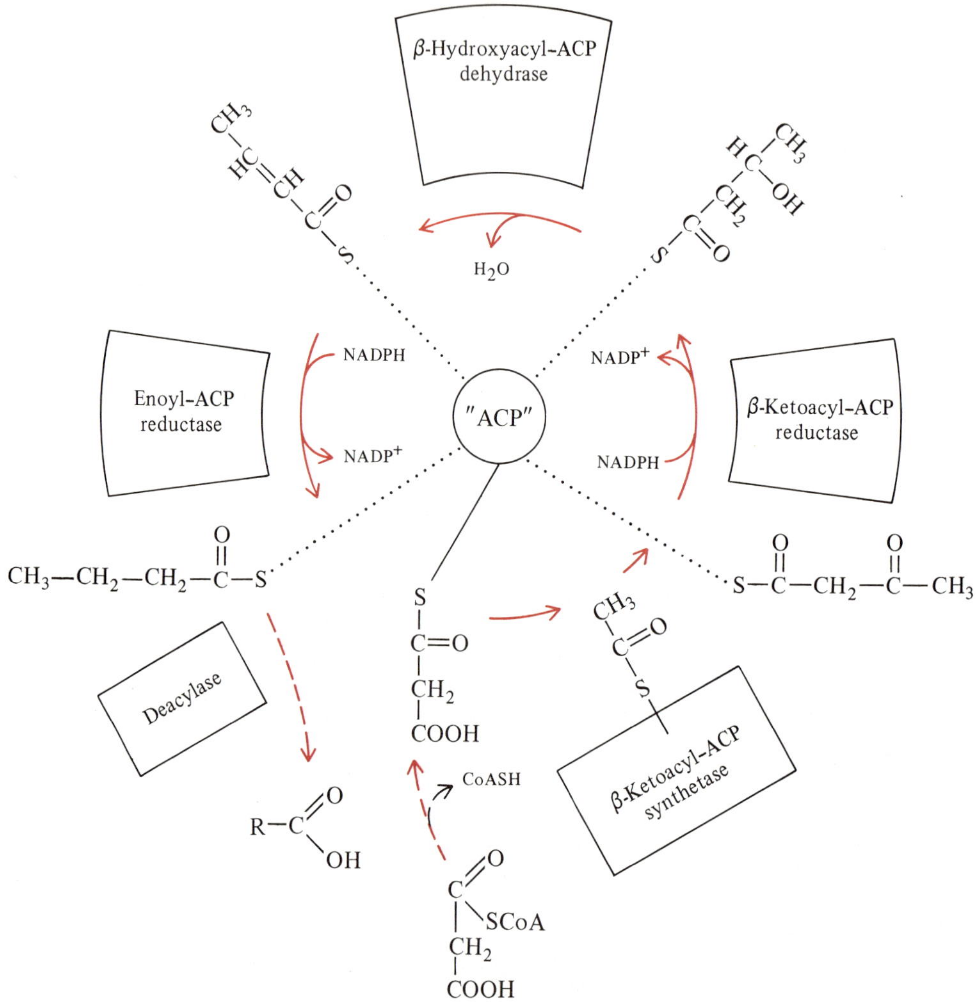

Figure 14-6
A schematic representation of the operation of a typical mammalian fatty acid synthetase complex. The deacylase will become active only when the chain length reaches an appropriate value.

inactive until the molecule reaches a certain size or size range. Very few synthetase systems are able to form fatty acids much longer than 16 carbons. In some tissues, further elongation of these products is carried out by very similar reactions in the microsomes. One of the interesting features of these microsomal enzymes is that they use the fatty acyl–CoA derivatives directly as substrates and do not use acyl carrier proteins.

Additional Readings

Scientific American article

D. E. Green, "The Metabolism of Fats," 190(1), 34 (January 1954).

A. I. Gurr and A. T. James, *Lipid Biochemistry: An Introduction*, 2nd edition (Halsted Press, New York, 1976).

S. J. Wakil (Ed.), *Lipid Metabolism* (Academic Press, New York, 1970). Articles comprising a thorough review of lipid metabolism at a sophisticated level.

Lipid Metabolism, Vol. 18 of *Comprehensive Biochemistry*, edited by M. Florkin and E. H. Stoltz (American Elsevier, New York, 1967). Advanced articles covering the metabolism of lipids in detail.

Study Questions

1. The free energy of hydrolysis for ATP → AMP + P–P$_i$ is given in Table 6-2. From this and the value of $\Delta G^{\circ\prime} = -6.3$ kJ/mole for the reaction of Figure 14-1, calculate the free energy of hydrolysis of the thioester bond of a fatty acyl–CoA derivative.

2. Calculate the energy yield in ATPs for complete oxidation of stearic acid (octadecanoic acid) to CO_2 and water. Why is this more than three times the yield for a 6-carbon fatty acid?

3. There is an old saying (often quoted in nutrition texts) that "Fats burn in the flame of the carbohydrates." Could you defend this statement on a metabolic basis today? Explain.

4. What compounds make up the common "ketone bodies"?

5. Production of acetyl–CoA from pyruvate occurs in the matrix of mitochondria; synthesis of fatty acids occurs outside the mitochondria. Suggest two ways in which the acetate units might be moved out through the inner membrane.

6. In Question 2, you will have found the yield in ATPs for conversion of 18-carbon stearic acid to nine molecules of acetyl–CoA. Compare this with the cost (in ATPs) of the reverse biosynthetic process (assuming each NADPH is worth three ATPs).

CHAPTER 15

Lipid Biosynthesis

Having examined the reactions of fatty acids, both degradative and synthetic, we now want to take a brief look at the processes by which several typical lipid structures are synthesized. We will deal primarily with the pathways of biosynthesis of glycerol lipids and of polyisoprenoid compounds, in particular the steroids. In addition, we will examine in a limited way one of the major uses of lipids in all cells, the formation of lipid-based cell membranes.

15.1 Biosynthesis of Glycerol Lipids

The fatty acids of the triacylglycerols present in the diet or within storage tissues, especially in animals, are capable of undergoing extensive transfer and rearrangement of position. In general, this occurs by a process in which one fatty acid is hydrolyzed off of the triacylglycerol and another then replaces it. The replacement, however, requires that the incoming fatty acid be in the form of the Coenzyme A ester. Thus, to transfer a fatty acid from one acyl glycerol to another, it would be necessary to have it hydrolyzed off the first, converted to the CoA ester by reactions we have already discussed, then transferred from CoA onto the accepting glyceride.

Transfers of this type work only as long as the acceptor molecule is either a mono- or diacylglycerol. Free glycerol is incapable of accepting a fatty acid from a fatty acyl–Coenzyme A ester.

Synthesis of the Glycerol Portion

We have seen how acetate can be derived from carbohydrates and made into fatty acids. The glycerol portion of the glycerol lipids is also derived from carbohydrate sources, being formed from dihydroxyacetone phosphate, one of the intermediates of glycolysis. This compound is reduced to a glycerol phosphate by a dehydrogenase of the cytosol using NADH as electron source. The resulting compound is optically active and can be described by any one of a number of systems for designating optical configuration:

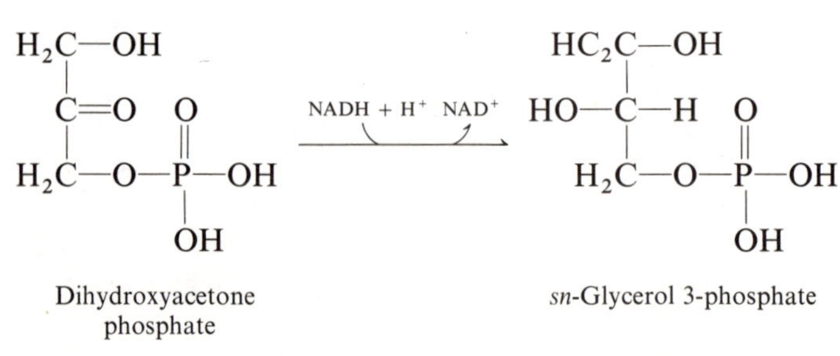

Dihydroxyacetone phosphate

sn-Glycerol 3-phosphate

At the present time it is most commonly called "*sn*-glycerol 3-phosphate" but the terms "L-α-glycerol phosphate" or "L-α-glycerolphosphoric acid" are well entrenched in the literature. All of the glycerol lipids that have structures showing optical asymmetry are based on this same configuration. You will recognize, of course, that some tri-, di-, and monoacylglycerols have symmetrical structures.

Free glycerol is converted to the same phosphate derivative by a specific kinase catalyzing the following reaction:

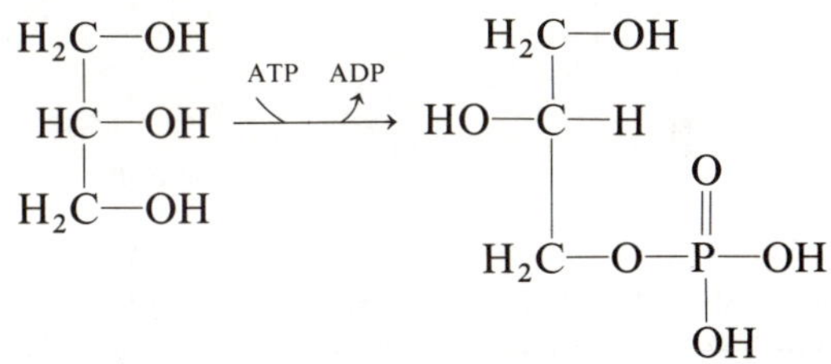

Glycerol *sn*-Glycerol 3-phosphate

Formation of Di- and Triacylglycerols

Once the glycerol phosphate is formed, it can react with two fatty acyl–Coenzyme A esters, sequentially, to form a **phosphatidic acid**:

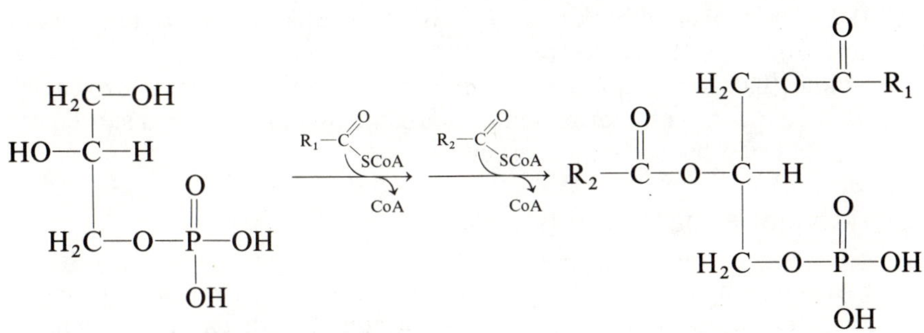

sn-Glycerol 3-phosphate A phosphatidic acid

In the presence of the ubiquitous phosphatases, these phosphatidic acids are rather rapidly hydrolyzed to diacylglycerols. Addition of a third fatty acid to convert the diacylglycerol to a triacylglycerol has already been discussed above:

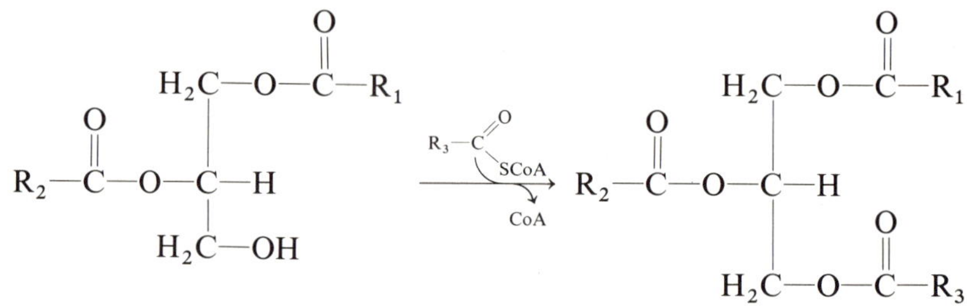

A diacylglycerol A triacylglycerol

Formation of Glycerol Phospholipids

The syntheses of the phosphatidyl cholines, serines, and ethanolamines are very closely related. We will use the synthesis of a phosphatidyl choline (a lecithin) to illustrate the reactions involved. In this case, the second alcohol is choline; it is first converted to a phosphate ester by a typical kinase enzyme using ATP. The choline phosphate thus formed then undergoes a type of anhydride exchange reaction with cytidine triphosphate (CTP) to yield cytidine-diphosphate–choline (CDP–choline) and inorganic pyrophosphate.

The CDP–choline then acts like a typical anhydride and reacts with the free hydroxyl group of a diacylglycerol which we have already synthesized. The result is formation of the phosphatidyl choline (lecithin) and release of cytidine monophosphate (CMP) as shown in Figure 15-1. The CMP can be converted back to CTP by the sequential transfer of phosphate groups from two molecules of ATP.

Note that this is another example of a synthetic process driven in the desired direction by formation of inorganic pyrophosphate, a substance subject to rapid hydrolysis in the cell.

Inositol Phospholipids

Also shown in Figure 15-1 is an alternative pathway by means of which synthesis of the *myo*-inositol derivatives is accomplished. This pathway may be used for synthesis of the other glycerol phospholipids in some tissues.

In this case, CTP reacts with the phosphatidic acid to form a cytidine diphosphate–diacylglycerol (CDP–diacylglycerol) as the reactive anhydride intermediate. This then reacts with the free alcohol of the inositol (or other group) to yield the phospholipid and CMP.

This variation in synthetic mechanism gets around the fact that no enzyme is known that is capable of forming *myo*-inositol phosphate (which would be required for the other biosynthetic pathway).

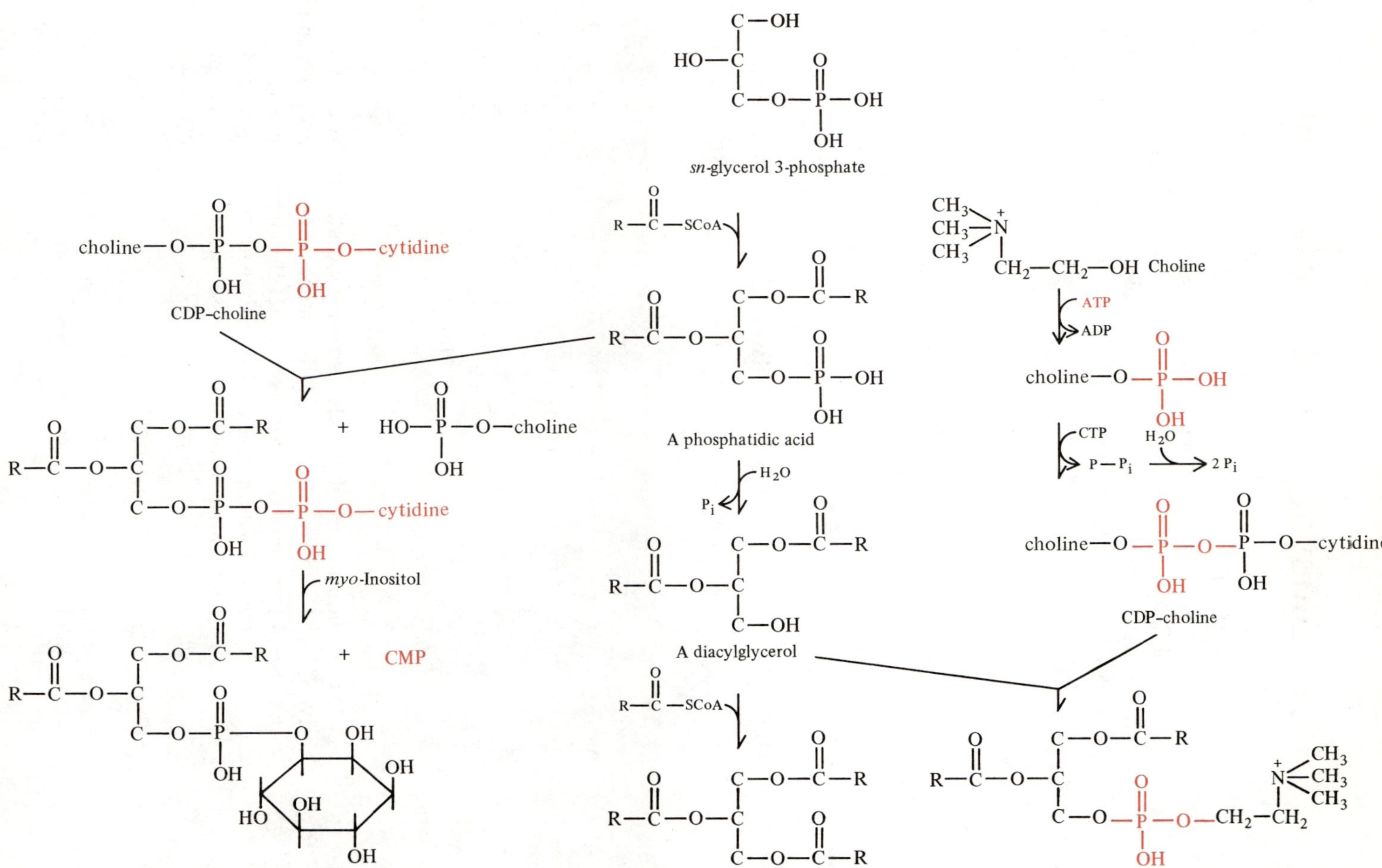

Figure 15-1
Pathways of biosynthesis of glycerol lipids. The central sequence of reactions shows formation of di- and triacylglycerols from *sn*-glycerol 3-phosphate. The reactions on either side outline the two pathways for formation of glycerol phospholipids. The hydrogens of the glycerol portion have been omitted for clarity.

15.2 Biosynthesis of Cholesterol

In animal systems, cholesterol occupies a central position in the pathway of biosynthesis of all of the steroids. As we will see shortly, essentially all of the other steroidal compounds found in animals can be formed from cholesterol. Thus, in animals, the pathway of *de novo* biosynthesis of steroids becomes first the pathway of synthesis of cholesterol. Very similar reaction sequences occur in plants and in some microorganisms, but the products in these cases are sterols of slightly different structures, such as ergosterol and stigmasterol[1]:

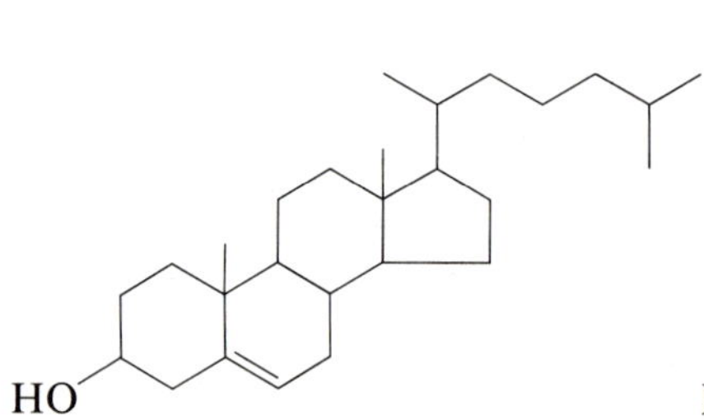

Cholesterol

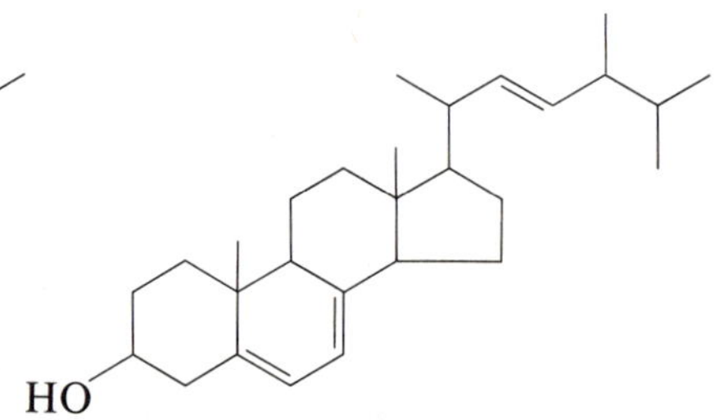

Ergosterol

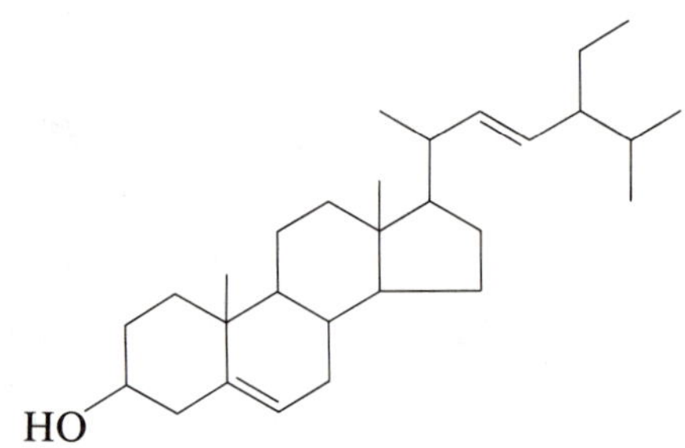

Stigmasterol

All of the carbons of the cholesterol molecule can be derived from acetyl–CoA. The early reactions of the condensation process are identical with those of fatty acid synthesis, except that the enzyme systems involved are cytoplasmic in location and deal directly with the Coenzyme A esters rather than an acyl carrier protein. Otherwise, the formation of a molecule of acetoacetyl–CoA from two molecules of acetyl CoA proceeds in the

1. Note again that in drawing steroid structures, it is customary to omit hydrogens and to indicate methyl groups simply by showing the bond leading to them. Substituents attached by dotted lines are below the plane of the ring system; those attached by solid lines are above.

same fashion, with formation of malonyl–CoA at the expense of ATP providing the driving force for the process.

At this point the pathways diverge sharply, with the next step in the direction of cholesterol being a condensation reaction in which the methyl group of another molecule of acetyl–CoA is added across the carbonyl double bond of the acetoacetyl–CoA. In the process, one molecule of Coenzyme A is lost, yielding as product 3-hydroxy-3-methylglutaryl–CoA. Reduction of the thioester group to the corresponding alcohol yields mevalonic acid. This portion of the biosynthetic pathway is shown in Figure 15-2.

The reduction reaction which produces mevalonic acid is catalyzed by a rather complex enzyme, hydroxymethylglutaryl–CoA reductase, and is the subject of much study because this is the rate-controlling step

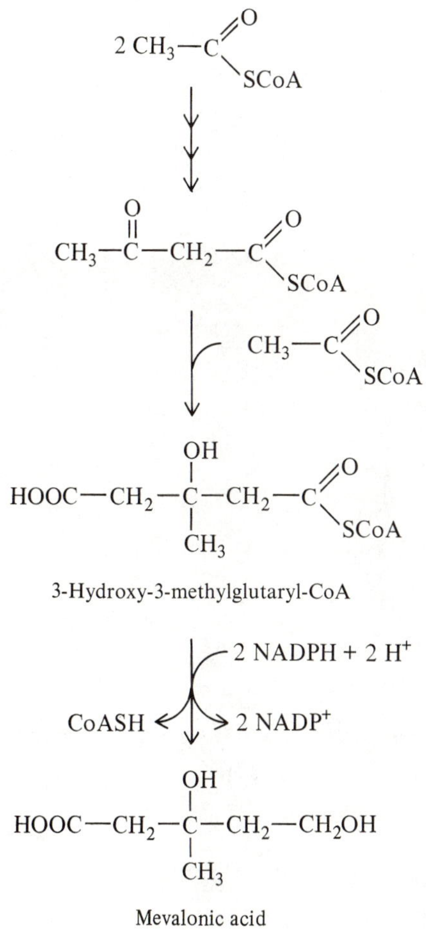

Figure 15-2 The pathway of biosynthesis of mevalonic acid from acetyl-Coenzyme A.

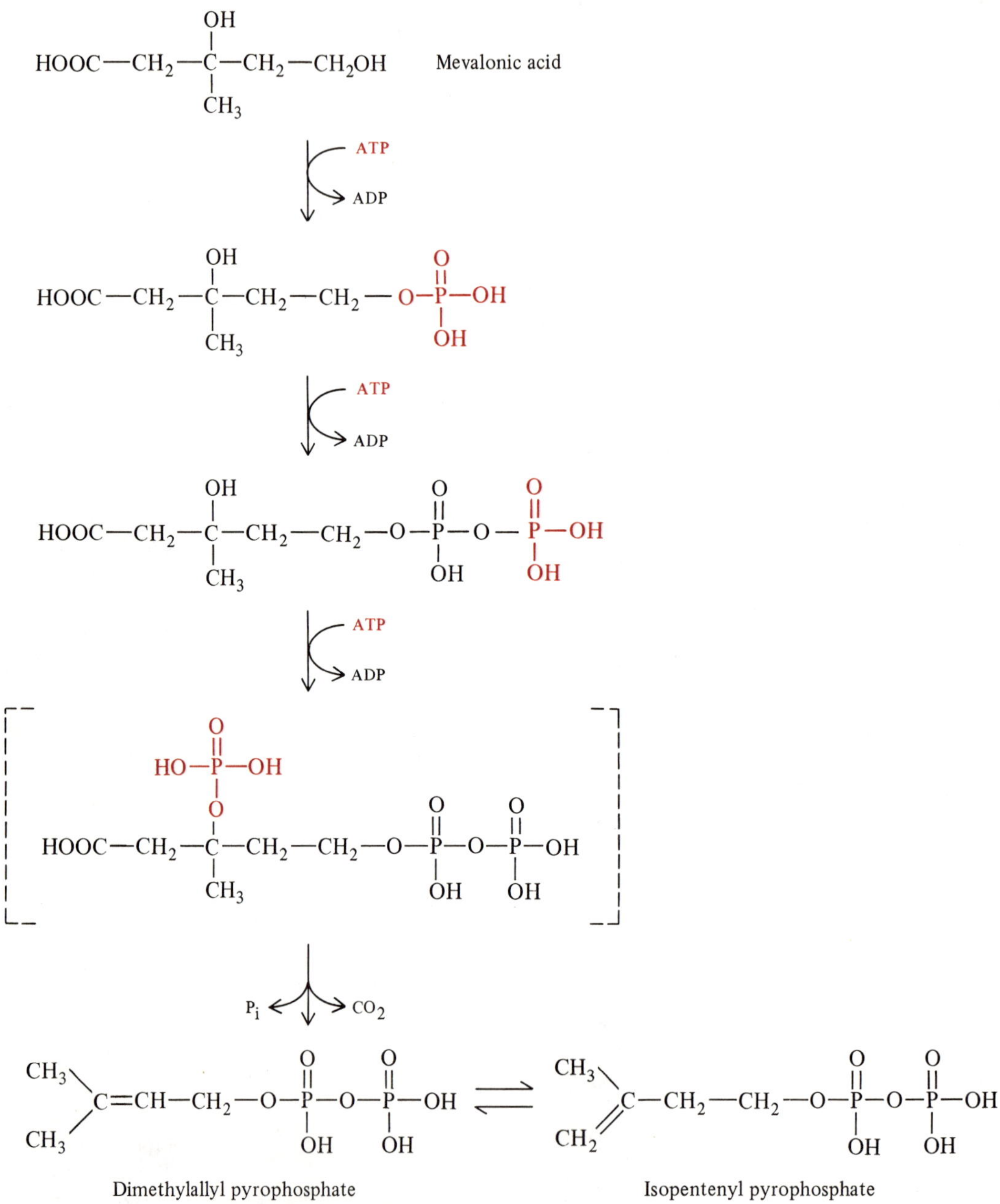

Figure 15-3
Conversion of mevalonic acid to the two "active isoprene units," dimethylallyl pyrophosphate and isopentenyl pyrophosphate.

in the synthesis of cholesterol. If a suitable means can be found for controlling the rate of this reaction it may be possible to reduce the synthetic rate in persons having abnormally high levels of cholesterol in the blood.

Formation of "Active Isoprene Units"

In the next portion of this pathway, the mevalonic acid is "activated" by the addition of a pyrophosphate group to the 5-position and a phosphate to the 3-position (Fig. 15-3). The formation of the pyrophosphate by sequential addition of two terminal phosphate groups from ATP is unusual and lacks a clear mechanistic justification.

Removal of the 3-phosphate group with formation of a double bond produces a β-unsaturated acid that easily undergoes a nonoxidative decarboxylation. The result is a close relative of isoprene called dimethylallyl pyrophosphate, which is readily isomerized to isopentenyl pyrophosphate. These two compounds represent the metabolically active branched, five-carbon units which are often referred to simply as **active isoprene units**. It is these compounds that are condensed together to form the wide variety of biological compounds that are called **polyisoprenoid** compounds, of which the steroids are one example. We will look at some other examples after we deal with the steroids.

The reactions involved in converting mevalonic acid to the active five-carbon pyrophosphates are shown in Figure 15-3.

Condensation to Form Cholesterol

The "active isoprene units" are capable of several types of condensation reactions. In the present case, three of them take part in a "head to tail" condensation, which results in a branched 15-carbon intermediate that still carries a pyrophosphate group. Two of these farnesyl pyrophosphates then come together "head to head" to form the terpinoid hydrocarbon **squalene**. This condensation process is complex and still not fully understood in a mechanistic sense; we will not attempt to deal with it here. These reactions are shown in Figure 15-4.

At first glance, we might appear to be still a long way from cholesterol. However, when we rewrite the squalene molecule with somewhat more realistic bond angles, we obtain the structure shown in Figure 15-5 as one possible conformation.

The squalene in this form undergoes a concerted shifting of electrons to form ring structures, the migration of one of the methyl groups to a new position, and the insertion of a hydroxyl group. The result is a compound (lanosterol) having the basic steroidal ring structure and both the hydroxyl group and the hydrocarbon side-chain characteristic of cholesterol. The excess methyl groups are oxidized to carboxyl groups and removed

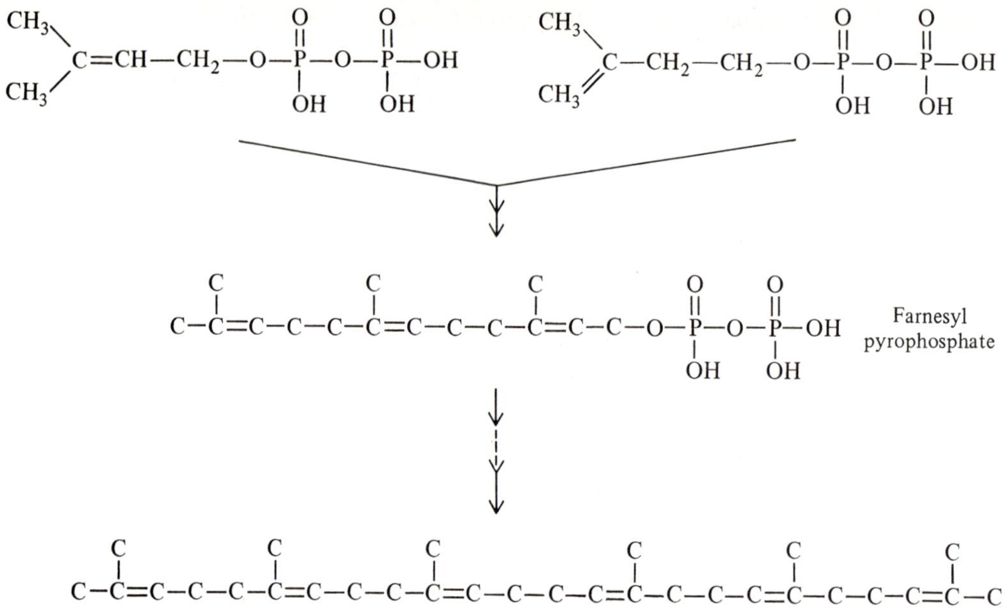

Squalene

Figure 15-4
An abbreviated version of the condensation of *active isoprene units* to form a molecule of squalene.

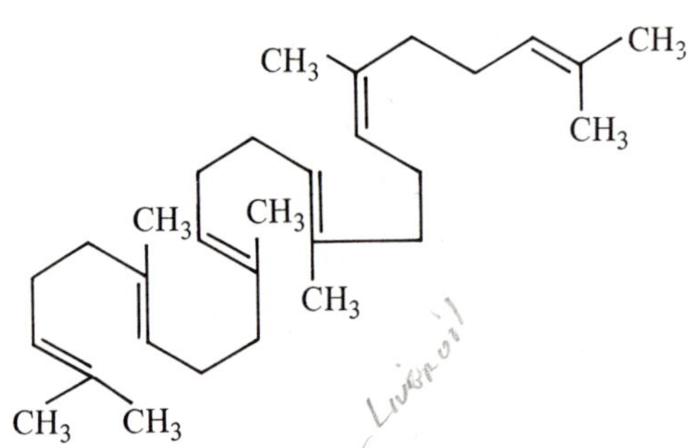

Squalene

Figure 15-5
The structure of squalene redrawn in one possible spacial conformation indicating its relationship to the ring structure of the steroids.

as CO_2. Shift of the position of the double bond completes the conversion to cholesterol as indicated in Figure 15-6.

It should be noted that much of our knowledge of this pathway from acetate to cholesterol came from studies using acetate and other compounds specifically labeled with ^{14}C and/or ^{13}C isotopes. Whole new systems of structural degradation had to be developed in order to be able to determine the amount of label in each carbon atom of the cholesterol molecule and in the other intermediates. The work of Bloch, Woodward, and others

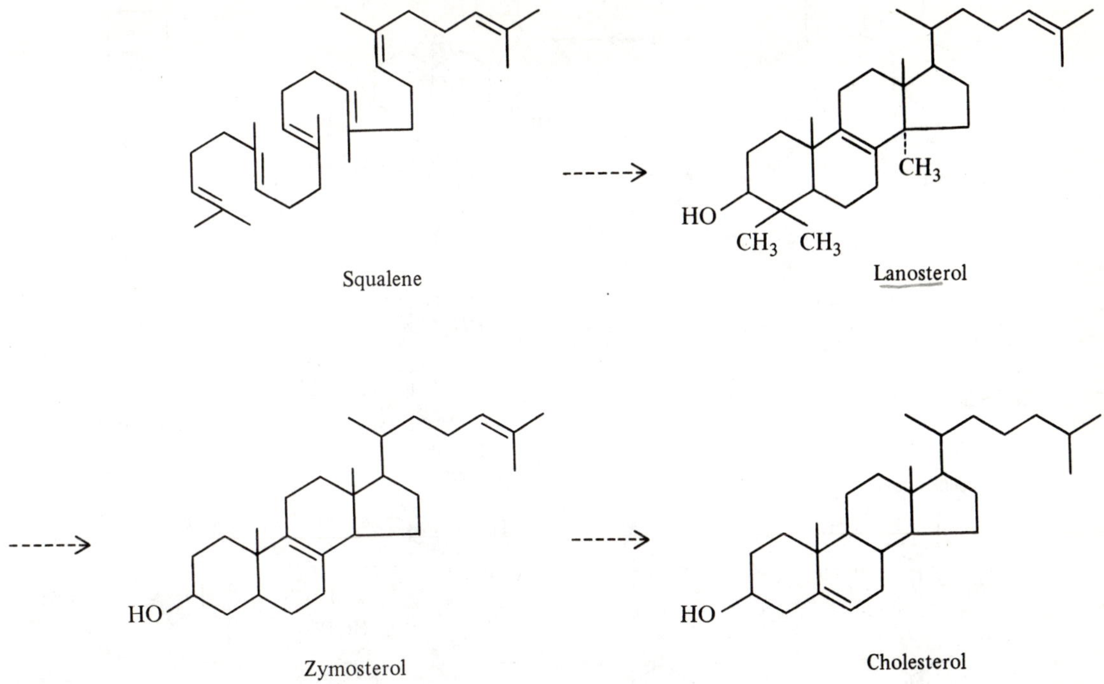

Figure 15-6
Some of the intermediates in the conversion of squalene to cholesterol.

added greatly to the ability of organic chemists to deal with structure analysis in such complex biologically important molecules.

A number of other metabolic intermediates are capable of entering this biosynthetic pathway without first being converted to acetyl–CoA. In particular, the normal degradative pathway for the carbon chain of the amino acid leucine is through the formation of 3-hydroxy-3-methylglutaryl –CoA. As a result, leucine is a very good precursor of cholesterol.

Metabolic Interconversions of Steroids

In Chapter 12 we introduced briefly some of the types of steroid structures. In Figure 15-7, some metabolic interrelationships between representatives of various classes of steroids are shown, demonstrating how they all can be formed from cholesterol.

The number of arrowheads shown in the figure represents approximately the number of enzymatic reactions needed for that conversion. The relatively small number of reactions necessary to convert one steroid hormone to another underscores the tremendous physiological significance of relatively minor changes in the steroid molecule. Thus progesterone (a hormone involved in maintaining pregnancy) can be converted to testosterone (one

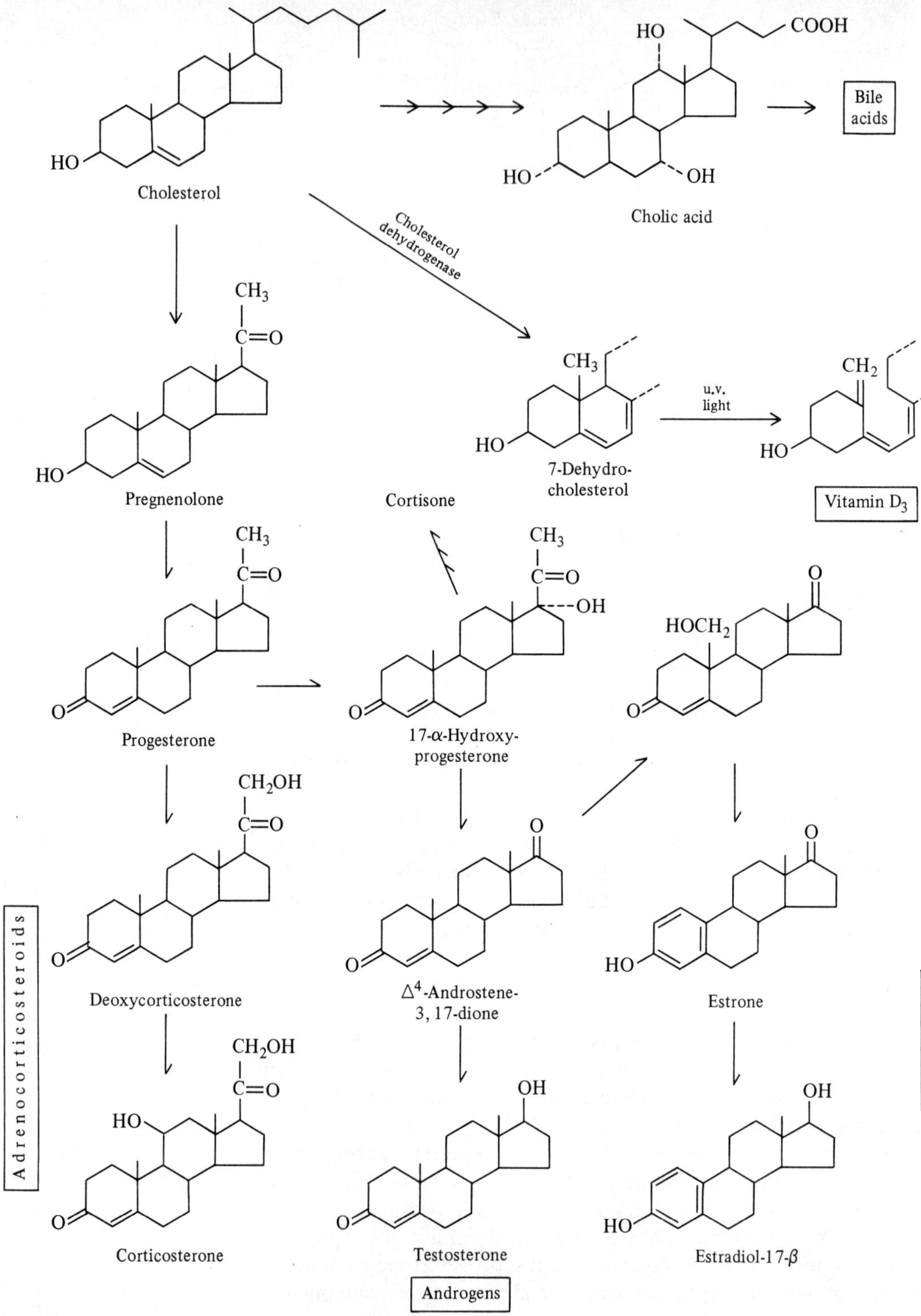

Cholesterol
Cholic acid
Bile acids
Cholesterol dehydrogenase
7-Dehydro-cholesterol
u.v. light
Vitamin D_3
Pregnenolone
Cortisone
Progesterone
17-α-Hydroxy-progesterone
Deoxycorticosterone
Δ^4-Androstene-3, 17-dione
Estrone
Corticosterone
Testosterone
Estradiol-17-β
Adrenocorticosteroids
Androgens
Estrogens

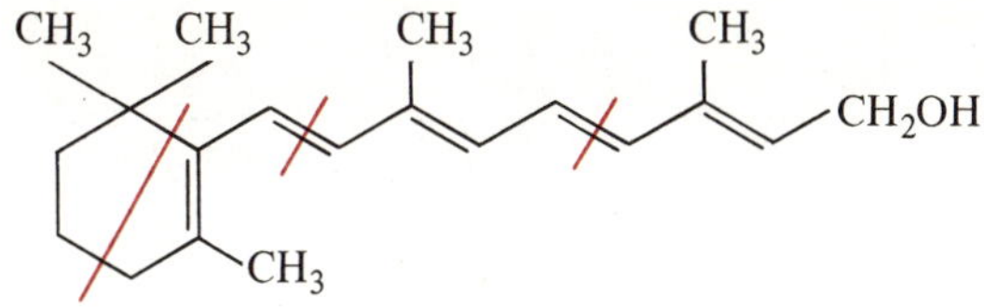

Vitamin A_1

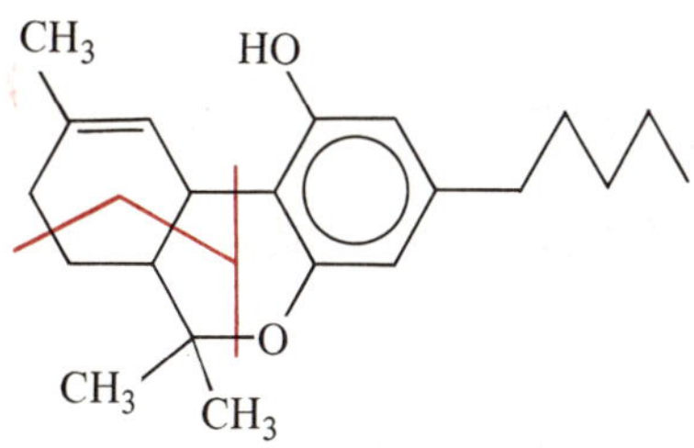

Δ′-3, 4-*trans*-Tetrahydrocannabinol (Δ′-THC)

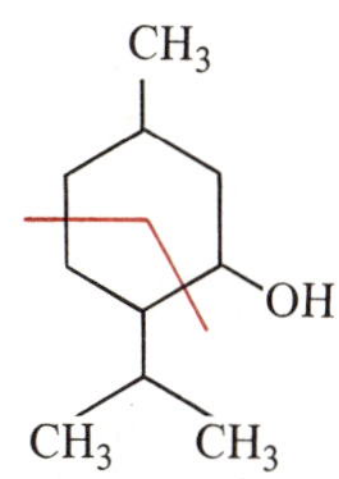

Menthol

Figure 15-8
Several molecules formed by condensation of isoprene units. Vitamin A_1 (retinol) and menthol (a natural constituent of peppermint oil and an unnatural ingredient in some cigarettes) are totally formed from isoprene units. The structure of Δ′-THC (an active ingredient of marijuana) is only partially derived from these compounds.

of the primary male sex hormones) in about three steps, or to cortisone in five or six reactions. While not all of these conversions are possible in all tissues, it is usually control of the balance among them, rather than the lack of one or more reactions, that governs the pattern of steroid hormone activity in an organism.

Other Polyisoprenoid Compounds

Many other compounds synthesized in biological systems arise from polymerization of the active isoprenoid molecules used above in cholesterol synthesis. We will not look at the individual synthetic pathways leading to these compounds, but several of them are shown in Figure 15-8 with markings to show how the isoprenoid units come together to form the final structure.

These polyisoprenoid compounds include simple hydrocarbons such as the terpene limonene (a constituent of pine extracts), Vitamin A and all of the carotenoids, the branched side chains of chlorophyll, ubiquinone, and of Vitamins E and K, and the complex structure of natural rubber latex, which is essentially a linear polymer of isoprene units.

Figure 15-7 (p. 242)
Some metabolic relationships among the steroids. Most of the reactions shown with single-headed arrows are reversible, leading to ready interconversions among the different types of steroids.

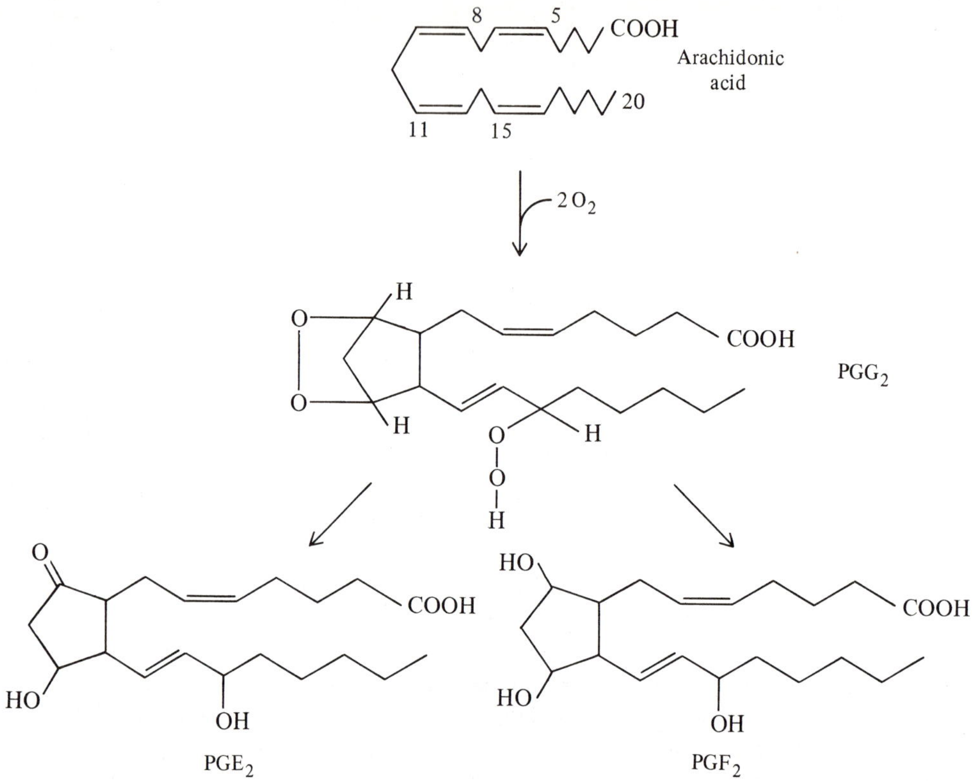

Figure 15-9
Biosynthesis of typical prostaglandins from arachidonic acid.

Formation of Prostaglandins

In recent years, a number of closely related compounds have been isolated that act in the manner of hormones in producing major physiological responses in animals. The exact nature of the metabolic means by which they exert their activities is not yet clear.

Biosynthesis of the prostaglandins generally starts with one of the unsaturated fatty acids, which is cyclized and modified to the particular form of prostaglandin being synthesized. The process is illustrated in Figure 15-9.

15.3 Structure of Membranes

It has been known for some time that the membranes that surround cells, and many of the subcellular organelles, contain substantial amounts

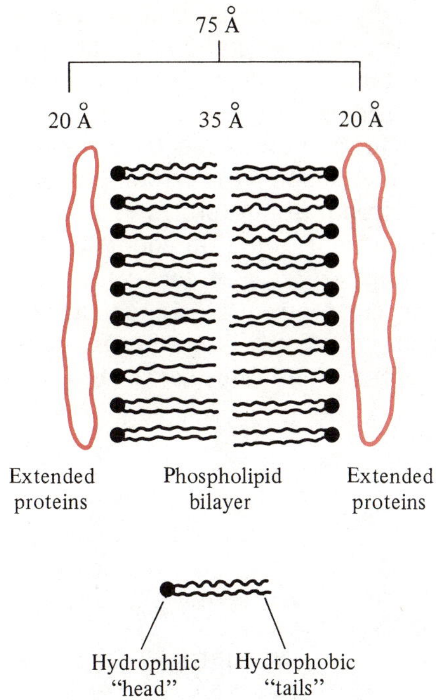

Figure 15-10
A model of the "unit membrane" concept of membrane structure.

of lipid materials. Early work on specialized membranes, such as that of red blood cells, combined with the results of electron microscopy, gave an overly simplified view of membranes as being composed of simple phospholipid bilayers coated on both sides with extended protein molecules. This view of a "unit" membrane structure is represented in Figure 15-10. A variety of discoveries in recent years have given rise to a more sophisticated view of the structure of membranes. The modern **fluid mosaic** model of the membrane's structure is illustrated in Figure 15-11. In this model, various protein molecules are imbedded in a bilayer of phospholipid molecules.

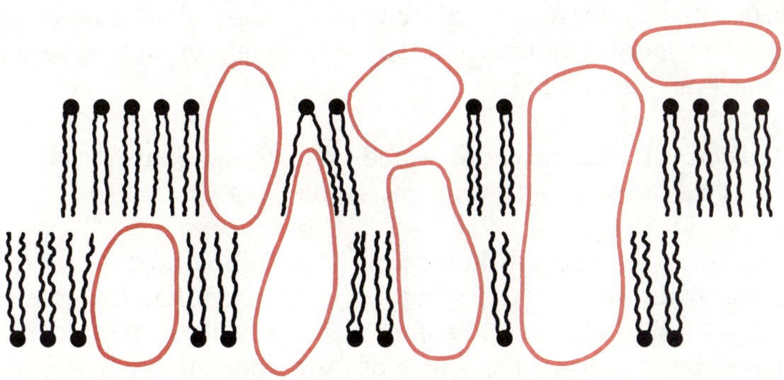

Figure 15-11
A model of one view of the fluid mosaic concept of membrane structure. Most of the protein molecules in this case either extend completely through a lipid bilayer or are embedded partially into it. These proteins are an integral part of the membrane and are more tenaciously held than the proteins that are only attached at the surface. Proteins in this system have some freedom to move laterally; hence the term "fluid" mosaic.

Even this depiction is an oversimplification, since a variety of other lipids are found in some membranes in addition to the phospholipids. Included in particular is a substantial amount of cholesterol and cholesteryl esters.

As we have learned more about the structures of membranes, we have also discovered more information concerning their critical role in many facets of cellular function. It is certainly clear that membranes are deeply involved in electron transport processes and the associated oxidative and photosynthetic phosphorylation. In addition, they have a central role in controlling many transport processes that determine the nature of cellular interactions with the environment. Research in the area of membrane structure and function is currently in a very productive phase and we can expect much more information in this key area in the next few years.

Additional Readings

Scientific American articles

R. A. Capaldi, "A Dynamic Model of Cell Membranes," 230(3), 26 (March 1974).

D. E. Luria, "Colicins and the Energetics of Cell Membranes," 233(6), 30 (December 1975).

S. Dagley and D. E. Nicholson, *An Introduction to Metabolic Pathways* (Wiley, New York, 1970).

A. I. Gurr and A. T. James, *Lipid Biochemistry: An Introduction*, 2nd edition (Halsted Press, New York, 1976).

S. J. Wakil (Ed.), *Lipid Metabolism* (Academic Press, New York, 1970).

Study Questions

1. A eucaryotic cell operating under aerobic conditions synthesizes one molecule of tristearoylglycerol from glucose molecules using reactions you have seen in this and previous chapters. What will be the *net* energetic cost in molecules of ATP (or equivalent nucleotide triphosphates) and in reducing power? (Assume that any NADH not needed for synthesis will be re-oxidized through the mitochondrial electron transport system, and that only the glycerol phosphate shuttle is functional in this cell.)

2. By means of what reactions would you suggest that the cell generate the needed reducing power for the synthesis of Question 1?

3. Draw the structure of cholesterol. Then, using an "X" to indicate carbon from the carboxyl group of acetate and "O" to represent the methyl carbon, label as many of the positions in the cholesterol molecule as you can to indicate the source of the carbon atoms.

4. Gossypol, a rather toxic phenolic constituent of cotton seeds, is a polyisoprenoid compound. Indicate which carbons you might expect to come from each of the six isoprenoid units involved:

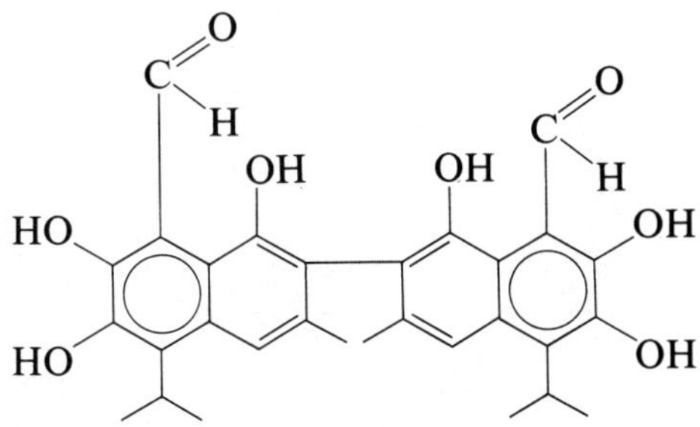

Gossypol

5. (a) Synthetic "membranes" composed only of a bilayer of phospholipid are relatively stable in aqueous environments. Comment on the forces tending to help stabilize this system. (b) The inner membranes of mitochondria contain about 70% by weight protein. On the basis of the fluid mosaic model for such a membrane, will the forces stabilizing this system likely be the same as those of part (a) or different? Explain.

CHAPTER 16

Amino Acid Metabolism I: General Reactions

The amino acids are central to the whole process of nitrogen metabolism, whatever the organism. First, of course, synthesis of the proteins that are critical to the life processes is dependent upon an adequate supply of the appropriate amino acids. But, beyond that requirement, there is a need in all biological materials for a remarkable array of nitrogenous compounds. These include all of the nucleotides, porphyrins, alkaloids, and many of the vitamins, hormones, and their relatives.

To a major degree, the biosynthesis of these more complex structures is based upon the use of amino acids as precursors. In some cases, the amino acids serve only as a source of nitrogen, but often the carbon chains of the amino acids accompany the nitrogen atoms as they are built into more complex structures. And then, in some cases, the amino acids serve a unique role in synthetic metabolism that has little or nothing to do with their nitrogen atoms.

We will deal in this chapter with some of the more general reactions that apply to many of the amino acids, including some of the processes by means of which cells dispose of nitrogen that is in excess of immediate needs. The next chapter will explore some of the special biosynthetic uses of certain specific amino acids. A discussion of protein synthesis will be left to Chapter 20.

16.1 The Concept of "Essential" Amino Acids

Essentially all plants and some microorganisms are capable of synthesizing all of the amino acids normally found in proteins, given an adequate source of carbon compounds and enough ammonia or an alternative form of nitrogen from which to make ammonia. On the other hand, animals and most of the other microorganisms have only a limited ability to synthesize the necessary range of amino acids. Usually, this reflects an inability to synthesize the required carbon skeleton, since the transfer of amino groups between the carbon chains of amino acids is very common, as we will see shortly. The amino acids needed for protein synthesis but not produced in a given organism must be supplied in its diet; thus the term "**essential amino acids**."

The list of essential amino acids varies greatly from one animal to another, and varies with the stage of the life cycle. For the immature chick, the demand for glycine (one of the simplest and most readily synthesized amino acids) outstrips the capacity to make it, and glycine becomes an essential amino acid. The adult chicken is capable of meeting the need by synthesis and does not need it in the diet. For an adult human being, the eight amino acids that are essential in the diet are

Valine	Phenylalanine
Leucine	Tryptophan
Isoleucine	Methionine
Lysine	Threonine

There is often some ambiguity in lists of essential amino acids. Arginine appears to be essential for growing children, and may be included for that reason. In lists referring to the needs of "mammals," histidine is usually included because it is required for rats, the common test subject, although histidine is not generally regarded as necessary for adult humans.

16.2 General Reactions of Amino Acids

A few reactions are common to a wide variety of amino acids; these include the transfer of amino groups from amino acid to keto acid (transamination), the related process of deamination to the alpha-keto acid, and nonoxidative decarboxylation.

Transamination

In the simplest overall sense, transamination involves the transfer of an amine group from one amino acid to an α-keto acid, creating a new amino acid and a new α-keto acid; thus;

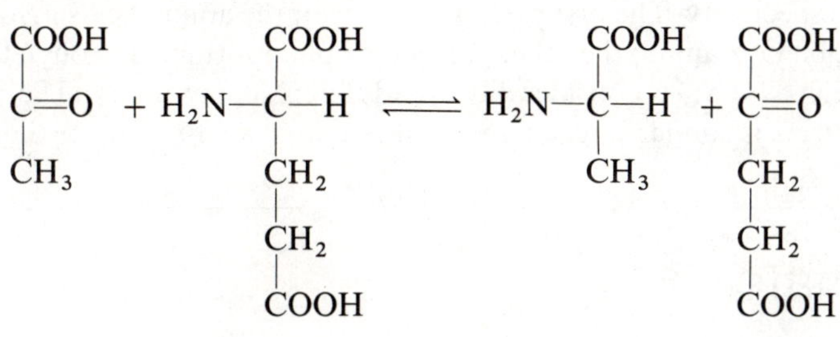

In practice, the reaction is a little more complex. First, it involves pyridoxal phosphate and a divalent metal ion as cofactors. The pyridoxal

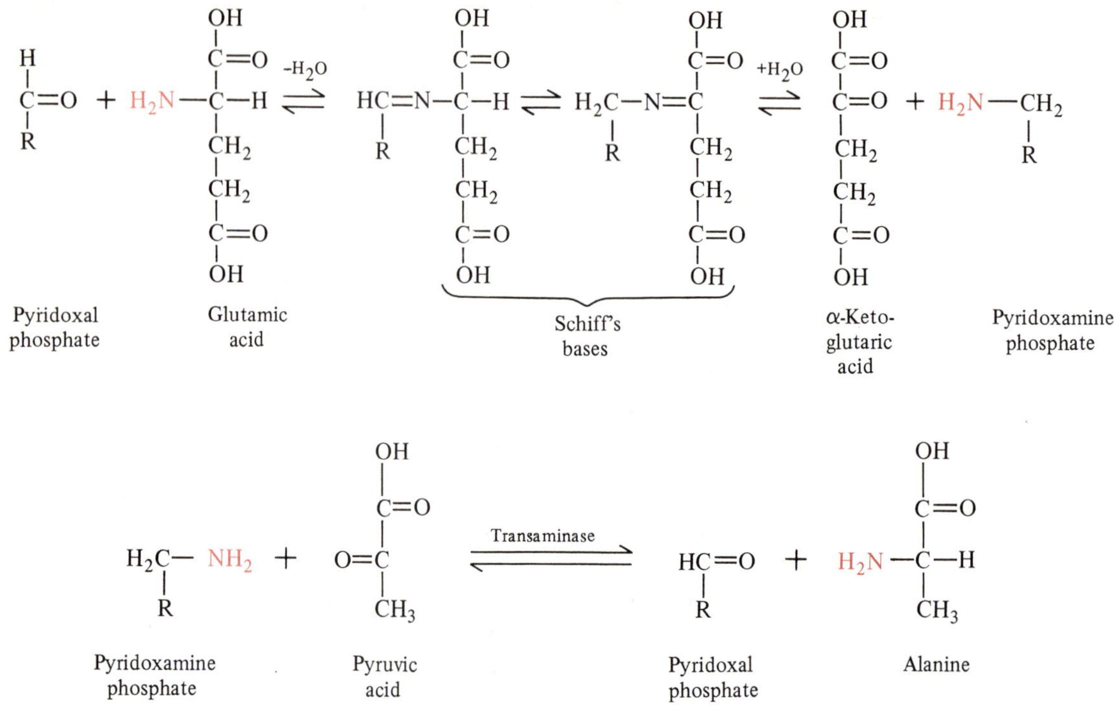

Figure 16-1
The mode of action of transaminase in transferring an amino group from glutamic acid to pyruvate to form alanine. The "R" symbol is used to represent most of the pyridoxal phosphate molecule.

phosphate serves as an intermediate acceptor for the amine group and functions in a pair of "push–pull" reactions as shown in Figure 16-1.

A number of transaminase enzymes are known; however, the predominant forms in a wide variety of tissues seem to have one of two types of substrate specificity. The first prefers that one of the amino acids involved in the reaction be alanine; the other partner is not important, although the rate of reaction will vary considerably for different compounds. The second transaminase is similar but requires that glutamic acid be one of the amino

Pyridoxal Phosphate

Any one of three compounds will meet the requirement for "Vitamin B_6" in humans. These compounds are pyridoxine, pyridoxamine, and pyridoxal. Whichever form of the vitamin is taken in, it must be converted to pyridoxal phosphate before it can be used as a coenzyme:

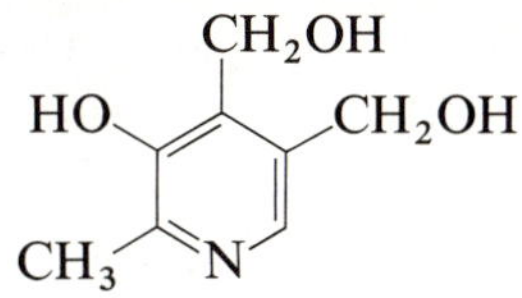

Pyridoxine

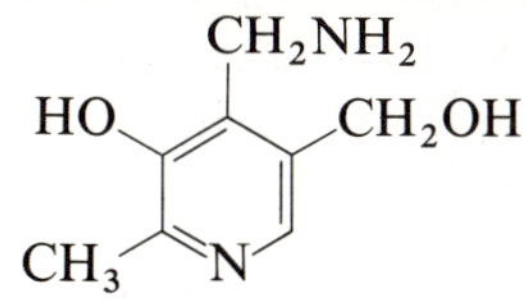

Pyridoxamine

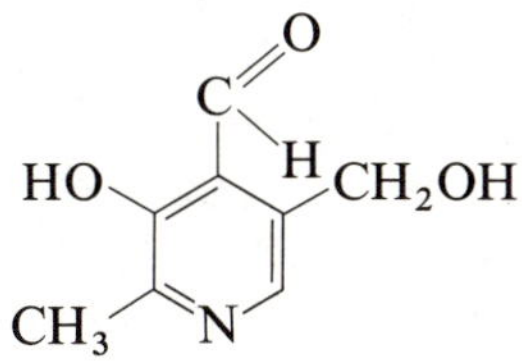

Pyridoxal

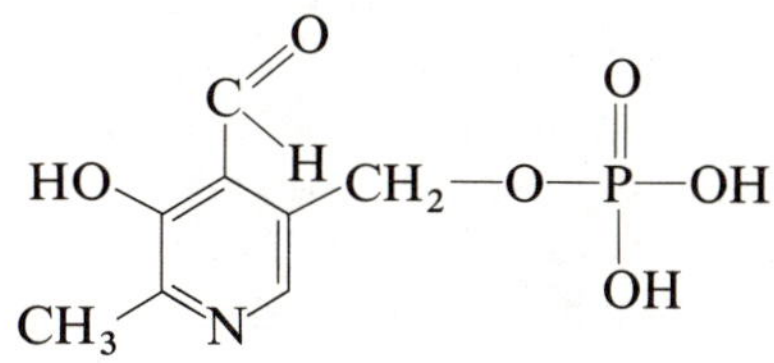

Pyridoxal phosphate

Pyridoxal phosphate is closely associated with amino acid metabolism since it is required for enzymes carrying out the general processes of transamination and decarboxylation, as well as a number of specific reactions of individual amino acids. In most cases, a divalent cation is also involved in forming the active enzyme–substrate complex.

acids involved in its reactions. Again, reaction rates will vary depending on the second amino acid, but alanine serves as an excellent cosubstrate:

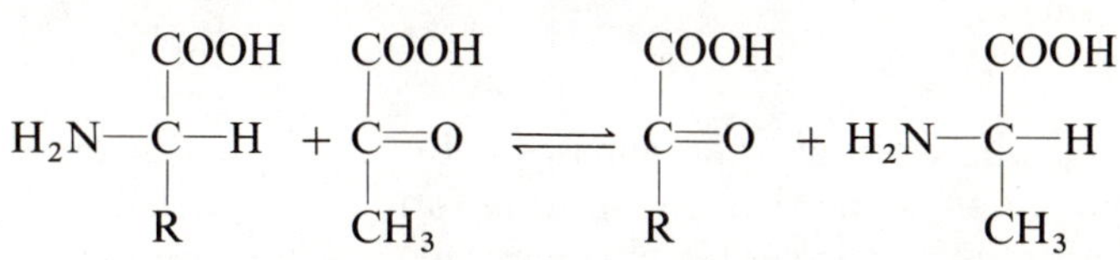

Pyruvic acid Alanine

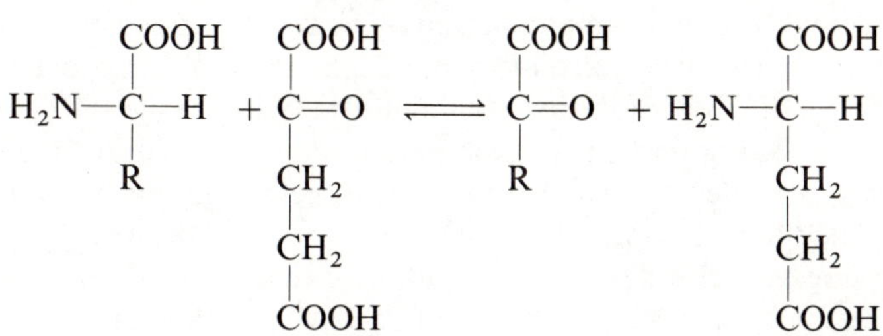

α-Ketoglutaric acid Glutamic acid

Thus, using these two transaminases, it is possible to transfer an amine group from any amino acid capable of donating one to almost any α-keto acid capable of forming an amino acid. We have already seen that the keto acids corresponding to the "essential" amino acids are not normally available, and proline cannot be formed by transamination because of its unusual structure.

Deamination

Amino acids that an organism ingests in excess of its immediate needs for protein synthesis and other specific biosyntheses are not stored for any period of time. Rather, they are processed in such a way that the nitrogen is removed and the carbon chains are used for the production of biologically useful energy. In most cases, carbon chains are converted into intermediates we have already identified in the various pathways for energy production or storage.

This process usually begins with the removal of the amine group to yield the α-keto acid. This can occur in several ways, the most common using the transaminases and an enzyme called glutamate dehydrogenase which catalyzes the following reaction:

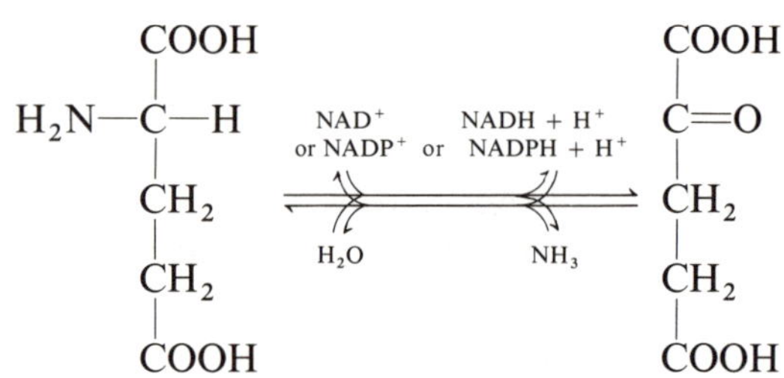

The net effect of the transaminases is to transfer amine groups from various amino acids to α-ketoglutarate to form glutamic acid. Glutamate dehydrogenase then oxidizes the glutamate and releases the amine group as ammonia. The overall result is the conversion of the original amino acid to the related α-keto acid and ammonia. This system is illustrated in Figure 16-2.

As mentioned earlier, proline cannot take part in this process and is degraded by a different pathway. Histidine and lysine, while they appear to be capable of taking part in transamination, do not do so in practice and also have special degradative reaction sequences. It should be noted that although they are not normally synthesized in animal systems by transamination due to lack of the appropriate keto acids, the other "essential" amino acids are in general capable of following the usual deamination process.

Deamination by Dehydration. Several specific amino acids are also capable of being deaminated to the keto acid by special reactions. As an example, serine and threonine can be dehydrated to an α,β-unsaturated amino acid which then rearranges to a form with a carbon-nitrogen double bond. Since this is exactly the same as a Schiff's base formed from the keto

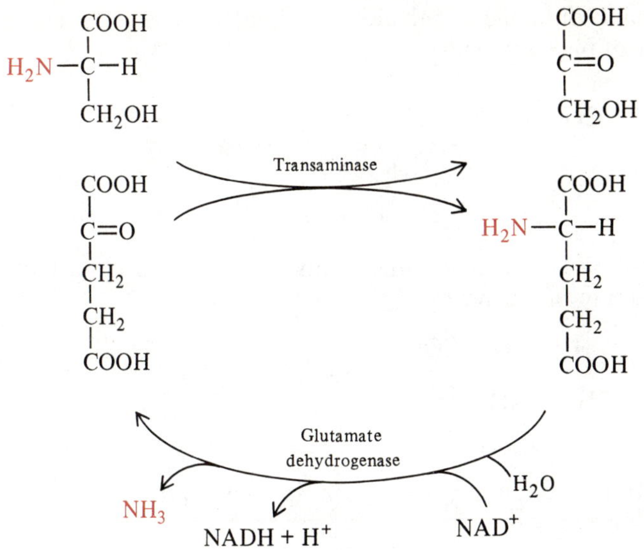

Figure 16-2
Deamination of serine resulting from the action of a transaminase and glutamate dehydrogenase.

acid and ammonia, it spontaneously hydrolyzes into these constituents. The sequence of reactions for serine is illustrated in Figure 16-3. Note that serine dehydrase is another pyridoxal phosphate enzyme.

Decarboxylation

Many of the amino acids are capable of undergoing a nonoxidative decarboxylation to the corresponding amine:

$$R-\underset{NH_2}{\overset{H}{\underset{|}{\overset{|}{C}}}}-COOH \longrightarrow R-\underset{NH_2}{\overset{H}{\underset{|}{\overset{|}{C}}}}-H + CO_2$$

The enzymes catalyzing these reactions are very specific, and a separate enzyme is required for each decarboxylation. All of these decarboxylases require pyridoxal phosphate as a coenzyme. Many of the amines resulting from decarboxylation serve important roles in the cells; we will consider only a few representative examples.

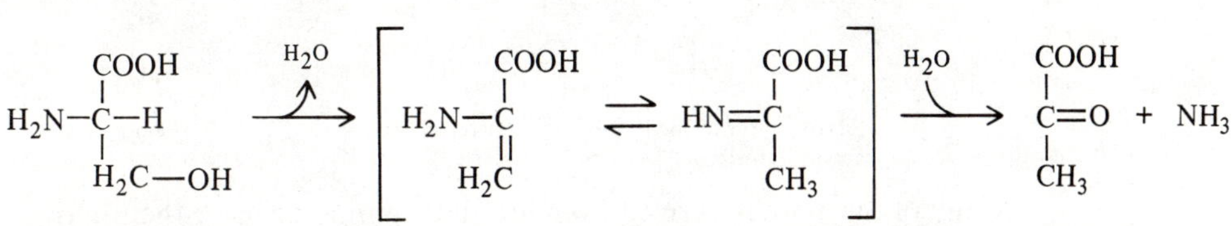

Figure 16-3
Deamination of serine resulting from the action of serine dehydrase.

1. Conversion of serine to ethanolamine and choline (required as components of phospholipids):

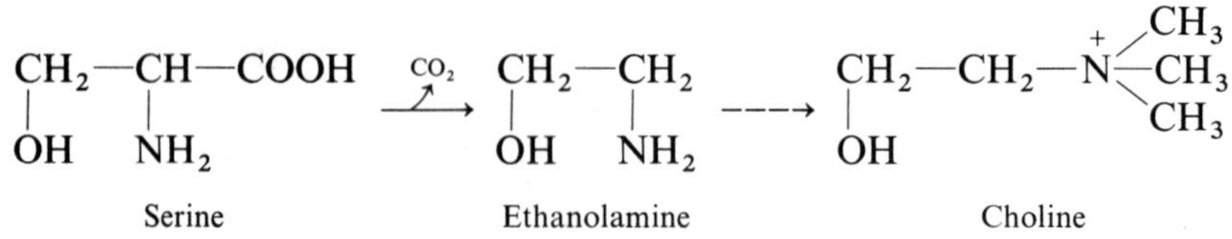

2. Conversion of cysteine to β-mercaptoethylamine (an important part of the structure of Coenzyme A):

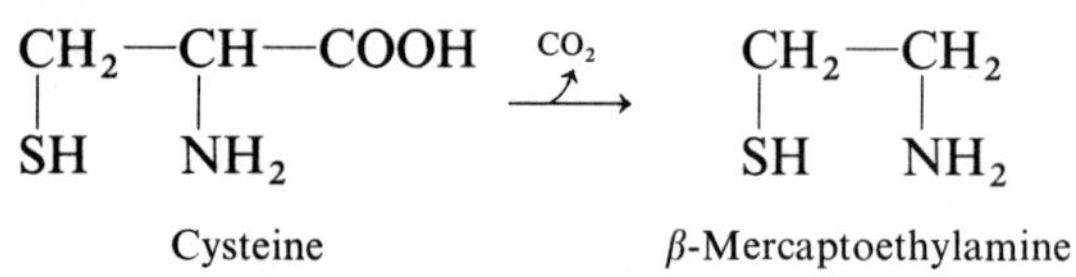

3. Production of tyramine, norepinephrine, and epinephrine (adrenaline) from tyrosine:

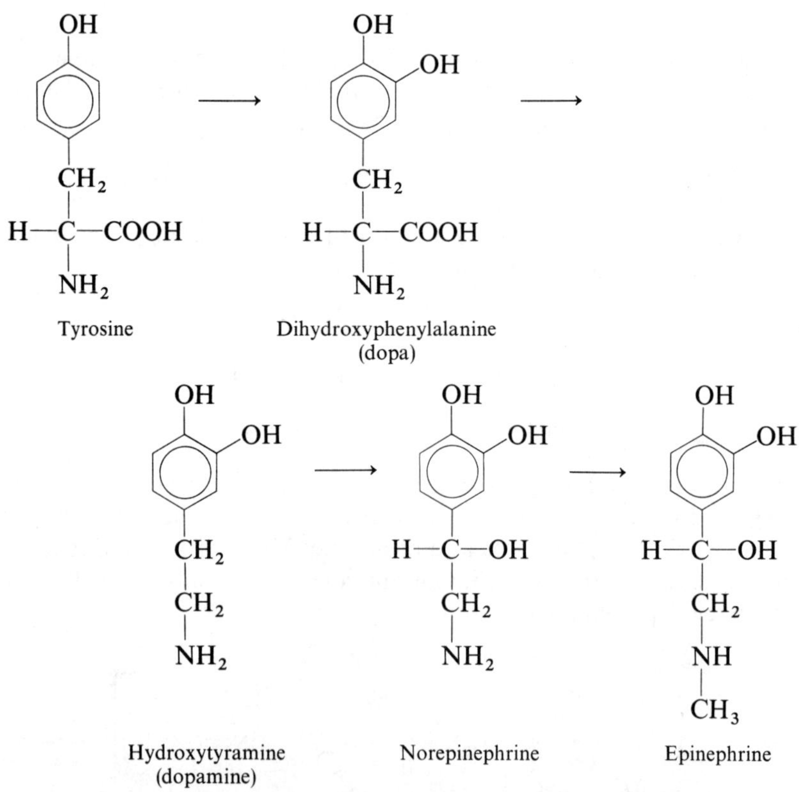

Many of the amines derived from aromatic amino acids or their derivatives act as hormones or as neurotransmitters or moderators of brain function in animals. These include tyramine, norepinephrine, epinephrine, tryptamine, and 5-hydroxytryptamine (both derived from tryptophan).

4. Production of histamine from histidine:

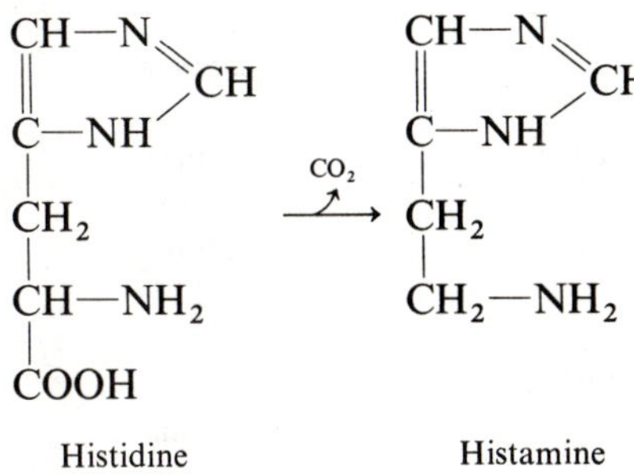

A considerable amount of the income of various pharmaceutical firms comes from products designed to reduce the effects of the histamine produced as a response to viral invasion in humans (especially from the common cold).

16.3 Amino Acid Degradation

Amino acids are not stored by cells in any significant quantity; in addition, there is no evidence of storage in the form of peptides or proteins which could yield amino acids to meet metabolic demands. Instead, amino acids that are in excess of the immediate cellular needs for protein synthesis and other specific metabolic requirements are utilized as sources of carbon atoms and of energy by means of reactions that eventually destroy the amino acid molecule.

For most of the amino acids, the first step on this road to destruction involves removal of the amine group as ammonia. We have seen earlier in this chapter how this process can be carried out using a combination of transminases and glutamic dehydrogenase. Some of the resulting α-keto acids are intermediates in oxidative pathways we examined as part of carbohydrate metabolism. These include pyruvic acid from alanine or serine, oxaloacetate from aspartate, and α-ketoglutarate from glutamate. Almost all of the other amino acids are converted into intermediates of carbohydrate or lipid metabolism by special degradative pathways.

We commented earlier on the impossibility, in animals, of producing the required blood glucose from fatty acids. Thus, under circumstances in which the carbohydrate intake is very restricted or impaired, production of glucose from amino acid carbon chains by reversal of glycolysis and related reactions is critical to survival, since blood glucose provides the primary source of energy for the brain.

TABLE 16-1

Metabolic fates of amino acids.

Product	Amino Acids
Pyruvate	alanine serine cysteine (glycine) (threonine)
α-Ketoglutarate	glutamic acid histidine arginine proline
Succinyl–CoA	valine isoleucine threonine methionine
Fumarate	phenylalanine tyrosine
Oxaloacetate	aspartic acid
Acetyl–CoA	leucine isoleucine (tryptophan)
Acetoacetate	leucine phenylalanine tyrosine

We cannot possibly deal here with all of the pathways of degradation of individual amino acids. We will consider only a few examples which give some view of the nature of the pathways involved and of their variety. For the rest, a chart is given in Table 16-1 of the products of degradative metabolism of the various amino acids.

Glucogenic and Ketogenic Amino Acids

All of the products of degradation of amino acids shown in Table 16-1 except acetyl–CoA and acetoacetate can undergo net conversion to glucose. The amino acids breaking down to these intermediates are thus called **glucogenic**. Those amino acids that break down through acetyl–CoA and/or acetoacetate cannot give rise to net formation of glucose. Instead, especially under conditions of carbohydrate deprivation, they are much more likely to give rise to ketone bodies. These amino acids are thus referred to as **ketogenic**.

Isoleucine: A Branched-Chain Amino Acid

The pattern of isoleucine breakdown should provide a comforting assurance to the student that similar compounds *do* tend to be dealt with by very similar sequences of reactions. We have already seen the similarity between oxidation of succinic acid in the citric acid cycle and of fatty acids during β-oxidation. In the pathway shown in Figure 16-4 we add another case to the list. First, the α-keto acid resulting from deamination is treated just as pyruvate and α-ketoglutarate were treated, by oxidative decarboxylation. The resulting acid is just another fatty acid with an attached methyl group and undergoes the reactions of β-oxidation, even to the point of a thiolase reaction. The propionyl–CoA is metabolized in several different ways in different tissues, but usually ends up as succinyl–CoA.

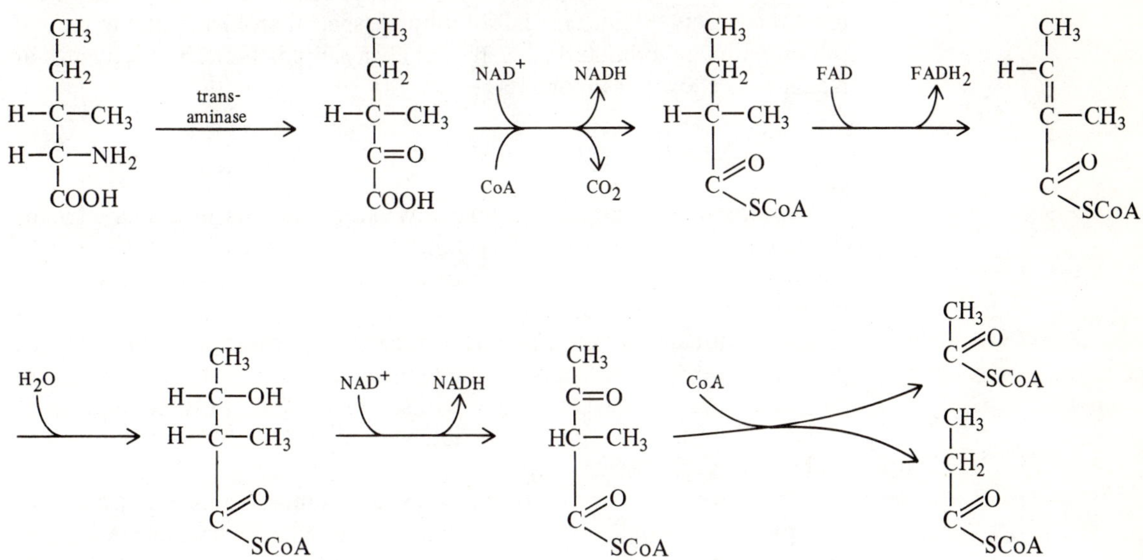

Figure 16-4
The pathway of metabolic degradation of isoleucine. By reactions not shown here, the propionyl–CoA can be converted to succinyl–CoA and enters the citric acid cycle in that form.

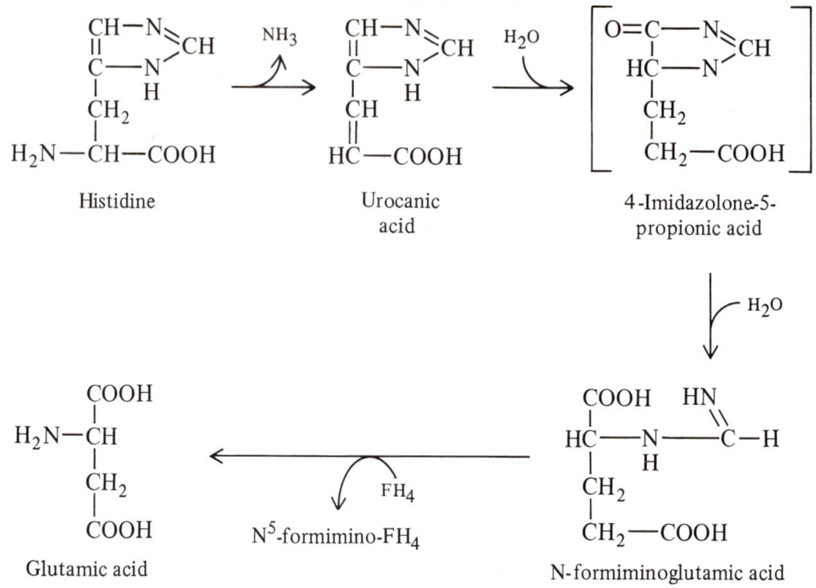

Figure 16-5
The pathway of metabolic conversion of histidine to glutamic acid. FH_4 is an abbreviation for tetrahydrofolic acid; N^5-formimino-FH_4 can take part in one-carbon metabolism through reactions to be considered in the next chapter.

Histidine—An Unusual Case

The metabolism of histidine is so unusual that it serves to indicate once more the variety of biochemical solutions to special problems, and to remind us that pathways of metabolism are not always predictable. The sequence of reactions is shown in Figure 16-5.

Nitrogen Excretion

Ammonia is produced in relatively large amounts in animals by the deamination of amino acids. The compound is quite toxic, and its accumulation in cells or in extracellular tissues in appreciable concentrations is incompatible with life.

A certain amount of ammonia is necessary for a variety of biosynthetic purposes, and some is "stored" safely in the form of the amide nitrogen of glutamine and asparagine. As we will see in Chapter 18, glutamine is especially important as a source of nitrogen in the synthesis of purine and pyrimidine nucleotides.

Aside from this storage capacity, excess ammonia must be disposed of as rapidly as possible. Organisms living in aqueous environments such as the oceans can simply excrete the ammonia directly. Land-based animals, however, have no such convenient mechanism for diluting the ammonia to nontoxic levels. For most of these, the solution lies in the formation of urea, a compound which can be tolerated in much higher concentrations than ammonia. Given a reasonable amount of water to dilute it and, preferably, a specialized storage site, the urea can be allowed to accumulate until it is convenient to excrete it.

Even urea is too toxic to permit its continued accumulation in the water-poor environment faced by developing embryos in the eggs of birds and reptiles. In this case, the waste nitrogen is made into purine nucleotides which are then oxidized to the end product, uric acid:

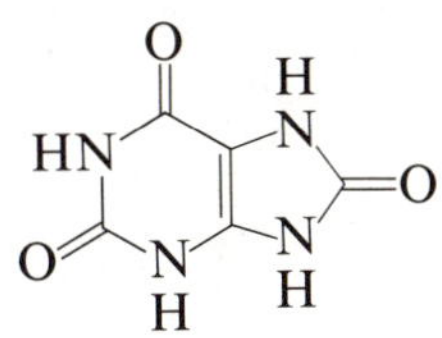

Uric acid

As we will see when we consider purine biosynthesis, this is a rather expensive process in terms of total energy, but it does solve the problem. Uric acid is a very weak acid and, more importantly, is remarkably insoluble. The crystals of this substance can be allowed to accumulate in an almost dry state without any disturbance to the osmotic pressure balance of the organism.

Urea Synthesis

The cyclic pathway by which urea is produced represents a minimal modification of the biosynthetic sequence by which arginine is generated from ornithine. The only additional reaction is the hydrolytic cleavage of the arginine to urea and ornithine.

The pathway as it operates to produce urea is variously known as the urea, ornithine, arginine, or Krebs–Henseleit cycle. None of these is very descriptive, so we will choose urea cycle, which has the virtue of being short. The reactions involved are shown in Figure 16-6.

The first step in the synthesis of urea is a complex reaction involving two molecules of ATP, and one each of CO_2, NH_3, and H_2O. The products are carbamoyl phosphate, two molecules of ADP, and one of inorganic phosphate:

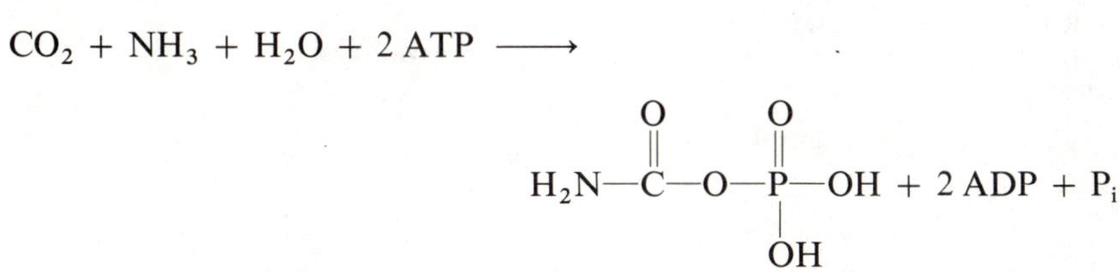

Carbamoyl phosphate

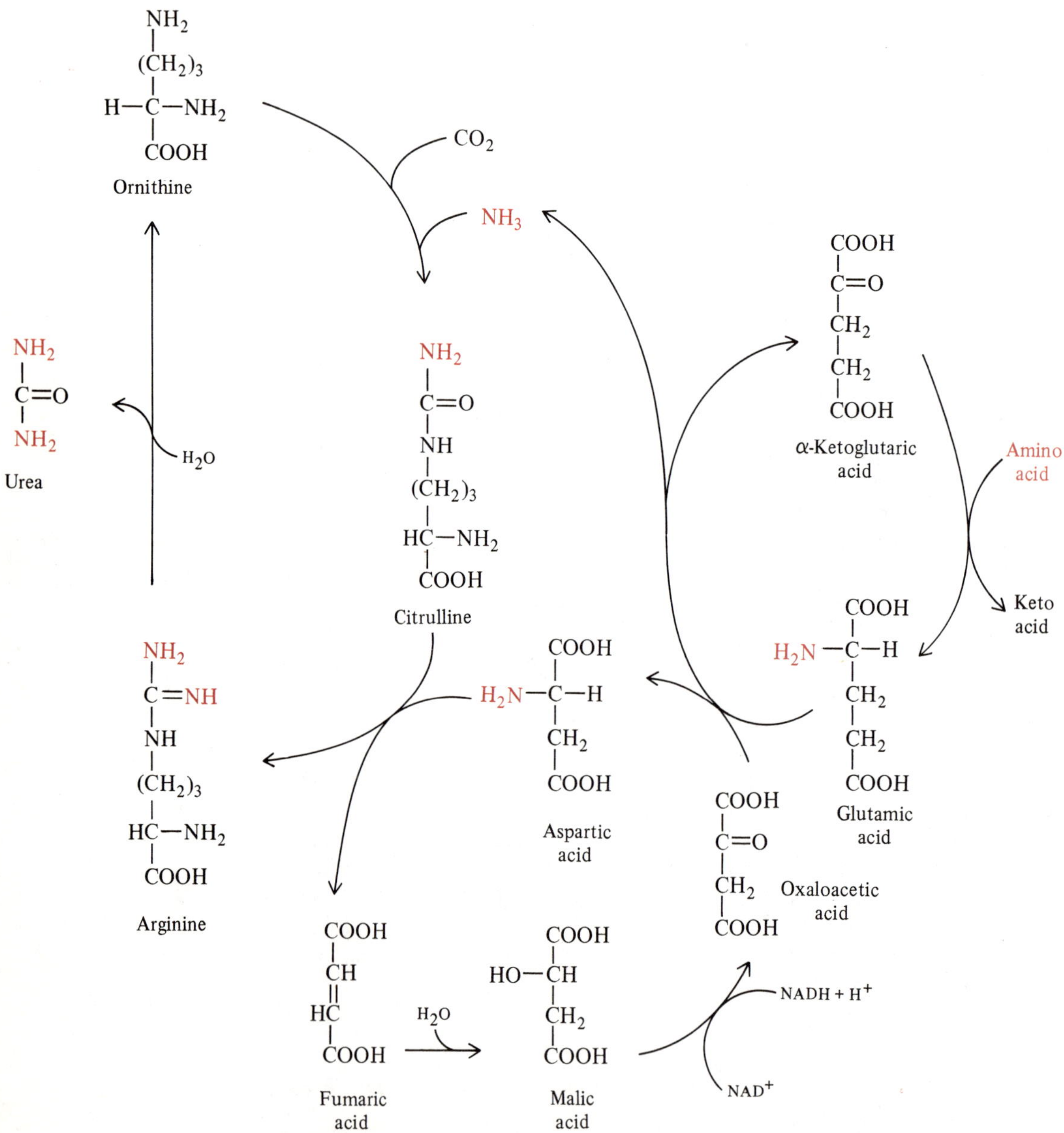

Figure 16-6
A schematic summary of the *urea cycle* and related reactions.

To make this reaction a little more understandable, it can be viewed as occurring in two steps. First, there is formation of an amide from carbonic acid and ammonia, using the energy of one molecule of ATP:

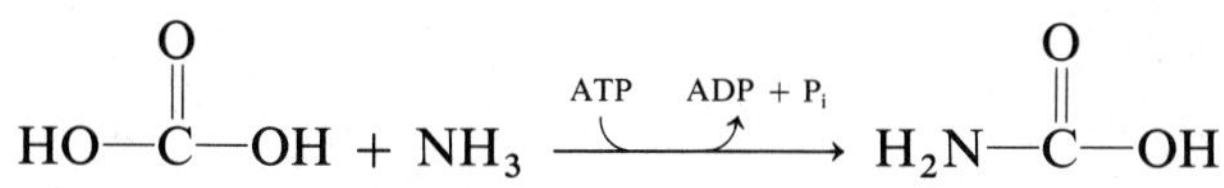

Carbamic acid

This is followed by the formation of the anhydride between the acid group of the carbamic acid and a phosphate group transferred from the second molecule of ATP:

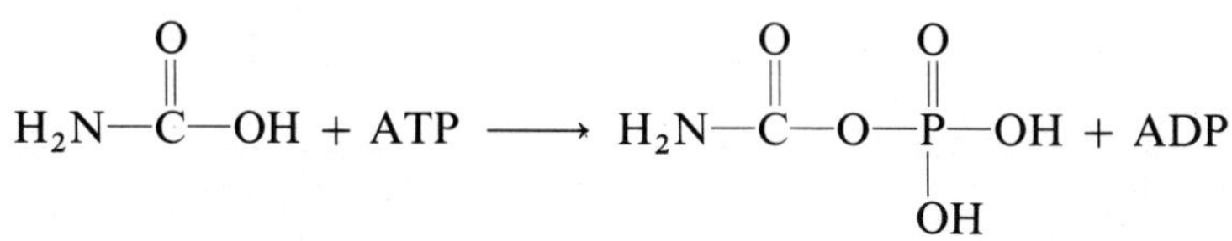

The carbamoyl phosphate is not very stable and must react immediately with ornithine to form citrulline with the release of inorganic phosphate:

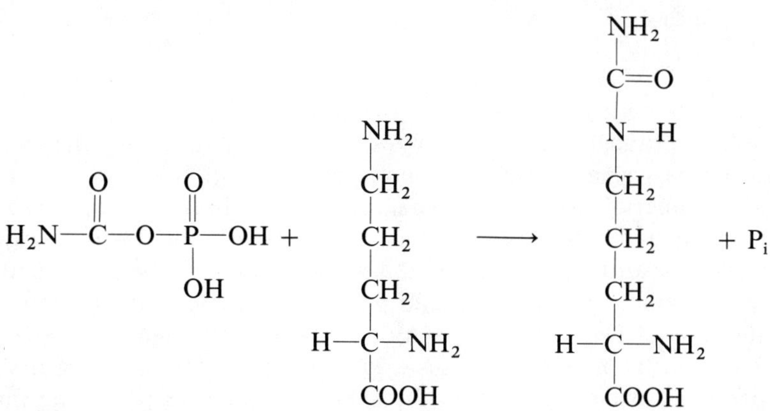

Carbamoyl phosphate Ornithine Citrulline

This portion of the urea cycle occurs within the matrix space of the mitochondria in eucaryotic cells. You will remember that glutamate moves easily through the inner membrane of the mitochondrion; thus, it can collect amine groups from amino acids in the cytosol, move into the mitochondrion, and deliver the ammonia by means of the mitochondrial glutamate dehydrogenase.

The product citrulline moves out of the mitochondrial matrix, and the remainder of the reactions take place in the cytosol. The second nitrogen arrives in the form of the amine group of aspartic acid and this compound reacts with citrulline, at the expense of ATP cleavage, to form argininosuccinic acid. This is another example of a reaction involving ATP that is nominally reversible but in practice is forced completely to the right by rapid hydrolysis of the inorganic pyrophosphate:

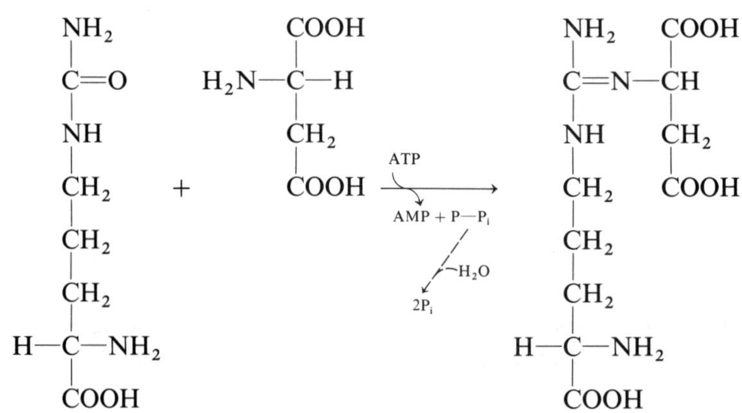

Citrulline Aspartic acid Argininosuccinic acid

An elimination reaction then produces arginine and releases as the other product fumaric acid. Hydrolysis of arginine catalyzed by the specialized enzyme arginase forms urea and regenerates ornithine.

The fumaric acid can of course be oxidized back to a molecule of oxaloacetate and then accept another amine group from glutamic acid to regenerate aspartic acid for the next turn of the cycle. This raises another interesting point. The energy cost for production of one molecule of urea is clearly four "ATPs worth" of energy. However, if the normal oxidative machinery of the mitochondrion is operating, the oxidation step necessary to regenerate oxaloacetate from fumarate is capable of producing three molecules of ATP by oxidative phosphorylation. The same is true for the NADH produced by the glutamate dehydrogenase reaction. Thus, the whole process as shown in Figure 16-6 would result in a net *production* of about 2 ATPs per molecule of urea formed (not counting any energy expended in membrane transport processes).

Additional Readings

R. W. McGilvery, *Biochemistry: A Functional Approach*, 2nd edition (Saunders, Philadelphia, Pa., 1979). This text contains a lucid account of amino acid metabolism with an emphasis on physiological implications.

C. R. Scriver and L. E. Rosenberg, *Amino Acid Metabolism and its Disorders* (Saunders, Philadelphia, Pa., 1973).

S. Grisolia, R. Baguena, and F. Mayor, *The Urea Cycle* (Wiley, New York, 1976). A rather advanced series of reviews.

D. Greenberg (Ed.), *Metabolic Pathways*, 3rd edition (Academic Press, New York, 1969). Volume 3 of this series contains an extensive description of the degradation and biosynthesis of the amino acids, as well as a number of articles on other aspects of nitrogen metabolism.

A. Meister, *Biochemistry of the Amino Acids*, 2nd edition (Academic Press, New York, 1965). A standard reference dealing in detail with amino acid metabolism and related processes.

Study Questions

1. When amino acids labeled with nitrogen-15 are fed to healthy human adults, the labeled nitrogen is eventually spread among all of the amino acids with three exceptions. Which three amino acids are the exceptions?

2. Write a balanced equation for each of the following types of reaction, showing complete structures for reactants and products (except ATP, NAD, and so on):
   (a) a transamination reaction
   (b) two different reactions involving ammonia
   (c) decarboxylation of an amino acid
   (d) a single-enzyme reaction which results in loss of the amine group of an amino acid as ammonia

3. What three compounds comprise the major forms in which excess nitrogen is excreted by animals?

4. Write out a series of complete and balanced equations for all of the reactions required for conversion of the amine groups of two molecules of alanine to a molecule of urea.

5. Given that the products of oxidative degradation of leucine are acetyl–CoA and acetoacetic acid, write out a series of reactions similar to those of Figure 16-4 to show how this process proceeds.

CHAPTER 17

Amino Acid Metabolism II: Specific Biosynthetic Pathways

We have already seen that the amino acids in general are involved in a wide variety of reactions affecting the metabolism of other cellular constituents. Now we want to examine briefly a few of the biosynthetic pathways for which single specific amino acids are required. We can deal here only with a few examples which illustrate the range of involvement of the amino acids in biosynthesis. In addition, in this chapter we will look at several selected examples of pathways of biosynthesis of "essential" amino acids, just to show how organisms that *can* synthesize them go about the job.

17.1 Glycine and the Synthesis of Heme and Other Polypyrroles

We have seen in Chapter 9 the iron–porphyrin structures characteristic of hemoglobin and the cytochromes; we have also seen the similar structure of the magnesium porphyrin we call chlorophyll. In connection with essential amino acids, we have commented on the great need for glycine in growing chicks. This need reflects in part the major role of glycine in the synthesis of the polypyrrole structures of the porphyrins.

The other major intermediate in the synthesis of the porphyrins is succinyl–CoA, a normal constituent of the Citric Acid Cycle. A simplified version of the biosynthetic pathway for formation of the structure of a heme molecule is shown in Figure 17-1. Other similar structures such as the cytochromes and chlorophyll are formed by closely related reaction sequences.

17.2 Single-Carbon Metabolism

The pathways of synthesis of many important biochemical compounds include one or more steps in which a single carbon atom is introduced into a molecule. This topic is introduced at this point because the metabolically active single-carbon units are largely formed from amino acids, and because many of the reactions in which they are used are closely involved with the metabolism of various amino acids.

We have seen several cases in which CO_2 is incorporated into an organic molecule. This obviously is metabolism of a single-carbon atom. However, the area of single-carbon metabolism is usually taken to exclude CO_2 and deal only with those one-carbon units that can be transferred readily from one compound to another and transformed from one oxidation state to another. As they appear when attached to compounds, these are the methyl

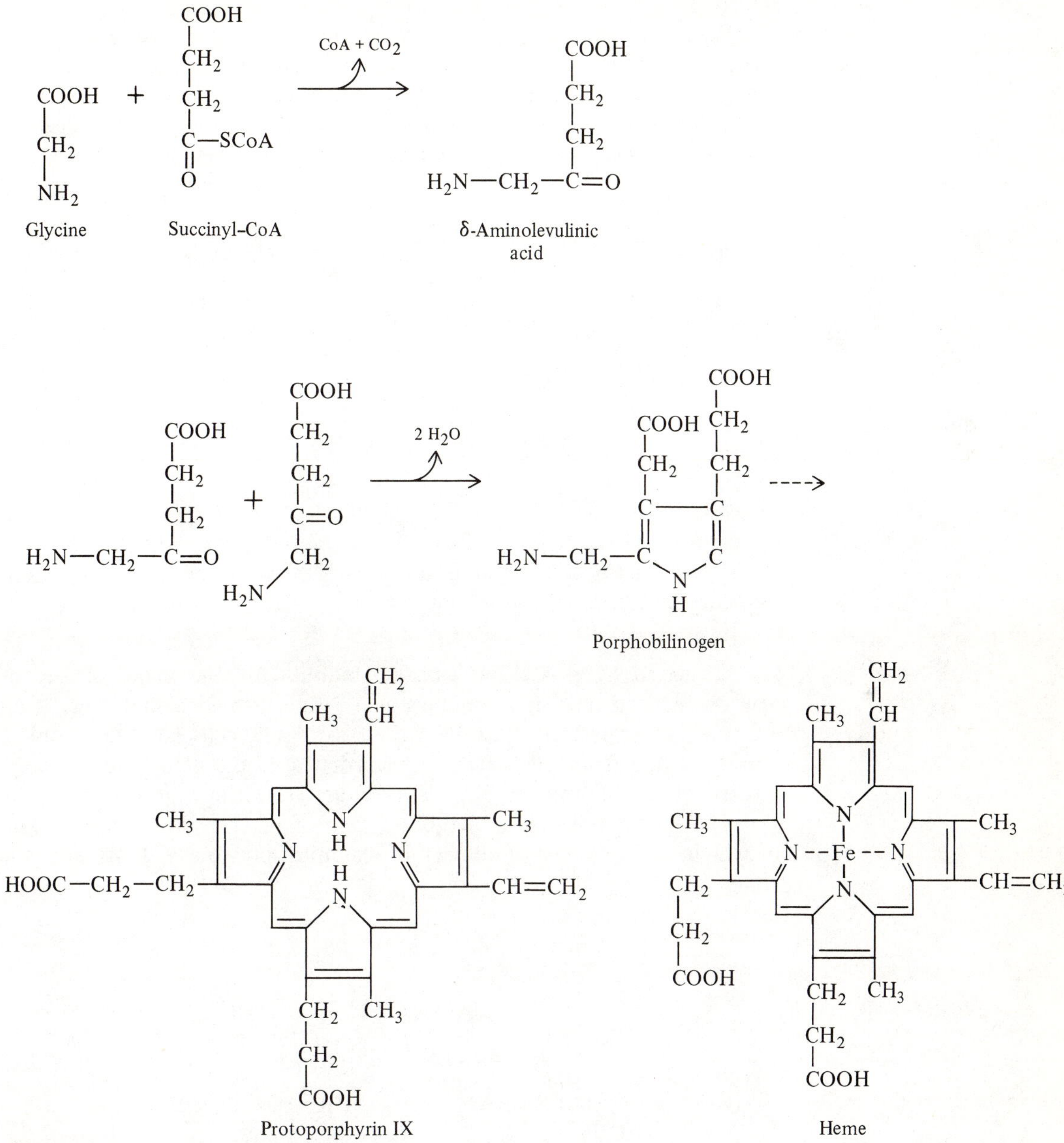

Figure 17-1
A simplified view of the biosynthesis of the heme molecule.

group —CH_3, the hydroxymethyl group —CH_2OH, and the formyl group —CHO.

When they are being carried by the coenzymes and other compounds involved in transfer to the ultimate acceptor molecules, the single-carbon units are usually attached either to nitrogen or sulfur atoms. If we look at the products that would result from hydrolysis of such a linkage, we can get a picture of the actual oxidation states of these groups:

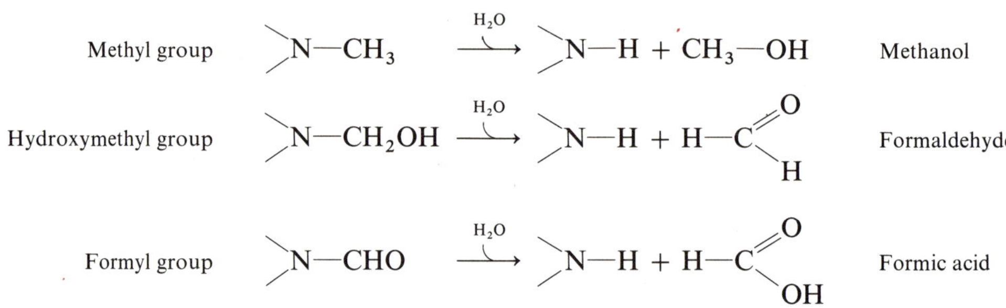

Considering the toxicity of all of these compounds for cells, it is easy to see why they are held in a bound state during the whole of their metabolism, from first production to final use.

Origin of Single-Carbon Units. Undoubtedly the major source of new one-carbon units is the amino acid serine. Provided that there is an adequate general level of amine groups available, cells have no problem making serine from one of the intermediates of the glycolytic pathway, glyceric acid 3-phosphate. The reactions are shown in Figure 17-2.

The serine molecule undergoes a type of reverse aldol condensation of which we have seen a number of examples previously. Normally, the

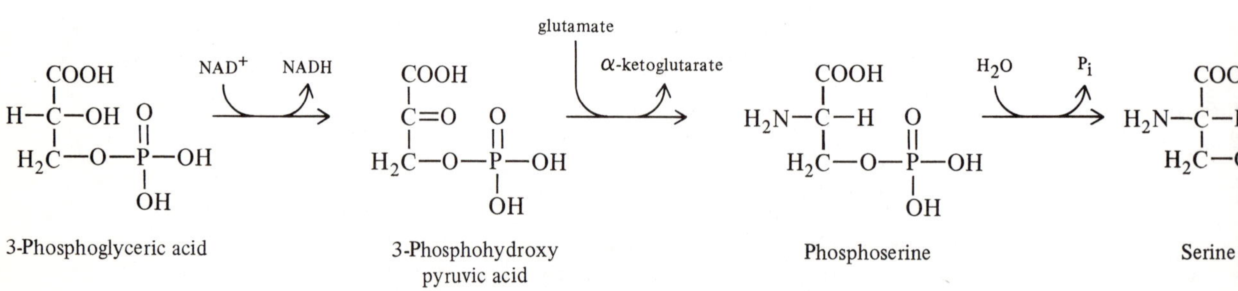

Figure 17-2
Formation of serine from 3-phosphoglycerate.

products we would expect would be free formaldehyde and glycine:

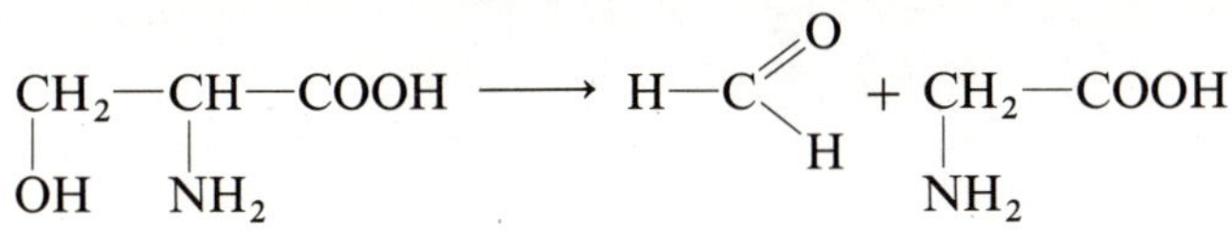

However, in this case the formaldehyde portion is transferred onto the nitrogen atoms of a coenzyme named **tetrahydrofolic acid** (which we will abbreviate FH_4). The result is a compound in which the carbon forms a bridge between the N^5 and N^{10} nitrogen atoms of the FH_4; this particular form is called N^5, N^{10}-*methylene*tetrahydrofolic acid. (It may make you feel better to think of this reaction as occurring by addition of one of the >N—H groups to formaldehyde, followed by a dehydration between the hydroxymethyl group and the other >N—H.)

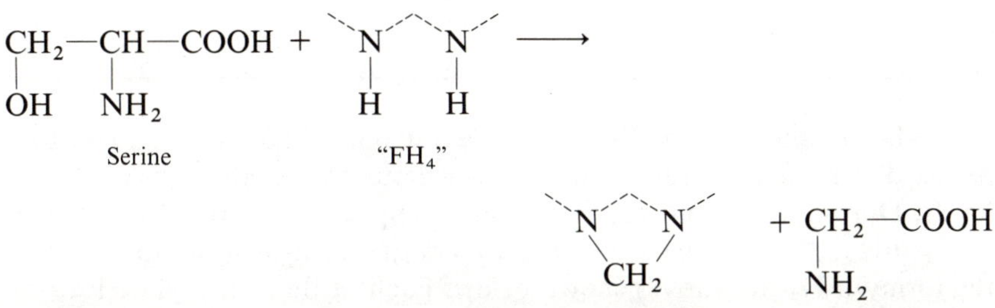

Serine "FH_4" N^5,N^{10}-methylene-FH_4 Glycine

Folic Acid and Derivatives

Folic acid is another of the B-group vitamins. It is composed of a pterin ring, a molecule of *p*-aminobenzoic acid, and a glutamic acid. In some organisms, several additional glutamic acids are attached in a short peptide chain:

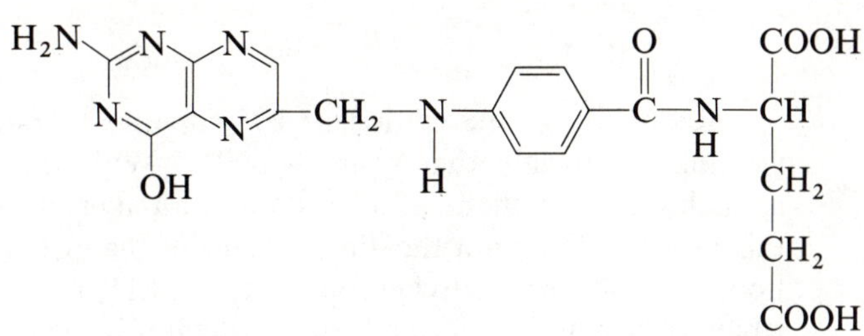

Folic acid

A number of microorganisms are capable of synthesizing folic acid but cannot make the *p*-aminobenzoic acid portion. For these organisms, *p*-aminobenzoic acid is a required vitamin.

Conversion of folic acid to the active coenzyme form involves a double reduction of the pterin ring to form tetrahydrofolic acid (FH_4):

The single-carbon units that are carried by tetrahydrofolic acid are attached to the N^5 or N^{10} position, or to both.

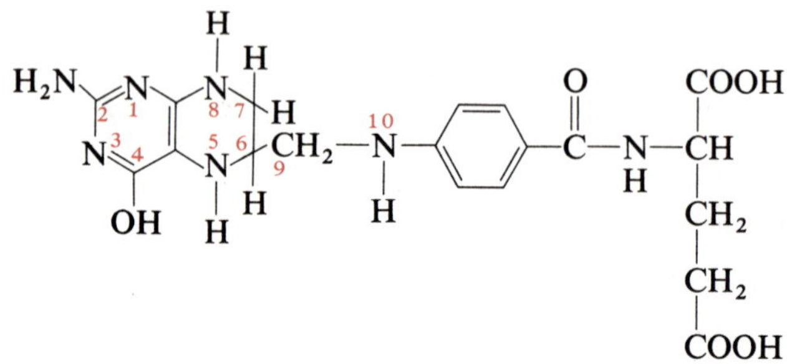

Tetrahydrofolic acid

While attached to the FH_4 molecule, the one-carbon unit can be either oxidized or reduced to form active intermediates at the other two oxidation levels. During oxidation, the first product appears to be the N^5, N^{10}-methenyl-FH_4; this can be converted by enzymatic hydrolysis to either of the formyl FH_4 derivatives shown below. Each of these formyl derivatives has its own role in specific biosynthetic processes:

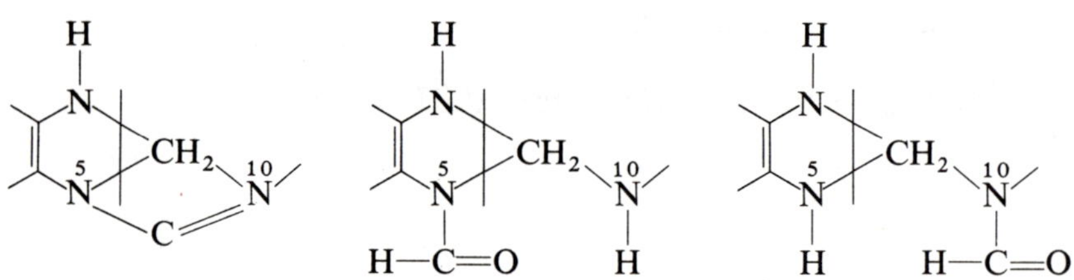

N^5,N^{10}-methenyl-FH_4 N^5-formyl-FH_4 N^{10}-formyl-FH_4

Formation of Methyl Groups

Reduction of the methylene FH_4 involves a rather complex reaction yielding specifically the N^5-methyl-FH_4. While methyl groups are used as such in the synthesis of a substantial number of important compounds, this methyl-FH_4 is not the direct source of the methyl group in those reactions. Instead, the methyl group of methyl-FH_4 is used to make methionine which is then activated to a form in which it is the primary source of labile methyl groups in cells. This role of methionine is indicated diagrammatically in Figure 17-3.

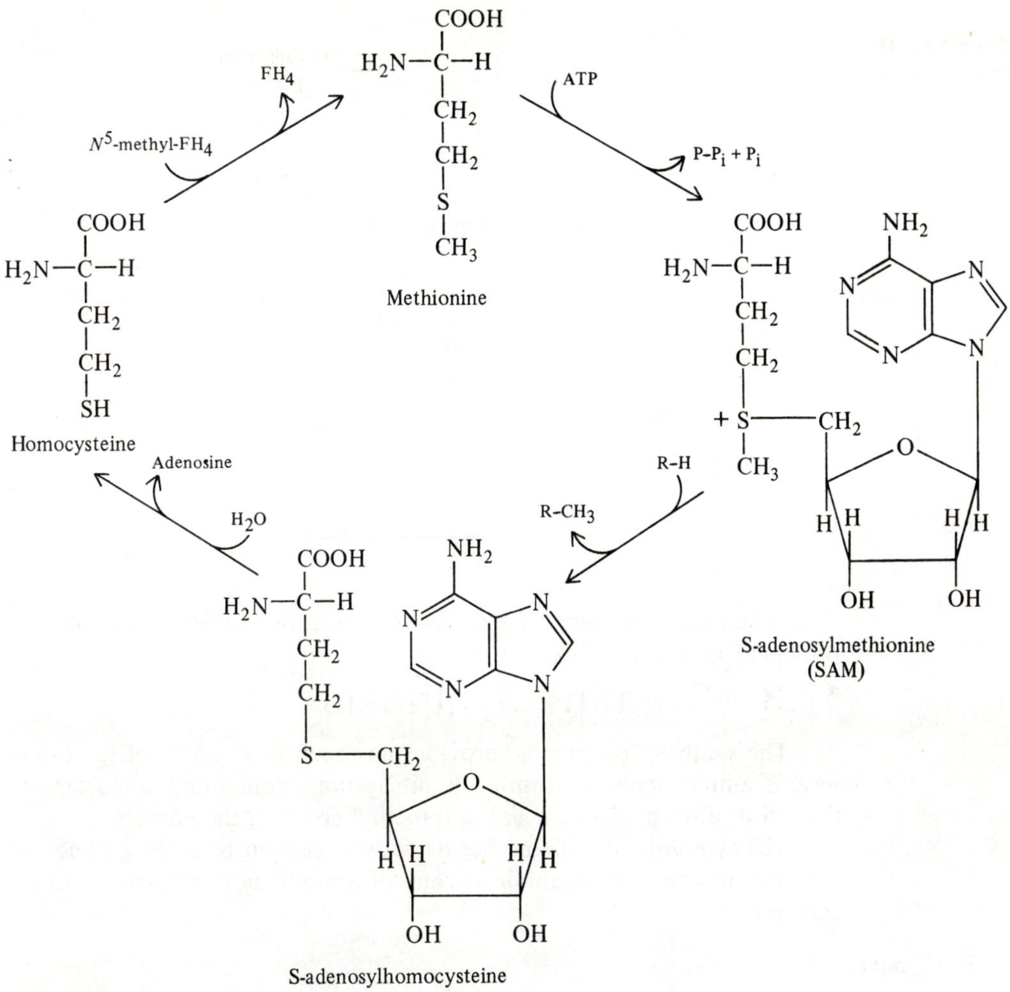

Figure 17-3
The role of methionine in methyl group transfer.

It must be emphasized here that this is only a brief and highly simplified look at an extremely complex process. A coenzyme form of Vitamin B_{12} is involved in the transfer of the methyl group from FH_4 to homocysteine in animals, many of the reactions are complex, and many aspects of metabolic control in these areas are still under study.

Animals are unable to synthesize homocysteine to take part in the synthesis of methionine, and thus methionine becomes an "essential" amino acid in the diet both because of its need for protein synthesis and to supply the intermediates needed to keep the cycle of Figure 17-3 in operation.

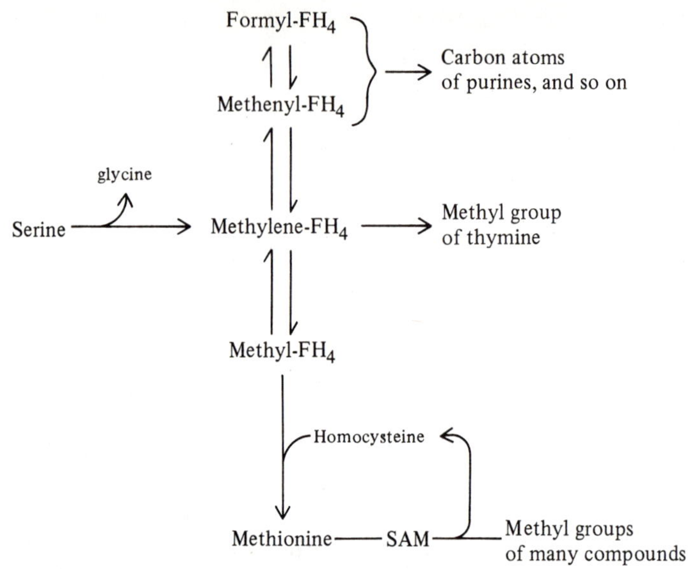

Figure 17-4
A summary of some of the reactions of single-carbon metabolism.

A summary of some of the reactions of single-carbon metabolism is presented in Figure 17-4.

17.3 Creatine Synthesis

The synthesis of creatine provides an excellent example of the use of several amino acids in forming a nitrogenous compound important to the cell. It also involves the use of a methyl group in the process.

The synthesis of creatine begins with a reaction between glycine and arginine in which the guanidino group of arginine is transferred onto the glycine:

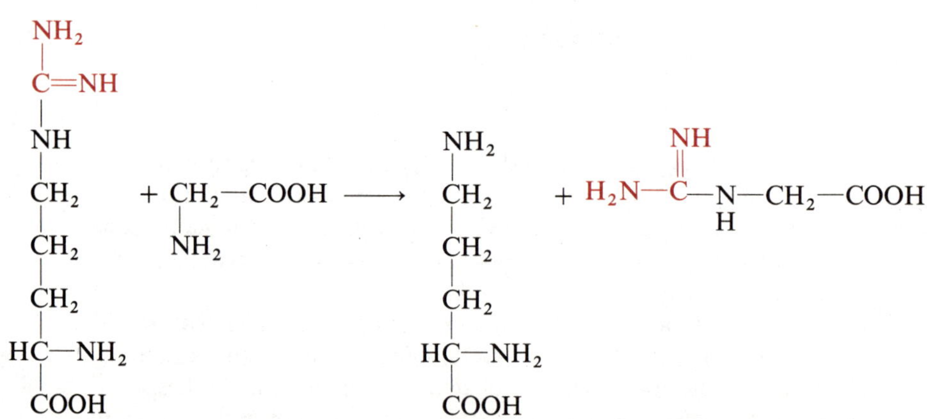

Arginine Glycine Ornithine Guanidoacetic acid

The next step requires the transfer of a methyl group from *S*-adenosylmethionine to the guanidoacetic acid. The products are creatine and *S*-adenosylhomocysteine:

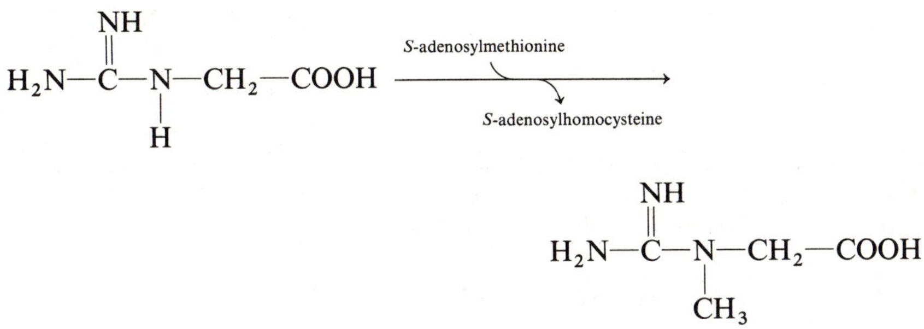

Creatine

The physiological significance of creatine merits some attention at this point. Its important function in muscle metabolism results from the following freely reversible reaction with ATP:

$$\text{creatine} + \text{ATP} \longrightarrow \text{ADP} + \text{creatine phosphate}$$

Because many reactions of metabolism are affected severely by any change in the ATP concentration, it is important that the level be kept within fairly narrow limits. At the same time, the muscle tissues should be capable of rapid response to a sudden need for action and should be capable of sustaining that action until glycolysis and other pathways of ATP production can be sufficiently stimulated to meet the needs. The use of creatine phosphate as a rapidly available storage form of phosphate-group energy permits the muscle cell to meet these requirements. In some organisms, such as many of the invertebrates, arginine phosphate serves this role instead of creatine phosphate.

17.4 Proline Biosynthesis

The synthesis of proline might be expected to be complex, in that the product is a cyclic compound and does not have a free amino group. In fact, proline is made rather simply from a molecule of glutamic acid. Reduction of the γ-carboxyl group of the glutamate to an aldehyde leads to formation of a Schiff base with the α-amino group. Reduction of this

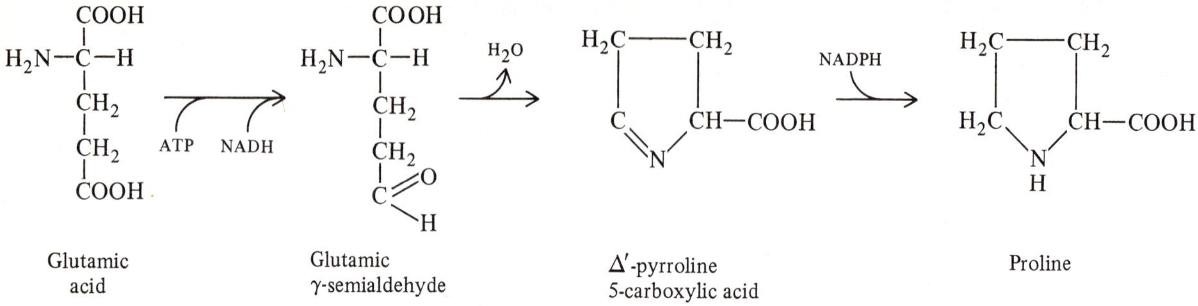

Figure 17-5
Biosynthetic formation of proline from glutamate.

intermediate at the expense of NADPH yields proline. This sequence of reactions is shown in Figure 17-5.

17.5 Biosynthesis of Essential Amino Acids

We have commented in the previous chapter on the inability of the human body, in common with many other animals and microorganisms, to synthesize certain amino acids. It is worth looking for a moment at several examples of the pathways by which plants (and some other organisms) are able to carry out these syntheses for us.

Biosynthesis of Phenylalanine

Most animals have only a very limited capacity to synthesize aromatic ring structures, in direct contrast to plants in which aromatic rings occur widely. Figure 17-6 is a brief outline of the pathway by which erythrose 4-phosphate and phosphoenolpyruvate, both intermediates of carbohydrate metabolism, can give rise to the carbon skeleton of phenylalanine. Note that the final step in this case is a simple transamination; the need for phenylalanine in humans arises from the inability to make the keto acid, not from a problem with amino group transfer.

Biosynthesis of Threonine and Lysine

Unlike phenylalanine and the remainder of the essential amino acids, these two compounds are not formed by direct amination of a keto acid. Threonine is formed from aspartic acid by a series of reactions which

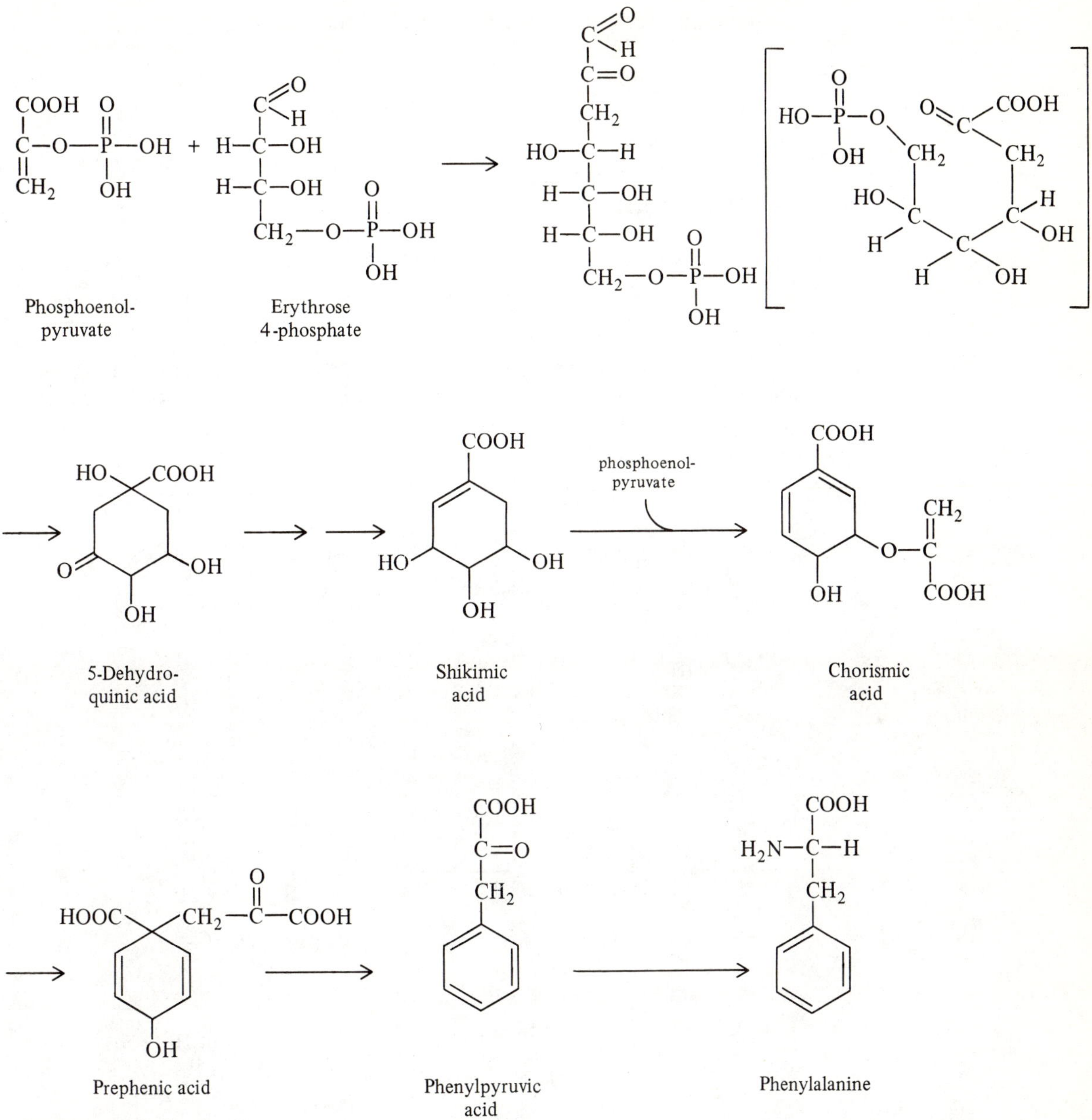

Figure 17-6
Pathway of biosynthesis of phenylalanine in higher plants.

preserve all four carbon atoms as well as the amino group:

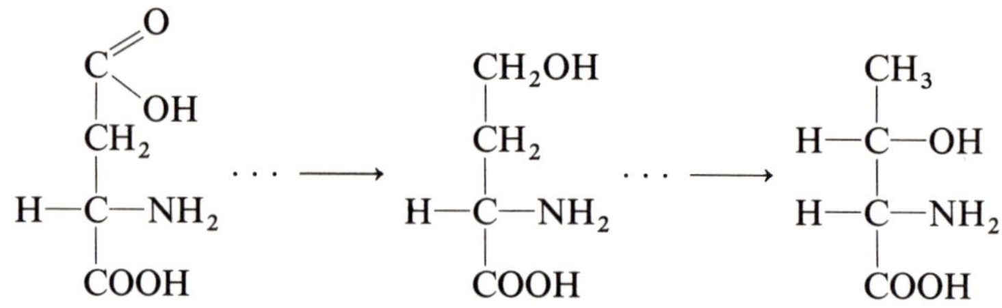

Aspartic acid Homoserine Threonine

Lysine synthesis seems to occur by different routes in molds, bacteria, and higher plants, but in each case the α-amino group is added before the structure is completed. An abbreviated version of one pathway of synthesis is shown in Figure 17-7.

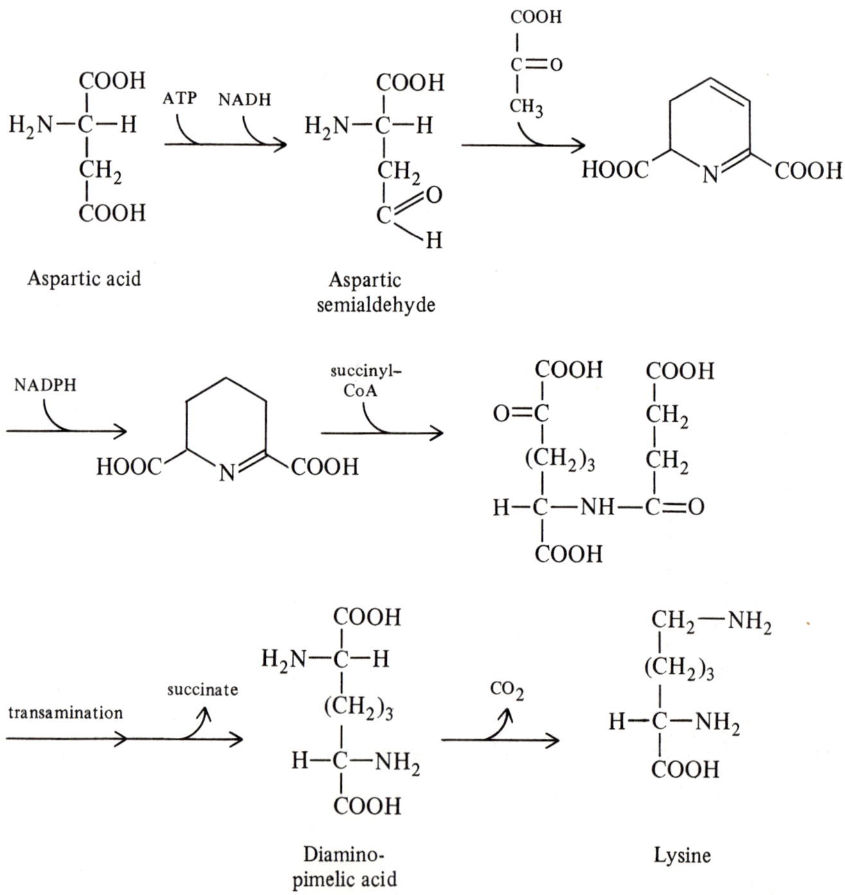

Figure 17-7
One of the pathways of biosynthesis of lysine. This pathway predominates in bacteria and the higher plants.

Additional Readings

The references suggested for Chapter 16 will serve equally well for this chapter. The student who wishes to pursue in detail the role of folic acid in metabolic processes might wish to consult the following additional source:

G. M. Brown, "Biogenesis and Metabolism of Folic Acid," in *Metabolic Pathways*, edited by D. M. Greenberg, 3rd edition, Vol. 4 (Academic Press, New York, 1971).

Study Questions

1. The immediate product of photosynthetic CO_2 fixation is glyceric acid-3-phosphate. Show how the carbons of this compound could give rise to all of the carbons of a molecule of δ-aminolevulinic acid (on the way to the synthesis of a chlorophyll molecule).

2. Write balanced equations for four different enzyme-catalyzed reactions in which serine is either a reactant or a product.

3. On the basis of the reactions of Figure 17-7, predict which of the carbons of lysine will be supplied by the various carbons of aspartic acid.

CHAPTER 18

The Nucleic Acids: Structure and Biosynthesis

The nucleic acids, another major group of biological polymers, have as their exclusive function the storage and utilization of the genetic inheritance of a cell. They also serve as a substantial portion of all viruses, in which capacity they are responsible for forcing the host cell to rearrange its metabolism to reproduce the viral constituents.

The monomeric unit of the nucleic acids is the **nucleotide**. In contrast to the amino acid monomers of proteins and the sugar monomers of the polysaccharides, nucleotides are capable of hydrolysis into simpler structural components. Each of the nucleotides is composed by definition of a nitrogen heterocyclic ring structure, a sugar unit joined to the heterocycle by a glycosidic bond, and at least one phosphate group attached to the sugar by an ester linkage. Two general classes of nucleic acids exist; those in which the sugar unit involved is D-ribose are called **ribonucleic acids** or **RNA** and those in which the sugar is D-2-deoxyribose are **deoxyribonucleic acids** or **DNA**.

18.1 Structures of the Nucleotide Monomers

The Nitrogen Heterocycles: Purines and Pyrimidines

Two different types of heterocycles (also often called "nitrogen bases") occur as part of the nucleotides of the nucleic acids. One type is based in the monocyclic structure of the parent compound **pyrimidine**, and the other on the bicyclic compound **purine**. The structures of the five "bases" that predominate in the nucleic acids are shown in Figure 18-1. Adenine, guanine,

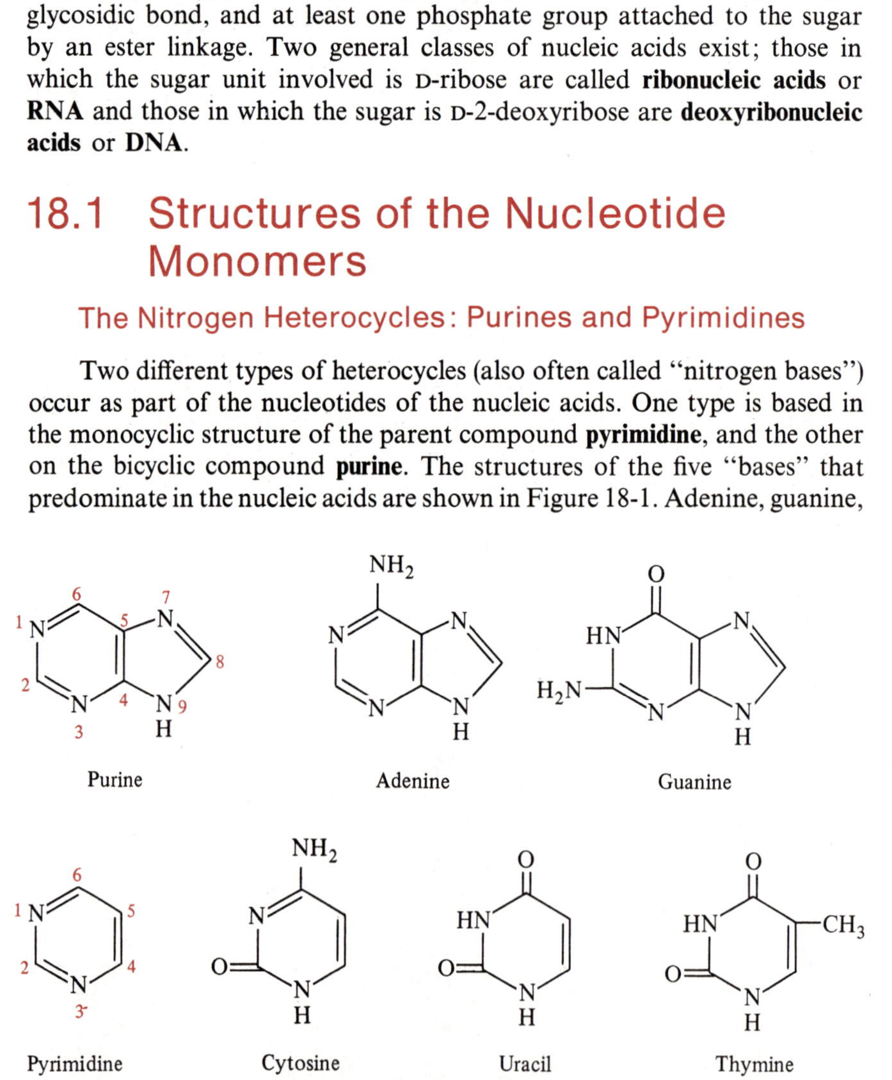

Figure 18-1
The structures of the major nitrogen heterocycles of the nucleic acids.

and cytosine occur in both DNA and RNA, while the uracil of RNA becomes the methylated form, thymine, in DNA.

A number of other purines and pyrimidines occur in some nucleic acids; they are largely methylated derivatives of the compounds listed above. Most of these occur in the small nucleic acid units called **transfer RNA**. We will look at the structures of the transfer RNA molecules and will examine their roles in protein synthesis in Chapter 20.

Nucleosides

A **nucleoside** is the name given to the structure that results when one of the nitrogen heterocycles is attached to a sugar through a glycosidic bonding to the appropriate nitrogen. All of the nucleosides involved in the nucleic acids are formed by linkage of either ribose or deoxyribose in a furanose form. All nucleosides (with the exception of one found in one of the Vitamin B_{12} coenzymes) have a β-glycosidic linkage:

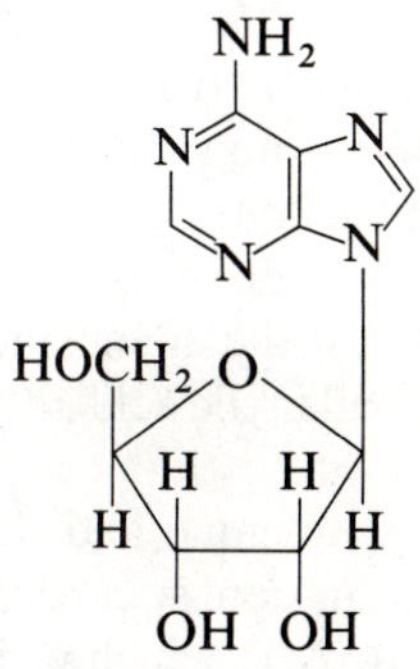

Adenine riboside (adenosine)

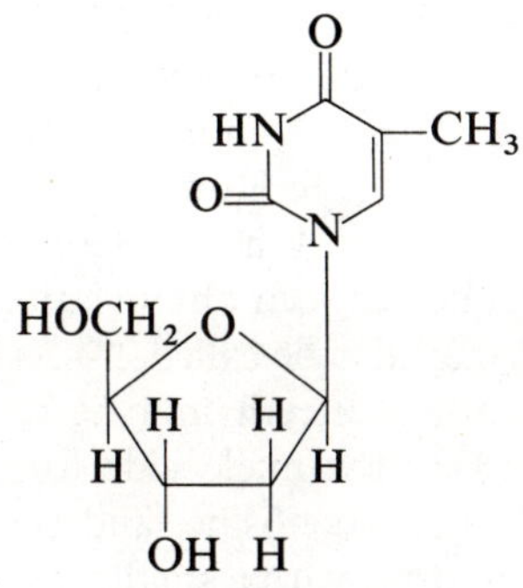

Thymine deoxyriboside (deoxythymidine)

The names of the nucleosides derived from each of the nitrogen heterocycles are as follows:

heterocycle	nucleoside(s)
adenine	adenosine deoxyadenosine
guanine	guanosine deoxyguanosine
cytosine	cytidine deoxycytidine
uracil	uridine deoxyuridine
thymine	deoxythymidine

There is only one naturally occurring nucleoside form of thymine, namely that formed from deoxyribose. While this nucleoside is sometimes called "thymidine," we will follow more common usage and refer to it as "deoxythymidine."

Forming the Nucleotides

All that is required to convert a nucleoside into a nucleotide is the addition of a phosphate ester to one of the hydroxyl groups of the sugar portion. A wide variety of compounds can be formed on this basic pattern. The simple nucleotides formed by adding phosphate to adenosine or to deoxythymidine are shown in Figure 18-2.

The existence of nucleoside triphosphates and diphosphates such as adenosine tri- and diphosphate was known well before the actual structures of the various nucleoside monophosphates were established. When it did become clear that the di- and triphosphates were attached to the 5′-position, it seemed reasonable to extend the abbreviations such as ATP and ADP to the 5′-phosphate compound and call it AMP (for adenosine monophosphate). The 2′- and 3′-monophosphates have no such abbreviation and must be identified in full. However, the nomenclature used in dealing with the various nucleosides and nucleotides is not completely standardized, nor are the frequently seen abbreviations. Thus, the adenosine nucleotides of Figure 18-2 may also be called 2′-, 3′- or 5′- adenylic acid and the thymine derivatives 3′- or 5′-deoxythymidylic acid.

Note also that the nucleoside formed by adenine and deoxyribose is usually called deoxyadenosine, and this designation is carried through in the abbreviated forms with a small "d" as in dAMP, which would represent deoxyadenosine 5′-phosphate.

Nucleoside Di- and Triphosphates

Each of the major nucleosides involved in the formation of nucleic acids exists in the forms of a 5′-nucleoside diphosphate and a 5′-nucleoside triphosphate, as we have already seen in Chapter 6. All of these are important in the process of formation of the nucleic acid polymers. In addition, each has other specific functions to carry out in the metabolism of the cell, usually in processes involved with energy transfer. The functions of ATP are of course pervasive; we have seen also the role of UTP in polysaccharide synthesis, of CTP in glycerol lipid biosynthesis, and of GTP in specific reactions of metabolism.

One additional nucleoside that has specific functions as the di- and triphosphate but does not form part of nucleic acid structures is **inosine**. The inosine nucleotides are the biosynthetic precursors of both adenine

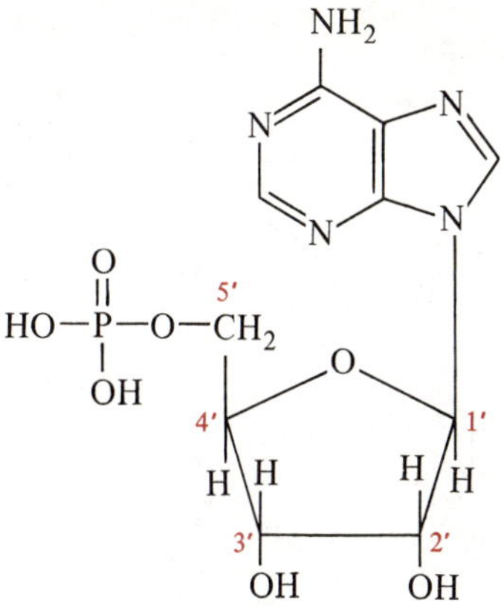

Adenosine 5′-phosphate
(AMP)

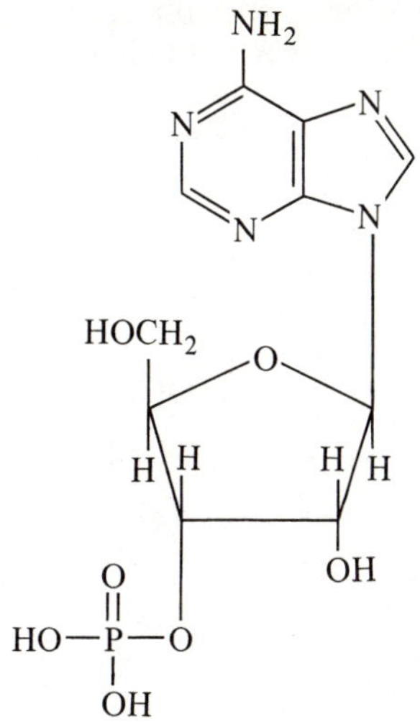

Adenosine 3′-phosphate

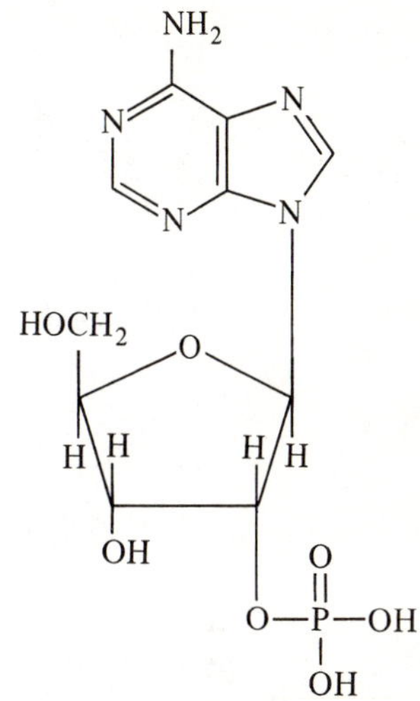

Adenosine 2′-phosphate

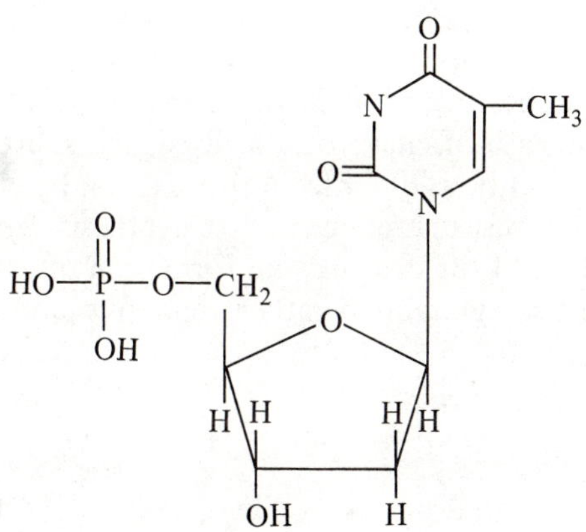

Deoxythymidine 5′-phosphate
(dTMP)

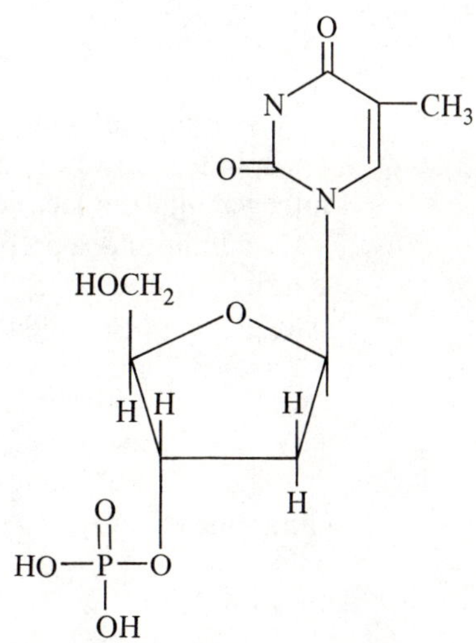

Deoxythymidine 3′-phosphate

Figure 18-2
The structures of the possible monophosphate nucleotides that can be formed from adenosine and deoxythymidine.

and guanine compounds, as well as serving as cosubstrates in the phosphoenolpyruvate carboxykinase reaction (Chapter 9):

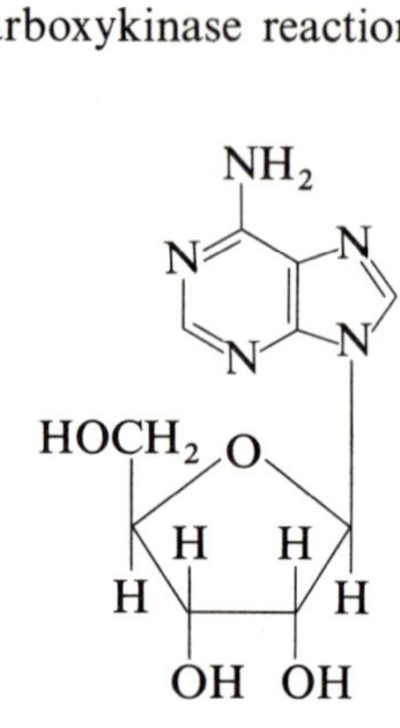

Adenosine

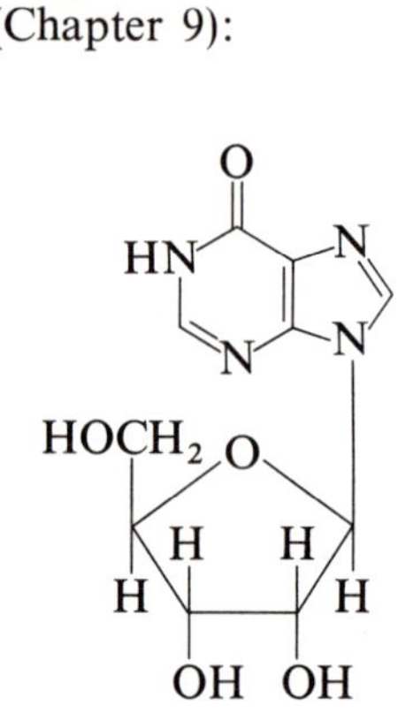

Inosine

18.2 Nucleotide Biosynthesis

Pyrimidine Nucleotides

Synthesis of the pyrimidines begins with the formation of carbamoyl phosphate. However, in the case of eucaryotes at least, the source of the nitrogen and the location of synthesis differ from the reaction by which the same compound is prepared for use in arginine or urea synthesis. Pyrimidine synthesis is entirely a function of the cytosol, and formation of carbamoyl phosphate occurring there uses glutamine rather than free ammonia as a nitrogen source:

$$\text{glutamine} + CO_2 + H_2O + 2\,\text{ATP} \longrightarrow H_2N{-}\overset{\overset{\displaystyle O}{\|}}{C}{-}O{-}\overset{\overset{\displaystyle O}{\|}}{\underset{\underset{\displaystyle OH}{|}}{P}}{-}OH + \text{glutamate} + 2\,\text{ADP} + P_i$$

Once the carbamoyl phosphate is formed, it condenses with aspartic acid and the pathway of pyrimidine synthesis proceeds as shown in Figure 18-3. Cyclization of the carbamoylaspartic acid to dihydroorotic acid is followed by an oxidation step catalyzed by a very complex flavoprotein enzyme.

Addition of the sugar-phosphate unit to orotic acid is worthy of some comment, since it is the first case we have seen of the use of ribose 5-phosphate 1-pyrophosphate (also called phosphoribosyl pyrophosphate and abbreviated PRPP). This compound is formed from ribose 5-phosphate in an

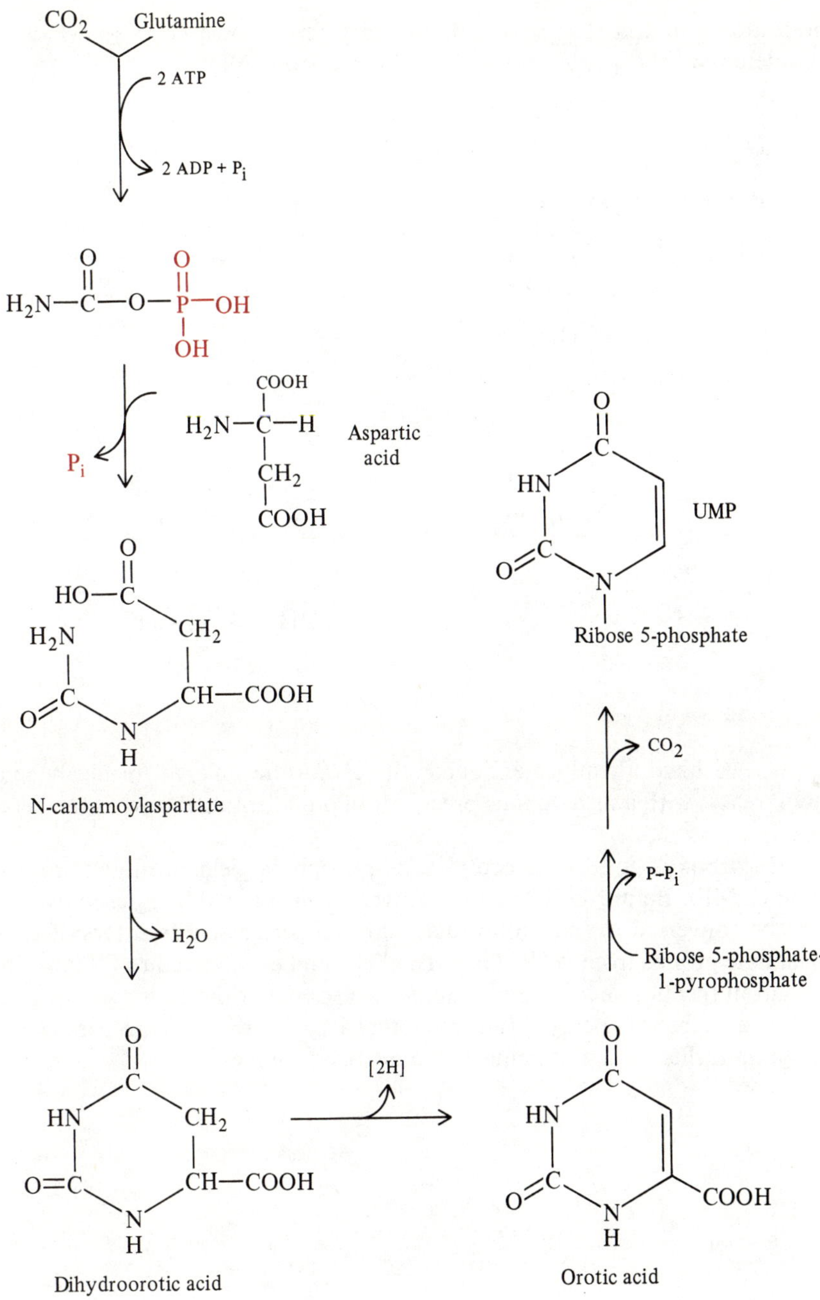

Figure 18-3
The pathway of biosynthesis of the pyrimidine nucleotides. Shown here is the synthesis of uridine 5′-phosphate (UMP), which serves as a precursor for the synthesis of the other pyrimidine nucleotides.

interesting reaction that is one of the very few known cases in which a complete pyrophosphate group is transferred from ATP:

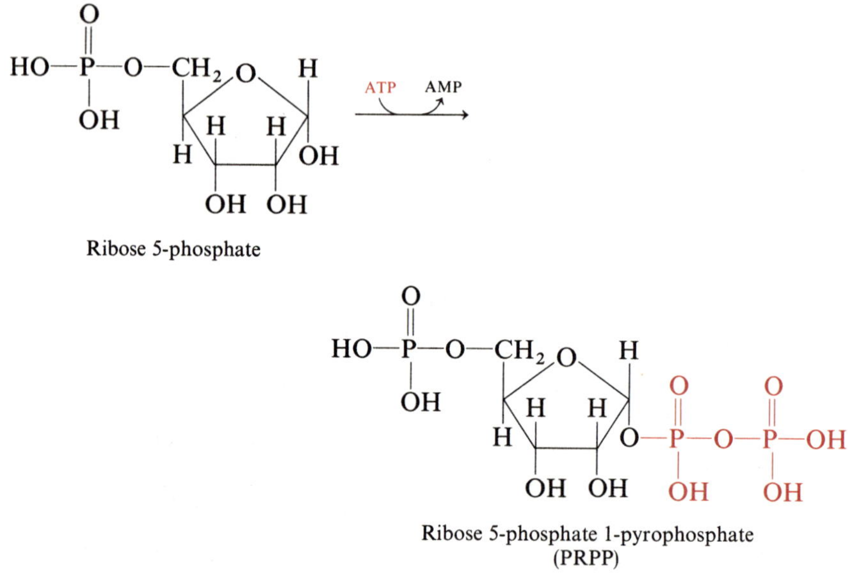

As we have already mentioned, the glycosidic linkage formed when PRPP reacts with a nitrogenous precursor of a nucleotide is always the beta form.

Decarboxylation of the orotidine 5′-phosphate yields uridine 5′-phosphate (UMP). Before UMP can be converted into a cytidine derivative, it must be converted to the triphosphate through successive transfers of two phosphate groups from ATP. Then the UTP can be converted to CTP by an amination reaction involving glutamine as a source of the amine group and ATP as a source of energy. Many bacteria carry out the same process using ammonia rather than glutamine as the nitrogen source:

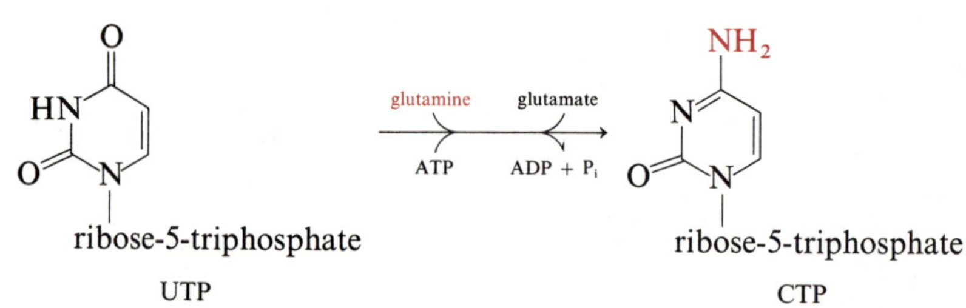

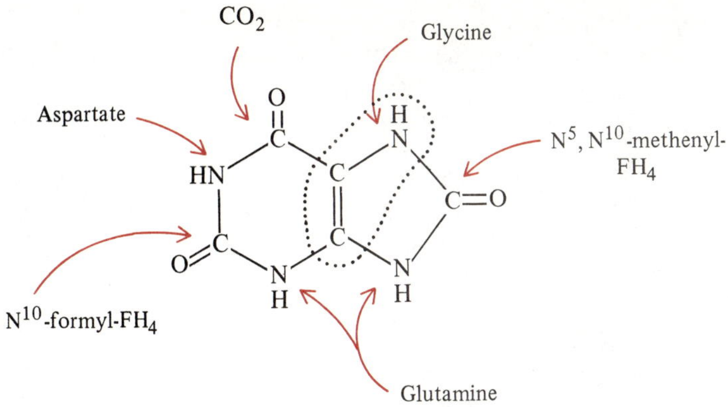

Figure 18-4
The metabolic sources of the various atoms that make up a purine ring. The compound shown is uric acid, the predominant degradative product of purine metabolism in humans.

Purine Nucleotides

Synthesis of the purine nucleotides follows a quite different approach. The pyrimidine heterocycle was formed by condensation of two preformed compounds, and then the sugar-phosphate group was added. Purines instead start with a small molecule with the sugar phosphate already attached and the ring is then constructed atom by atom in what gives the appearance of being a laborious process.

Even before there was much information about the pathway of synthesis of the purines, it was possible to label various compounds with isotopes and look for their appearance in the final purine ring. It was in that way that the major precursors of the purines were first identified. These are shown in Figure 18-4.

The reactions involved in the synthesis of inosine 5′-phosphate, the parent compound of all the other purine nucleotides, are shown in Figure 18-5. Most of the reactions are self-explanatory; we will comment on only a few of the more interesting ones.

The process begins this time with the displacement of the pyrophosphate from 5-phosphoribosyl 1-pyrophosphate by an amine group from glutamine. An amide is then formed between the 5-phosphoribosylamine and glycine, followed by transfer of a one-carbon unit from N^5,N^{10}-methenyl-FH_4 to produce the *N*-formylglycinamide structure. Transfer of another atom of nitrogen from glutamine is followed by closure of the imidazole ring, both reactions requiring the assistance of energy from cleavage of ATP. Addition of CO_2 prepares the compound for introduction of the final ring nitrogen by reaction with aspartic acid. This formation of an amide followed by release of fumarate is exactly analogous to the reactions converting citrulline to arginine in the Urea Cycle (Chapter 16).

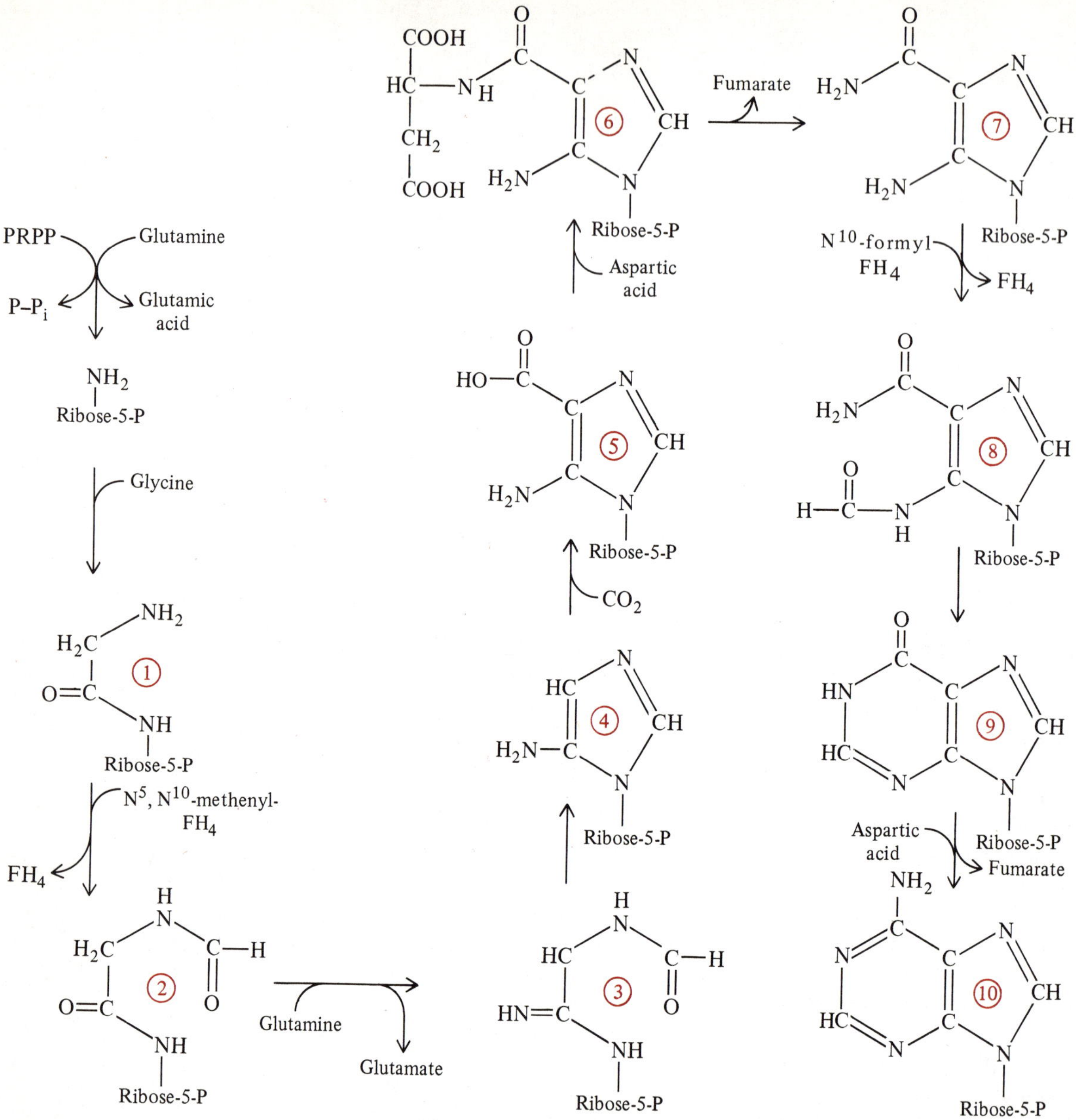

(1) Glycineamide nucleotide
(2) Formylglycineamide nucleotide
(3) Formylglycineamidine nucleotide
(4) 5-Aminoimidazole nucleotide
(5) 5-Aminoimidazole carboxylate nucleotide
(6) 5-Aminoimidazole 4-succinylcarboxamide nucleotide
(7) 5-Aminoimidazole 4-carboxamide nucleotide
(8) 5-Formamidoimidazole 4-carboxamide nucleotide
(9) Inosine 5′-phosphate (IMP)
(10) Adenosine 5′-phosphate (AMP)

Figure 18-5
An abbreviated version of the biosynthetic pathway of formation of purine nucleotides. The involvements of ATP or other nucleotide triphosphates as a source of energy are not included in the figure but are discussed in the text.

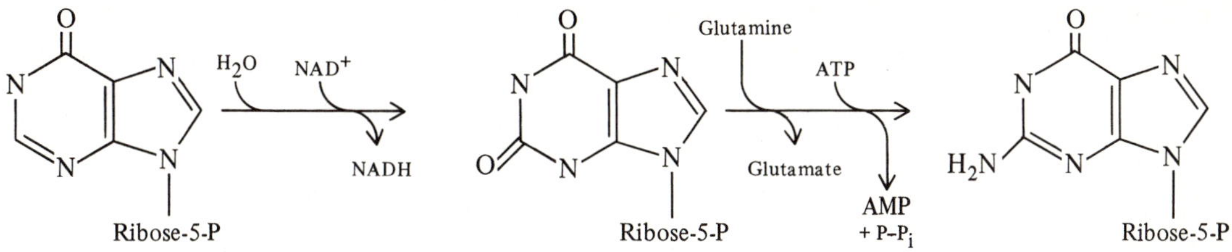

Figure 18-6
Synthesis of guanosine 5′-phosphate (GMP) from IMP.

In view of the use of N^5, N^{10}-methenyl-FH_4 as a source of a formyl group in the reaction mentioned above, it is interesting that the final carbon atom of the synthesis is supplied in the form of a formyl group donated very specifically by N^{10}-formyl-FH_4. Closure of the ring at this point completes the synthesis of inosine 5′-phosphate (IMP). Formation of adenosine 5′-phosphate (AMP) is then possible through another transfer of an amine group from aspartic acid. In this case, GTP rather than ATP supplies the energy for attachment of aspartate.

Synthesis of guanosine 5′-phosphate (GMP) from IMP requires oxidation at carbon-2 of the purine ring, followed by amination to form GMP, as shown in Figure 18-6. The source of the amine group in this case is glutamine, and again the energy of ATP is needed for the process.

The pathways of purine and pyrimidine biosynthesis are under close metabolic control. We will look at some of the mechanisms of this control in Chapter 21

18.3 Structures of the Nucleic Acids

We have called the nucleic acids polymers based upon nucleotides as a monomeric unit. Whatever the kind of nucleic acid, be it DNA or RNA of whatever type, the structure of the basic polynucleotide chain is the same. This structure is created by linking the 5′-phosphate of one nucleotide to the 3′-position of the next in a phosphodiester bonding.

The backbone of the polynucleotide chain thus is formed by alternating pentose and phosphate groups. The various bases are free to project from the chain much as do the side-chains of the amino acids of a polypeptide. A representation of a portion of a polynucleotide chain illustrating these features appears as Figure 18-7.

It should be emphasized that DNA and RNA represent classes of nucleic acids. Within each of these classes are many individual kinds of

Figure 18-7
A typical portion of a DNA chain.

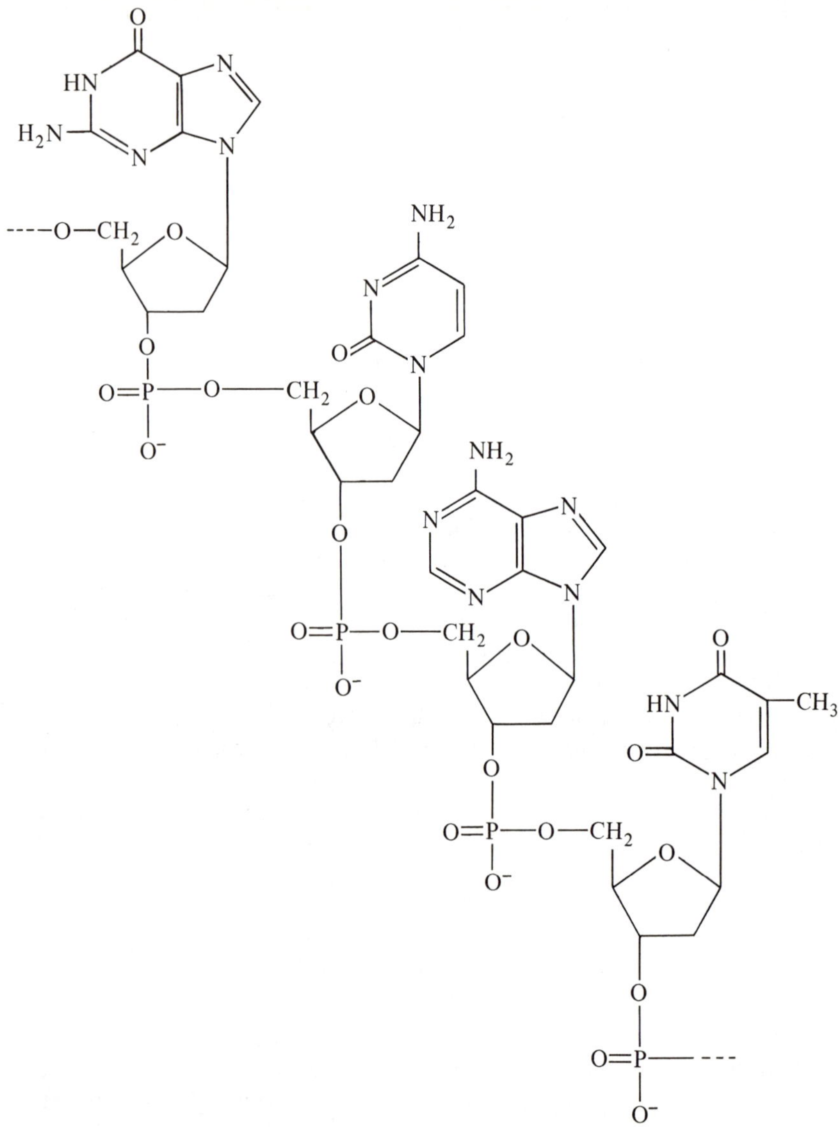

nucleic acid molecules, differing in the proportions of the constituent nucleotides and in the sequences in which the nucleotides are linked together. The nucleotide sequence of a nucleic acid molecule is its primary characteristic, just as the amino acid sequence is the primary characteristic of a protein. We shall see in the next chapter how these two sequences are related to one another and how the sequences of bases in the nucleic acids are able to serve as the genetic reservoir of a cell.

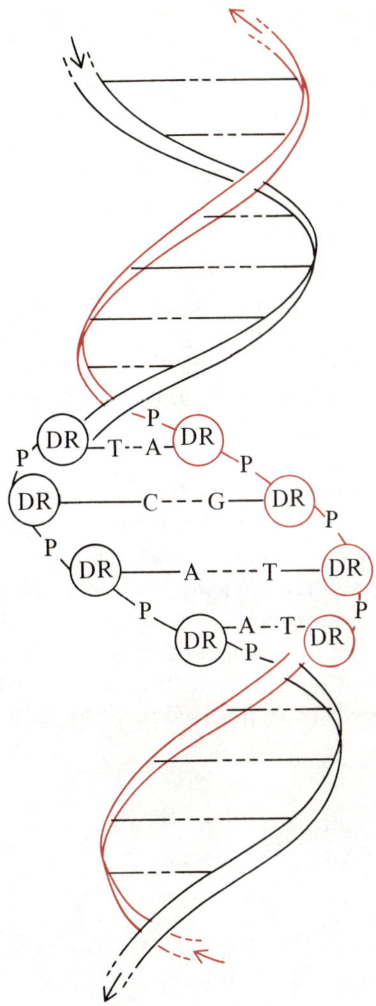

Figure 18-8
A simplified representation of the double helix structure of DNA. A, T, G, and C represent adenine, thymine, guanine, and cytosine, respectively. The deoxyribose sugar units are represented by a circled "DR."

The DNA Double Helix

With only a few exceptions, the DNA of cells exists in a coiled form involving two separate chains or strands of polynucleotide. The basic structure was first proposed by Watson and Crick in 1953. The major features of this structure are depicted in Figure 18-8.

Briefly, two polynucleotide chains are coiled together around the same axis. The nitrogen heterocycles of each of the constituent nucleotides of both chains project inward toward the axis of the helix. These nitrogen bases are all planar or nearly planar molecules. In the double-strand DNA structure, they are aligned so that they are parallel to one another and perpendicular to the axis of the helix.

When the nitrogen bases are thus arranged, it is possible for them to form multiple hydrogen bonds between the bases of the two chains. This bonding, and electronic interactions between the stacked heterocycles, is responsible for the great stability of this double-strand structure. Temperatures on the order of 70–100°C are required to "melt" this bonding and cause dissociation of the two strands.

Two other aspects of the Watson–Crick structure for DNA are very important to our understanding. First is a very serious limitation imposed on the base pairs which may form hydrogen bonds. The dimensions of the helix and the limits imposed by the three-dimensional structures of the heterocyclic bases themselves permit multiple hydrogen bonds only between adenine and thymine or between guanine and cytosine. The nature of the hydrogen bonding is shown in Figure 18-9.

This limitation in the allowable base-pairs in the opposite chains tells us that the structures of the two chains must be very closely related. Each nucleotide unit of one chain must be matched by its base-pair counterpart in the other chain. Given the nucleotide sequence of one of the chains, you could readily write the sequence of the other. The two chains are thus said to be **complementary** with respect to their nucleotide sequences.

The second interesting feature to bear in mind is that the two chains of nucleotides are running in opposite directions. Since the phosphodiester

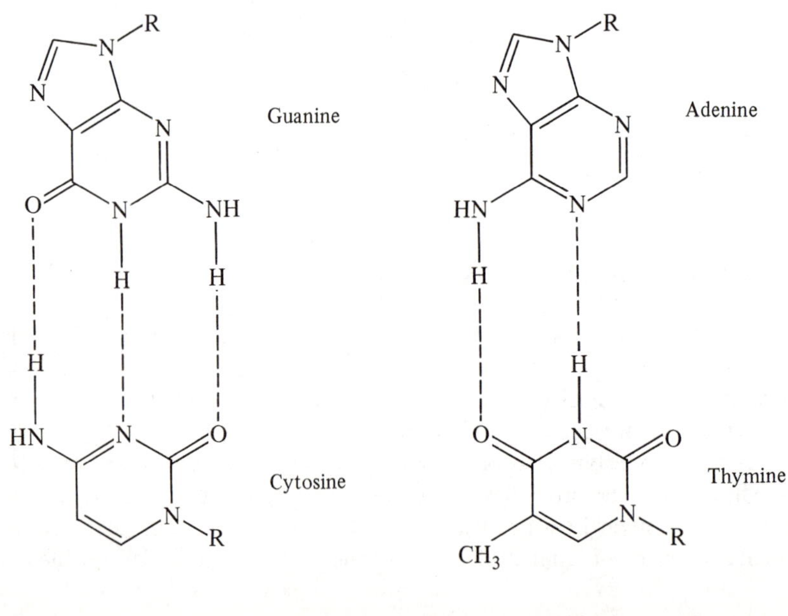

Figure 18-9
A more detailed picture of the hydrogen bonding between nitrogen base pairs of DNA.

bonds go from the 3′-position of one sugar to the 5′-position of the other, there is a definite directionality to the bonding, just as there is in a polypeptide chain. If one of the strands shown in Figure 18-8 has a free 5′-hydroxyl on the left and a free 3′-position on the right, then the other will have its free 5′-hydroxyl on the right and the free 3′-group on the left. This is indicated in the figure by the arrows going in different directions on the two chains.

Secondary Structure of the RNAs

Most of the RNA molecules lack a simple secondary structure comparable to that of DNA. However, they often are extensively folded and occur in close association with protein molecules in complexes that may have well defined structures. The high-molecular-weight RNA of the ribosomes or of a virus such as Tobacco Mosaic Virus would seem to fit in this category.

Messenger-RNA molecules are formed using as a template a portion of the DNA of the cell. They then proceed to the ribosomes, where they serve as a pattern for the synthesis of polypeptide chains: hence the name "messenger-RNA". These molecules appear to have no defined secondary structure, and are not normally associated with specific proteins except while they are interacting with ribosomes in the process of protein synthesis.

The smaller **transfer-RNA** molecules serve as a means by which specific amino acids can be recognized and brought into position as specified by the message of the messenger-RNA in order to produce the desired sequence of amino acids during protein synthesis. These transfer-RNA molecules have rather well defined secondary structures involving localized hydrogen bonding to form several loops. We must emphasize here, however, that all of the RNAs occur only in single-strand form, and that the hydrogen bond "pairing" occurring within the transfer-RNA molecule is between segments of the single polynucleotide chain. A schematic picture of this limited bonding, and of the three-dimensional structure that results, is shown in Figure 18-10. We will look at this structure in more detail when we examine the role of these molecules in the protein synthesis process.

Sizes of Nucleic Acids

The nucleic acids are truly *macro*molecules. The entire genetic information for a procaryotic bacterium such as *E. coli* is present in the form of a single circular double-stranded chain of DNA. The molecular weight of this DNA has been estimated as being on the order of 10^9 daltons. It is difficult to determine the size of intact DNA molecules in eucaryotic cells, with their more elaborately organized chromosomal structures. The average molecular weight of DNA isolated from such cells is on the order of 10^7 daltons, but that probably represents degraded material.

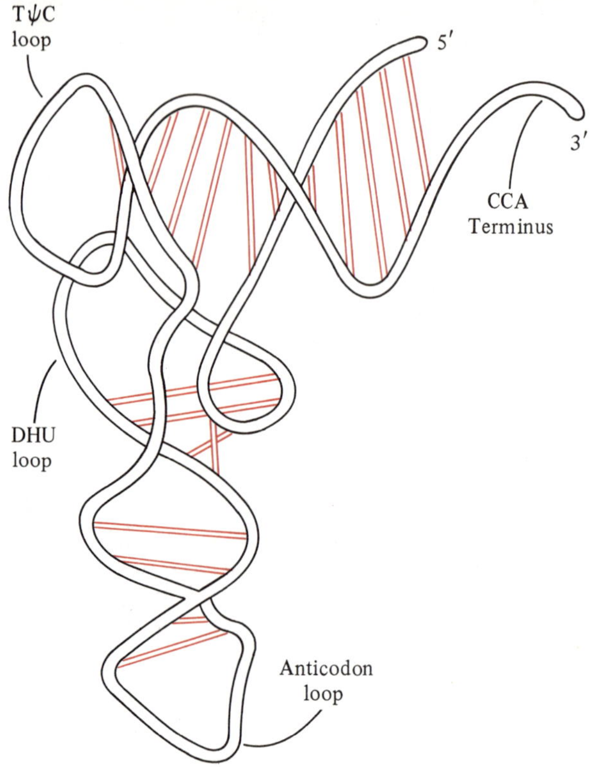

Figure 18-10
A simplified view of the three-dimensional structure of a typical transfer RNA molecule. The shaded lines represent the internal hydrogen bonding between segments of the molecule that stabilizes the structure shown. (From S. H. Kim, et al, *Science*, 185:435–440 (1974). Copyright © 1974 by the American Association for the Advancement of Science.)

The RNA chain of a Tobacco Mosaic Virus is of single-chain form with a molecular weight of about 2 million; this is about the same size range as found in ribosomal-RNA. The smallest of the functional varieties of RNA are the transfer-RNAs, with an average molecular weight of about 25,000. In between these and the real giants mentioned above are a large number of messenger-RNAs and other molecules of a wide range of sizes.

18.4 Sequencing of Nucleic Acids

Within the last few years, two methods have been developed which permit rapid determination of the nucleotide sequences of quite large polynucleotide segments. The first method was that of Sanger, who also pioneered the sequencing of proteins. His procedure involves the use of a series of highly specific enzymes, and will not be described here. The procedure of Maxam and Gilbert, reported in 1977, has the advantages of involving primarily chemical procedures and in practice is somewhat simpler and more rapid. We will take a brief look at the steps involved in this process and the bases on which it is founded.

TABLE 18-1
Specificities of typical restriction enzymes.

<table>
<tr><th>Enzyme</th><th>Required Structure and Points of Cleavage</th></tr>
<tr><td>EcoRI
(E. coli)</td><td>--- G|A A T T C ---
--- C T T A A|G ---</td></tr>
<tr><td>HaeIII
(H. aegyptius)</td><td>--- G G|C C ---
--- C C|G G ---</td></tr>
<tr><td>HindIII
(H. influenzae)</td><td>--- A|A G C T T ---
--- T T C G A|A ---</td></tr>
</table>

Preparation of Polynucleotide Fragments

Just as we saw that proteins had to be cut up into workable pieces for amino acid sequencing, nucleic acids must be cleaved into smaller units before any sequencing method can be used. In the case of DNA molecules, this is usually accomplished using **restriction enzymes** isolated from various bacterial sources. These enzymes catalyze hydrolysis only when very specific sequences occur in a chain. Several examples of this specificity are shown in Table 18-1. Using a series of such enzymes, the original DNA molecule can be reduced to segments suitable for sequencing. However, before proceeding any further several stages of separation and purification are necessary. The individual segments must be separated from one another and, if they are double-strand segments, must be converted to the single-strand form by treatment with a reagent that will disrupt the hydrogen bonding. Purification may involve separation on gel permeation columns, gel electrophoresis, or other similar techniques.

Labeling the Chain

The purified fragment to be analyzed must be "tagged" by the addition of ^{32}P in the form of a phosphate group at either the 5′- or the 3′-end of the chain. If there is already a phosphate group at that position, it is removed by the action of a phosphomonoesterase before the molecule is reacted with ^{32}P-labeled ATP as indicated in Figure 18-11.

The Chemical Reactions

The sample is now divided into four portions and each portion is subjected to a different specific chemical procedure. The intent of each of these procedures is to bring about a specific removal of one base from the

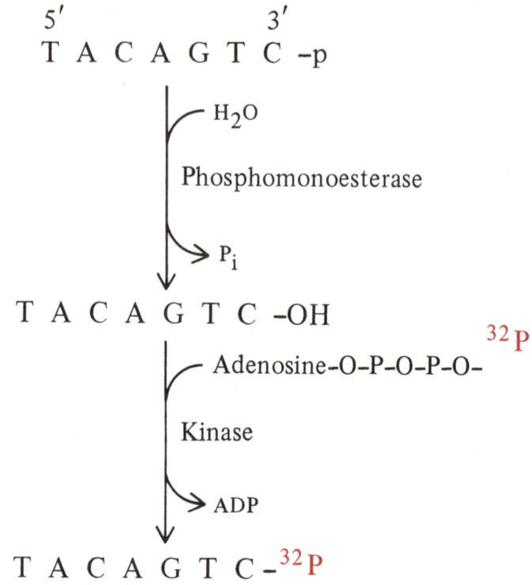

Figure 18-11
An example of the reactions used to label a DNA strand with a ^{32}P phosphate group prior to sequencing.

chain, leaving a point at which the chain will then undergo alkaline hydrolysis readily. In every case, a very limited amount of reagent is used in order to try to obtain only one break in each of the molecules in the sample. The chemical treatments are as follows:

1. Treatment with dimethyl sulfate followed by heating at about 90°C. This procedure results in methylation of guanine and adenine residues, followed by preferential removal of the methylated bases. Under these conditions, much more G than A is methylated and they are removed at about the same rate, so that cleavage at the point of a guanine residue will predominate ($G > A$).
2. Methylation is again carried out with dimethylsulfate, but this time the product is treated with acid under cold conditions. The result is essentially complete removal of methylated adenines and only a limited removal of methylated guanines ($A > G$).
3. Treatment with hydrazine and piperidine removes both cytosine and thymine residues ($C + T$).
4. Carrying out the same hydrazine treatment in the presence of salt (NaCl) suppresses the removal of thymine groups and permits a preferential removal of cytosine (C only).

The final chemical treatment in each case is an alkaline hydrolysis which cleaves the strand at the point where a base has been removed. Bear in mind that only one end of the sample strand is labeled with ^{32}P, so any fragments that are split off which do not contain that end will not be noticed in the later analyses (and thus will not interfere with the interpretation of the results).

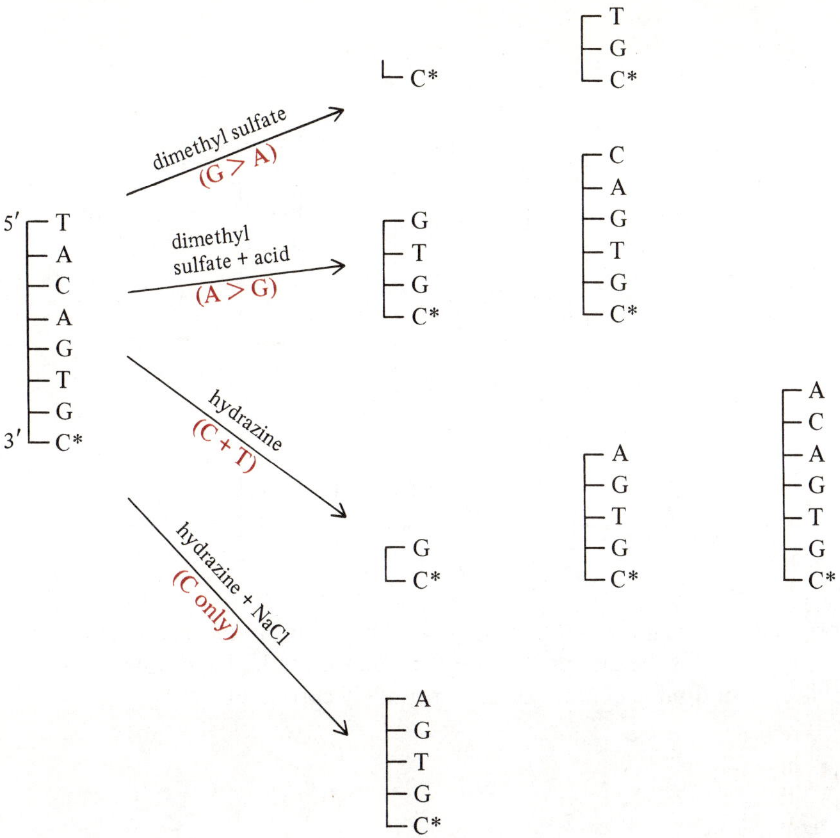

Figure 18-12
The products of action of the four Maxam and Gilbert procedures on a hypothetical segment of DNA. The asterisk in each case indicates the nucleotide containing the ^{32}P label. Because of the limited amounts of reagents used, there will always remain some unreacted material as well as the products shown.

The results of chemical treatment for a hypothetical short polynucleotide sample are shown in Figure 18-12.

Gel Electrophoresis

The key to both of the common methods of sequencing of nuclei acids is the ability to separate and identify polynucleotides differing by only one nucleotide residue. The development of very precise methods of separation using electrophoresis in thin slabs of polyacrylamide gel allow the separation and analysis of polynucleotides up to 150 units in length in a single gel. A simplified version of the results of such a gel separation is shown in Figure 18-13, using the products of the chemical action shown in Figure 18-12.

The order of the nucleotides can be read directly from the patterns of the four gel samples, as indicated by the letters on the right of the figure. Note that you cannot determine the nature of the one labeled nucleotide in this system. However, it is of course possible to label the other end of the

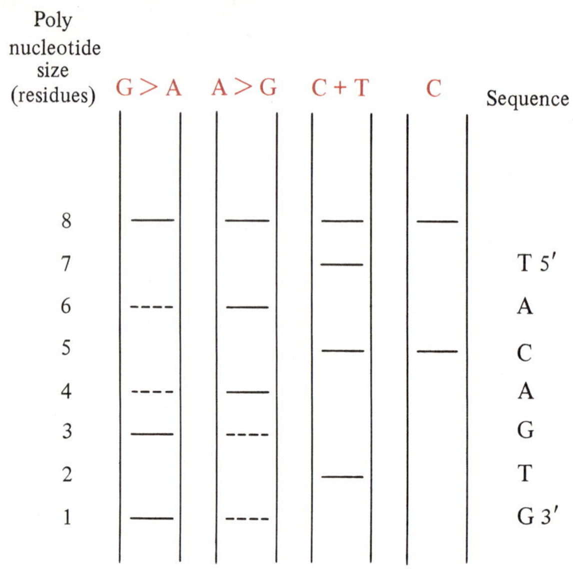

Figure 18-13
The pattern of radioactive bands obtained on gel electrophoresis of the products of the four chemical procedures of Figure 18-12. The dotted lines indicate lighter bands obtained by partial reaction of purine bases. The bands shown with eight residues reflect unreacted material.

original strand and repeat the analysis to get this information and to confirm the accuracy of the first procedure. Because the method is simple and rapid, this type of double checking can be routinely carried out.

Additional Readings

Scientific American articles

F. H. C. Crick, "Nucleic Acids," 197(3), 188 (September 1957).

A. E. Mirsky, "The Discovery of DNA," 218(6), 78 (June 1968).

J. C. Fiddes, "The Nucleotide Sequence of a Viral DNA," 237(6), 54 (December 1977). This article contains a description of the methods used by the Sanger group for sequencing DNA.

A. Rich and S. H. Kim "The Three-Dimensional Structure of Transfer RNA," 238(1), 52 (January 1978).

R. L. P. Adams, R. H. Burdon, A. M. Campbell and R. M. S. Smellie, *The Biochemistry of Nucleic Acids*, 8th edition (Academic Press, New York, 1976). An excellent source of information on many aspects of nucleic acid structure and function presented at an introductory level.

F. Wold, *Macromolecules: Structure and Function* (Prentice-Hall, Englewood Cliffs, N.J., 1971). The chapters dealing with the nucleic acids provide much useful information in a very organized and condensed form.

S. Dagley and D. E. Nicholson, *An Introduction to Metabolic Pathways* (John Wiley, New York, 1970). A good source for more detailed discussion of the metabolic pathways involved in nucleotide syntheses and interconversions.

D. Freifelder (Ed.), *The DNA Molecule: Structure and Properties* (W. H. Freeman, San Francisco, 1978). A collection of original research papers that made important contributions to elucidation of the structure and properties of DNA.

Study Questions

1. Draw the complete structures for each of the following nucleotides:
   (a) adenosine-3′-phosphate
   (b) deoxythymidine-5′-phosphate
   (c) cytidine-5′-phosphate
   (d) 3′-guanylic acid
   (e) UMP
   (f) dAMP

2. Write balanced reactions to illustrate the role of each of the following compounds in purine biosynthesis:
   (a) glycine
   (b) N^{10}-formyltetrahydrofolic acid
   (c) aspartic acid
   (d) 5-phosphoribosyl-1-pyrophosphate
   (e) ATP
   (f) GTP

3. You have a short segment of double-stranded DNA and one of the strands has the sequence A–C–G–T–A–A–C–T–G (reading from the 5′ to the 3′ end).
   (a) Draw a sufficient portion of this chain to show the essential structural features of the molecule.
   (b) Again reading from the 5′ to 3′ end of the chain, what would be the sequence of the other strand?

4. Averaging the molecular weights of AMP and CMP will give you a rough estimate of the contribution of one nucleotide to the molecular weight of a nucleic acid. Then, if a particular transfer-RNA molecule contains 77 nucleotide residues, estimate its approximate molecular weight.

5. Assume that the hypothetical DNA segment of Figure 18-12 is labeled at the 5′-end and again subjected to Maxam and Gilbert sequencing.
   (a) Indicate the labeled fragments that would result for each of the four chemical treatments.
   (b) Draw a pattern similar to that of Figure 18-13 to show the results you would expect on gel electrophoresis of each of the treated samples.

CHAPTER 19

DNA: The Genetic Inheritance of Cells

In the first chapter we commented on the need for every reaction in a living cell to be catalyzed. Inherent in that need is the result that a cell can only produce those compounds and carry out those activities for which it has enzymes able to catalyze the necessary reactions. In short, the nature of the cell and of its activities will be largely specified by the nature of the protein molecules it can synthesize.

Thus we find that the genetic information carried forward from one generation to another is simply the information the new cell will need in order to make all of the proteins peculiar to that particular organism or cell strain. At any given time, of course, the cell may choose not to use all of the available data that are in its inheritance. This matter of the control of expression of genetic information is of great importance, especially in the differentiation of cell function that takes place in multicellular organisms. We will deal with this question in Chapter 21.

19.1 A Note on Nomenclature

The fields of genetic expression, protein synthesis, cellular differentiation, and related areas of exploration have developed almost explosively in the years since about 1960. In the process, they have developed a great deal of new, and at times rather esoteric, terminology. Fortunately, we do not need to deal with too much of it in an introductory discussion, but there are some things you should be aware of at this point. First, the name **molecular biology** has somehow become associated with this branch of biochemistry and is commonly used in spite of its questionable etymology. Second, there are certain major terms that we need to mention before going any further. A **chromosome** is an organized structure that houses the DNA that is the genetic inheritance of a given cell; a cell may have one chromosome (common for procaryotic cells) or a number of them (characteristic of eucaryotic cells). A **gene** is an amount of hereditary information (that is, an amount of a DNA chain) sufficient to supply the data necessary for the synthesis of one polypeptide chain, and the **genome** is the total collection of genes possessed by any one cell.

Further, in describing the events that occur during the overall process that leads to protein synthesis, we make several important distinctions. Thus the information contained in the DNA molecule is first **transcribed** into the information carried by a messenger RNA molecule. This information is then **translated** into a certain amino acid sequence by the action of the transfer RNAs and the various components of the ribosomes.

19.2 Role of DNA

The DNA of cells serves as the primary reservoir of genetic information. In the simpler procaryotic cells, this DNA usually occurs as one or more macromolecules distributed randomly within the cell. In eucaryotes, the bulk of the DNA occurs within the nucleus, although some DNA is found within both mitochondria and chloroplasts.

When we were talking about proteins, we decided that the structure of a functional protein is determined almost entirely on the basis of its primary structure; that is, its amino acid sequence. In order to synthesize a certain specific protein, it is necessary only to somehow specify its particular amino acid sequence. By extension, then, the information that must be present in the genetic record is a listing of all of the amino acid sequences of all of the possible proteins that that particular cell could possibly produce. Since the nucleic acids are made up of nucleotides and the proteins of amino acids, there must clearly be some system by which a series of nucleotides can be made to represent a series of amino acids; in short, the message must be expressed in a coded form.

19.3 The Genetic Code

A series of ingenious experiments in chemistry and biochemistry in the late 1950s and early '60s resulted in complete elucidation of the **genetic code**. This code requires a specific sequence of three nucleotides in DNA in order to specify one amino acid in the finished protein. These nucleotide triplets occur sequentially along the DNA chain, with no "overlapping" in the reading of the code.

With four different nucleotides being used three at a time, a total of 64 different triplets are possible. Since this is considerably in excess of the 20 amino acids to be represented, it is not surprising to find that some amino acids can be represented by more than one nucleotide sequence. In these cases, the first nucleotide is nearly always constant, with variation possible in the second unit and common in the third. The code is shown in condensed form in Table 19-1.

The nature of the code was primarily determined by using synthetic polyribonucleotides to replace messenger-RNA during protein synthesis and observing which amino acids were then incorporated into the polypeptide chain. The code in Table 19-1 is expressed in terms of the triplet sequences as they would occur in one of these synthetic polymers or in a natural messenger-RNA molecule. One of these triplets is generally called a **codon**. The sequence of each codon is read from left to right starting with the 5′-end and going toward the 3′-end of the polynucleotide. Thus, for example, the two codons for phenylalanine are UUU and UUC.

Note that the three codons UAA, UAG, and UGA are used to indicate the end of a polypeptide chain. No separate code is needed to start a chain, since the first amino acid is always methionine, or a methionine derivative using the same code.

TABLE 19-1
The Genetic Code: Each possible codon in mRNA represents either a specific amino acid or an instruction to end the polypeptide chain. The latter cases are shown as "C.T." indicating chain termination.

First Position (5′ end)	SECOND POSITION				*Third Position (3′ end)*
	U	C	A	G	
U	Phe	Ser	Tyr	Cys	U
	Phe	Ser	Tyr	Cys	C
	Leu	Ser	C.T.	C.T.	A
	Leu	Ser	C.T.	Trp	G
C	Leu	Pro	His	Arg	U
	Leu	Pro	His	Arg	C
	Leu	Pro	Gln	Arg	A
	Leu	Pro	Gln	Arg	G
A	Ile	Thr	Asn	Ser	U
	Ile	Thr	Asn	Ser	C
	Ile	Thr	Lys	Arg	A
	Met	Thr	Lys	Arg	G
G	Val	Ala	Asp	Gly	U
	Val	Ala	Asp	Gly	C
	Val	Ala	Glu	Gly	A
	Val	Ala	Glu	Gly	G

The Genetic Code is often referred to as a "universal" code, since all of the living materials thus far tested use the same code for protein synthesis. The complete universe remains, for the moment, beyond our grasp.

19.4 DNA Replication

When a cell divides in the normal growth process, it becomes necessary to produce a copy of the genetic material for the new cell. While the details of this process are by no means clear, the basic approach involved has been known since the elegant experiments of Meselson and Stahl in 1958. Using ^{15}N-labeled nucleotides, they were able to show that the double-stranded DNA was being separated and a new complementary strand was being manufactured for each of the original strands. Their experimental results are diagrammed in Figure 19-1. They were able to obtain these results because the DNA strands involved had different masses depending upon which isotope of nitrogen they contained. The original parent DNA had two strands made entirely of nucleotides containing nitrogen of mass 15. The first generation product DNA had one strand made up of ^{15}N nucleotides

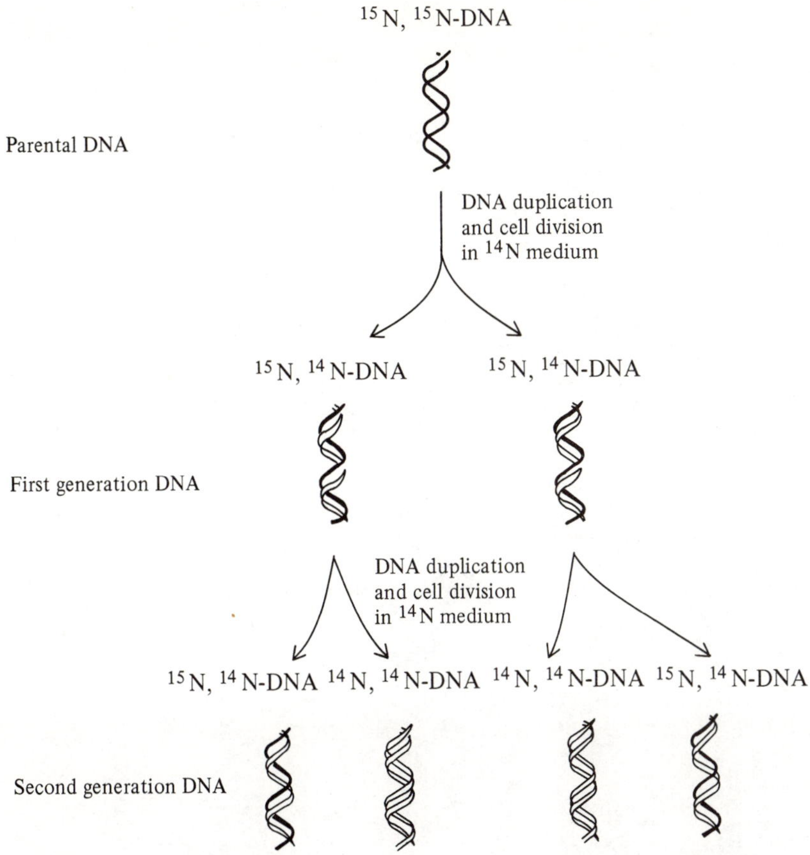

Figure 19-1
A diagrammatic representation of the Meselson–Stahl experiments on DNA replication. The dark helices represent strands of DNA which were present in the original cell before DNA duplication and contain ^{15}N nucleotides. The light helices represent DNA strands formed in the process of replication in the presence of ^{14}N nucleotides only.

and one of ^{14}N nucleotides, and was thus a lighter molecule which could be separated from the parent by high-speed centrifugation. Similarly, some of the second generation molecules were lighter still, being composed only of ^{14}N nucleotides, and thus could be separated readily from the other two varieties.

Much of our information of DNA replication comes from the study of procaryotic organisms such as *E. coli*; we will deal primarily with the nature of these systems.

In order that a portion of a DNA chain may be made available for replication, it is necessary that the two strands be separated sufficiently to allow access of the necessary enzymes and the new strands to be formed. This creates certain problems, particularly in the case of the circular DNA of procaryotes, since this separation must involve a substantial degree of uncoiling of the helix. This situation has been resolved with the discovery of an enzyme complex (originally called a "swivelase" and now usually referred to as a "gyrase") which makes a break in one strand, allows it to

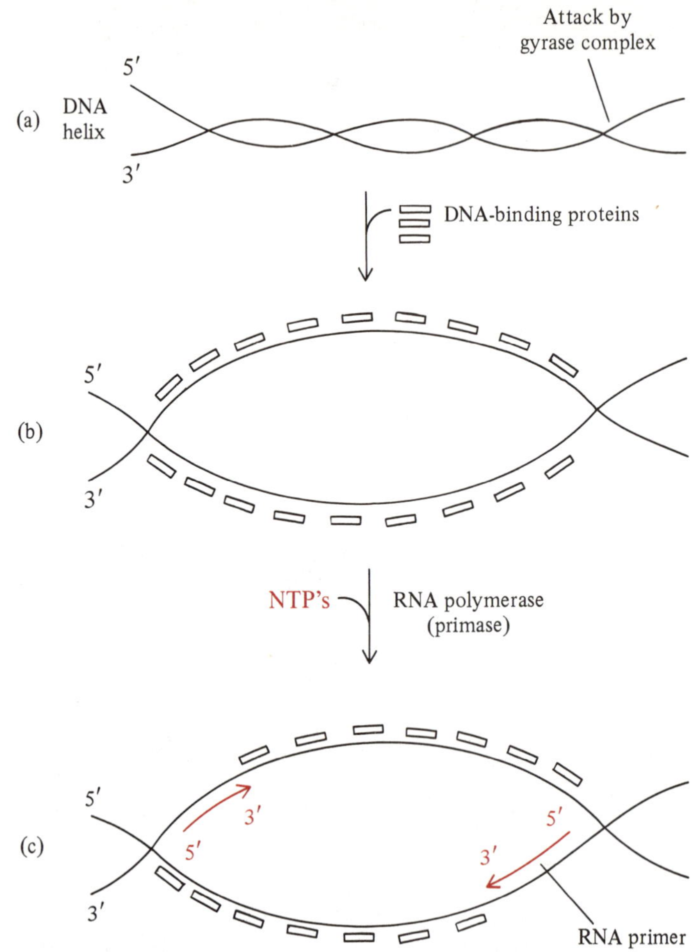

Figure 19-2
The beginning of the process of DNA replication. Attack of the gyrase complex opens, uncoils, and reseals the DNA into a bubble form stabilized by DNA-binding proteins. A special **DNA-directed RNA polymerase (primase)** then begins the formation of primer strands of RNA.

unwind for a distance, then reseals the original break. To prevent refolding of the single-strand portions thus produced, a group of protein molecules called **DNA-binding proteins** become attached to these sections, producing a "bubble" in the DNA chain as indicated in Figure 19-2.

The Need for RNA Primers

There are in *E. coli* three different enzymes that act as **DNA polymerases**: these are often referred to as *pol* I, *pol* II, and *pol* III. While any one of these can extend a polynucleotide chain with a free 3′-OH group, none is capable of **starting** such a chain. The ability to start a chain is, however, common to the **DNA-dependent RNA polymerases** which carry out the synthesis of RNA molecules on the basis of the information in DNA chains. Replication then begins with attachment on one of the single strands of the bubble of a protein called (for want of a better name) **preprimase**. This protein binds and permits a special RNA polymerase (**primase**) to begin the manufacture of a short (about 100 bases) segment of RNA complementary to the DNA chain. This will occur on both strands of the bubble, giving the situation shown in Figure 19-2(c).

Extension of the Chain

As mentioned above, the DNA polymerases must work from the free 3′-OH of a polynucleotide chain. The reaction involved is shown in Figure 19-3. Thus, with an appropriate supply of deoxyribonucleoside triphosphates (dNTPs), polymerase III (*pol* III) can extend a new DNA strand from the 3′-end of the RNA primer. As this process takes place, the DNA-binding proteins are removed and are re-used to stabilize the newly exposed chains as the opening bubble moves along the DNA. If we consider that the gyrase complex moves in only one direction along the DNA, a new problem becomes evident. While *pol* III can continue to extend the one new chain from the growing 3′-end as new lengths of single strand open up, it is unable to do so with the other strand since it runs in the opposite direction. Thus, it is necessary to have new units of primer inserted so that the intervening space can be filled from the 3′-end of the primer. These intermediate segments of DNA can be isolated under certain circumstances, and have lengths on the order of 1000 base pairs. The picture at this point is shown in Figure 19-4.

The cell now takes advantage of the special properties of DNA polymerase I (*pol* I) which allow it to hydrolyze off nucleotides of the RNA primer one at a time from the 5′-end (**5′-exonuclease** activity) and replace them with the appropriate deoxynucleotides by adding on to the 3′-OH of the new DNA strand. The final reaction linking the free 3′-OH of this section of replaced primer to the 5′-phosphate of the *pol* III product is carried out by a separate enzyme called **DNA ligase**. These processes are shown schematically in Figure 19-5.

A Note on Nucleotide-Pairing Specificity

It is generally believed that the specificity of the interchain A–T and G–C pairings is as much determined by the spatial requirements of the active sites of the DNA polymerases as by the specific hydrogen bonding between the bases themselves. It is important in a case such as this to distinguish between the fact that these pairings are critical to the stability of the double-stranded DNA helix and the means by which the pairings are brought about during synthesis.

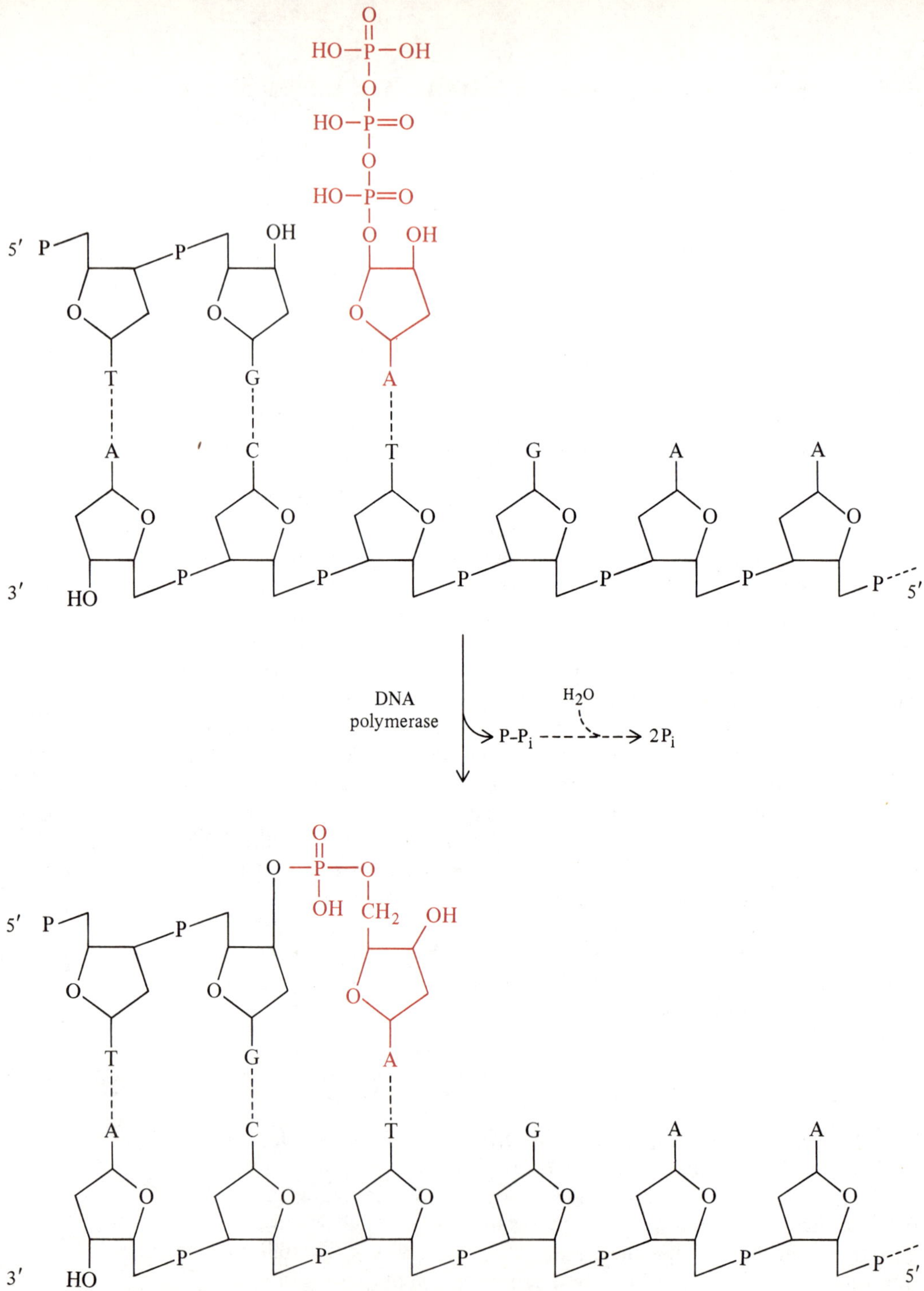

Figure 19-3
The nature of the reaction catalyzed by a DNA polymerase.

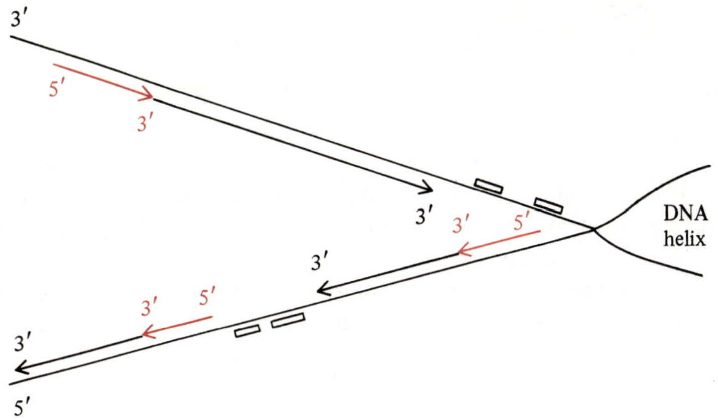

Figure 19-4
A replicating DNA molecule in the process of synthesis of DNA chains from the 3′-ends of RNA primers. Note that one strand can be extended continuously but the other must be formed in segments as new DNA is opened up.

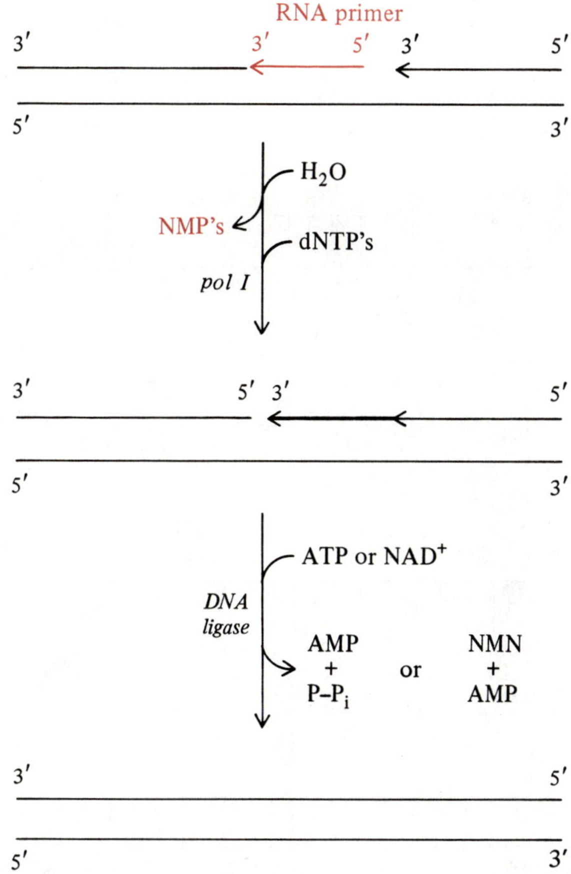

Figure 19-5
Removal of RNA primers and joining of DNA segments to complete replication of a section of DNA.

19.5 DNA Repair

Work with *E. coli* damaged by ultraviolet radiation has shown that some strains have the capability to carry out repair of the damaged DNA, at least when only one strand has been damaged. The major effect of UV radiation is the covalent fusing of two pyrimidine bases, which distorts their normal alignment and interferes with the structure of the DNA chain at that point; a situation shown in Figure 19-6.

The suggested mechanism for the repair of such a defect involves first the recognition by an **endonuclease** (an enzyme that hydrolyzes a bond some-

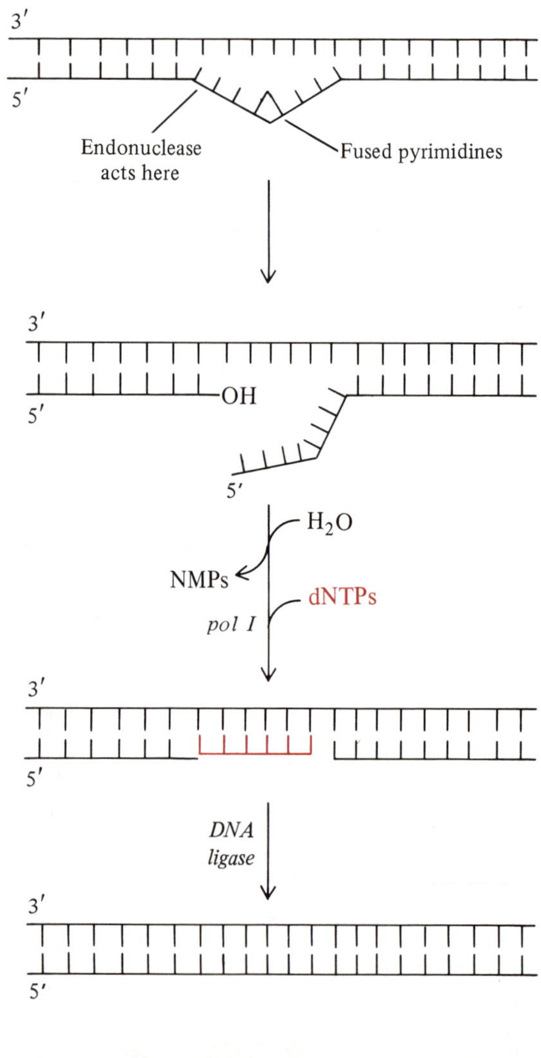

Figure 19-6
A mechanism for repair of UV radiation induced damage to bacterial DNA.

where inside a polynucleotide chain) that there is a structural defect. This enzyme then creates a *nick* (a single bond cleavage of the one chain) to one side of the problem area, leaving a free 3′-OH exposed. DNA polymerase I (*pol* I) then makes use of its polymerase activity to extend the chain from the free 3′-position and makes use of its ability to sequentially remove nucleotides from the 5′-end of a chain to remove the damaged segment. Just as in normal replication, final closure of the gap in the chains depends upon the action of DNA ligase. These steps are also illustrated in Figure 19-6.

19.6 A Protection Mechanism for Bacteria

In looking at methods of preparation of DNA segments suitable for sequencing (in Chapter 18), we mentioned the use of very specific enzymes, called **restriction enzymes**, isolated from bacterial sources. These enzymes are normally part of an ingenious mechanism possessed by some strains of bacteria to protect them from invasion by foreign DNA (such as that of a bacterial virus). The process involves first the action of another enzyme, a **modification methylase**, which recognizes the same nucleotide sequence as the restriction enzyme with which it is paired. This methylase uses the usual methyl group donor, *S*-adenosyl methionine, to methylate one specific nucleotide of the sequence, as shown in Figure 19-7. Once the sequence is methylated, it is no longer recognized by the appropriate restriction endonuclease. Foreign DNA, lacking the methyl groups, is rapidly cleaved in both strands and rendered ineffective. This "restriction" on the DNA that is allowed to be present in the cell is the source of the descriptive name for these enzymes.

19.7 Palindromic Sequences in DNA

A palindrome is simply any statement that reads the same in either direction. An example is the quote that it has been suggested Napoleon might have made had he thought of it, "Able was I ere I saw Elba." The concept has been stretched to describe sections of double-strand DNA in which the two strands have the same sequence but running in reverse directions. You will note that the enzyme recognition site shown in Figure 19-7, and the additional restriction enzyme sites shown in Table 18-1, are all palindromic in nature. As research progresses, there is increasing evidence that these palindromic regions serve as attachment sites for a number of proteins in their interactions with DNA, including many of those involved in the control of expression of genetic information.

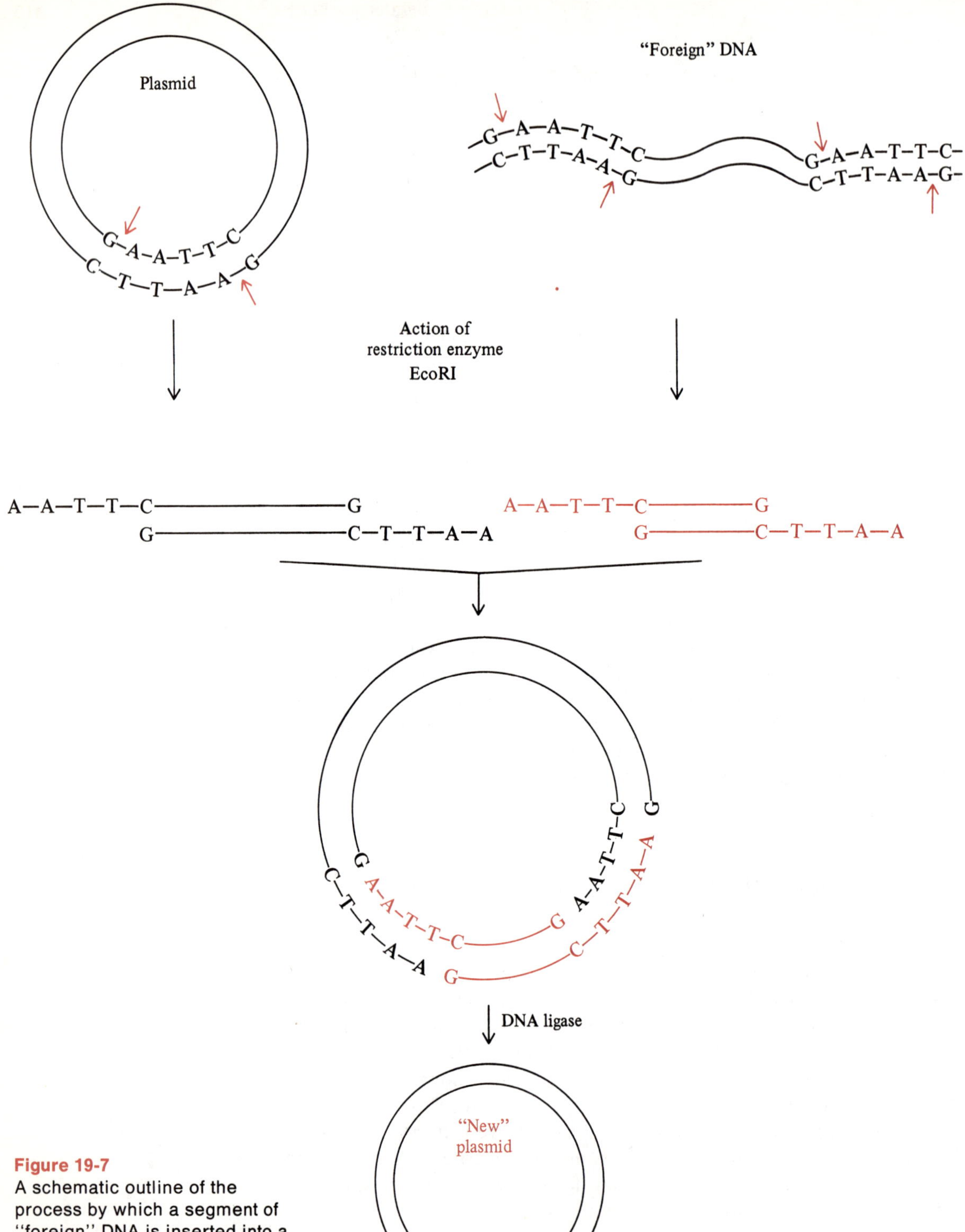

Figure 19-7
A schematic outline of the process by which a segment of "foreign" DNA is inserted into a plasmid capable of replication and translation within a bacterial cell.

19.8 Recombinant DNA

The discovery of the various restriction enzymes and the nature of their action has led to a whole new area of research sometimes referred to as "genetic engineering." Among other things, it is possible to combine portions of DNA from two quite different sources, and then to insert the resulting **recombinant DNA** molecule into an appropriate cell where it will be reproduced and have its genetic messages expressed. We will look at the basic methods involved before saying a word about the implications and the controversy involved in this kind of work.

Many microorganisms contain, in addition to their own "chromosomal" DNA, other small circular molecules of DNA called **plasmids**. During the process of cell division, these plasmids are reproduced and passed on just as is the chromosomal DNA. Since various strains of *E. coli* contain a number of plasmids that are quite well defined, this is usually the organism of choice for recombinant DNA work. The process begins with isolation of one of the plasmids and its treatment with a restriction endonuclease of the type that makes "staggered" cuts in the two DNA chains of the plasmid (for example, *Eco*RI or *Hind*III as shown in Table 18-1). As a result of the double cuts, the molecule will tend to come apart and spend a substantial proportion of its time in the open-chain form. However, it does have "sticky" ends on the chains that will tend to associate with one another. This property is important to the next step in the process.

The second sample of DNA can be taken from any double-stranded DNA that has an appropriate site for action of the restriction enzyme being used and that is not methylated at that site. When dealing with a typical eucaryotic source, one of the difficult problems can be separation of the desired segment of DNA from all of the other fragments produced by action of the endonuclease on the very large DNA structures characteristic of these cells. In some cases, unseparated mixtures of these pieces have been used, which has added to concerns about the biological hazards of the process, which we will address shortly. In any case, the "foreign" DNA segment will now have sticky ends identical with those of the plasmid DNA. If the two are mixed at this point, some of the chains will associate in such a fashion as to create a new circular form including both the plasmid and foreign DNAs, as shown in Figure 19-7. Treatment at this point with a DNA ligase will heal the breaks in both chains and create a new pseudoplasmid. Of course, in the process of ligase action, you will also get some reformed original plasmid and some circular foreign DNA. The latter molecule is generally incapable of replication.

When the recombinant molecule is allowed to enter an *E. coli* cell, it carries with it the original plasmid mechanisms that cause its replication through future generations of these cells. At the same time, both the original DNA of the plasmid and the foreign DNA that has been inserted can be used to make messenger-RNA and subsequently the appropriate protein molecules.

Concern and Controversy

As has been true of so many major discoveries, the ability to insert functional foreign genes into a bacterial system provides both great promise and great potential hazards. The possible advantages include using bacterial systems to produce protein molecules otherwise available only in small amounts from natural sources, such as blood clotting factors for treating hemophilia, polypeptide hormones, specific globulin fractions, and enzymes of limited distribution. Some of the other suggestions that have been made lie farther down the road but would lead to even more major changes in the biological world. This include giving microorganisms associated with plants other than clover and the legumes the ability to fix nitrogen to increase crop yields of staple food products.

The dangers of course arise from our lack of knowledge of the exact effects of incorporating various foreign DNA units into bacterial cells. This is particularly true when the material being inserted is of eucaryotic origin. The possibilities of producing materials hazardous to animal hosts or of producing antibiotic-resistant strains of bacteria are among the very real concerns felt in the scientific community. These concerns have given rise to a set of guidelines that are generally adhered to by workers in this field. These include the use of *E. coli* strains that cannot live in humans or other animals, and the use of high-security laboratories for work with pathogenic organisms. No doubt these measures will be refined and modified as increasing knowledge permits a better estimate of the dangers and the means of avoiding them. The public debate that has taken place on this topic has been a major force in speeding establishment of protective measures that do not at the same time unnecessarily impede the attainment of whatever benefits this line of research may eventually provide. It is also evidence of the increasing recognition that scientists must be prepared to take part of the responsibility for dealing with the wider implications of their new-found abilities.

Additional Readings

Scientific American articles

R. C. Clowes, "The Molecule of Infectious Drug Resistance," 228(4), 18 (April 1973). A discussion of the role of bacterial plasmids as carriers of drug resistance.

S. N. Cohen, "The Manipulation of Genes," 233(1), 24 (July 1975).

C. Grobstein, "The Recombinant-DNA Debate," 237 (1), 22 (July 1977).

J. D. Watson, *Molecular Biology of the Gene*, 3rd edition (W. A. Benjamin, New York, 1976). A remarkable book covering a wide range of topics in molecular biology in depth but in a form that is highly readable for the beginner. Chapters 9, 10, and 13, which deal with the role of DNA and the nature of the genetic code, are particularly appropriate to this chapter.

R. L. P. Adams, R. H. Burdon, A. M. Campbell, and R. M. S. Smellie, *The Biochemistry of Nucleic Acids,* 8th edition (Academic Press, New York, 1976). An excellent source at the introductory level.

A. Kornberg, *DNA Replication* (W. H. Freeman, San Francisco, 1980). An advanced book that provides an outstanding source of data on the current state of the field; not easy reading for the novice.

Study Questions

1. In terms of genetic expression, define the nature of the terms "replication," "transcription," and "translation."

2. Write out all of the codons that represent the amino acid serine. For each, write the 5′- to 3′-sequence of nucleotides in the corresponding complimentary nucleotide triplet of DNA.

3. One strand of a palindromic region of DNA has the sequence 5′–A–T–T–C–G–A–T–G–C–3′; draw a diagram showing the sequences of both strands in this region.

4. The enzyme ribonuclease contains 124 amino acid residues per molecule. What is the minimum number of nucleotides that you would expect to find in the gene supplying the information for synthesis of this protein?

5. The type of replication mechanism found by Meselson and Stahl is usually called a "semiconservative" process. Can you suggest the nature of the products that would have been found in the first generation offspring cells if the mechanism had been a "conservative" one? Further, is it theoretically possible to have a "nonconservative" mechanism for replication?

6. The DNA ligase of bacterial systems usually relies on the hydrolysis of NAD to AMP and NMN (nicotinamide mononucleotide) as an immediate source of energy. Use your growing knowledge of biochemical processes to suggest a reaction by which the NAD could be regenerated from NMN.

CHAPTER 20

Genetic Expression and Protein Synthesis

It is tempting to compare the processes of protein synthesis with the operation of a large clothing factory. The factory must have somewhere a pattern from which each dress, suit, and so on in their line can be made. However, it would be impractical at best to send that original pattern out to the cutting floor, and even more ridiculous to send all of the patterns out to the cutters at the same time, since only some of the styles are needed at any given time. As a result, the original patterns are kept in a vault; when a particular style is needed, templates are made from the original and sent out to be used in cutting the fabric. As you might expect, these working templates have a very limited lifetime.

In exactly the same sense, the DNA of a cell is generally held in a tight association with protein molecules and, in eucaryotes at least, is isolated away from the other activities of the cell. When a protein is to be synthesized, a working template in the form of a molecule of RNA is prepared and sent to the site of synthesis. In this process, the code specified in the DNA is "transcribed" into the RNA molecule. The ribosomes then "translate" the information from this template into a specific polypeptide sequence.

20.1 Transcription

We have already noted the complementary nature of the two strands of DNA, and noted that the two strands are not likely to be identical. Whether the RNA to be synthesized is messenger RNA, ribosomal RNA, or transfer RNA, only one of the DNA strands carries the code for that product, and it is only that strand that serves as a pattern for transcription. Since the RNA is made by aligning complementary bases along the parent DNA chain in much the same fashion as in DNA replication, the resulting RNA will have a sequence that is complementary to that of the one operative strand of the DNA. Of course, all of the units that would be thymine in DNA will now be uracil in the RNA. The result of a typical transcription is shown in Figure 20-1.

The process of transcription is carried out under the direction of a complex enzyme, **DNA directed RNA polymerase**, which has also been called

3′-A–T–G–C–A–A–T–G-5′ One strand of DNA

ATP, UTP, CTP, GTP → (RNA polymerase)

5′-U–A–C–G–U–U–A–C-3′ Corresponding section of RNA

Figure 20-1 Transcription of a segment of a DNA chain into a corresponding portion of RNA.

"transcriptase." This enzyme as normally isolated from a bacterial source such as *E. coli* contains five polypeptide subunits (of four different structures) associated together. The entire structure is needed to initiate transcription, a process that requires the recognition of a suitable sequence as a starting point, followed by the opening of about a six-nucleotide stretch of the double-strand DNA to permit access of the enzyme units. Formation of the chain then begins in much the same fashion mentioned in Chapter 19 for the formation of RNA primer molecules (Figure 20-2). The chemical reaction involved is essentially that shown in Figure 19-3 except that the nucleotides involved are ribonucleotides.

Shortly after the chain is started, one of the five subunits of the enzyme dissociates and remains free until transcription is complete. This specific subunit is apparently needed either for recognition of the starting point or for the process of opening the double helix. Elongation of the RNA chain continues by addition to the free 3′-end until a termination signal is received. This can occur through the binding of another protein factor which causes

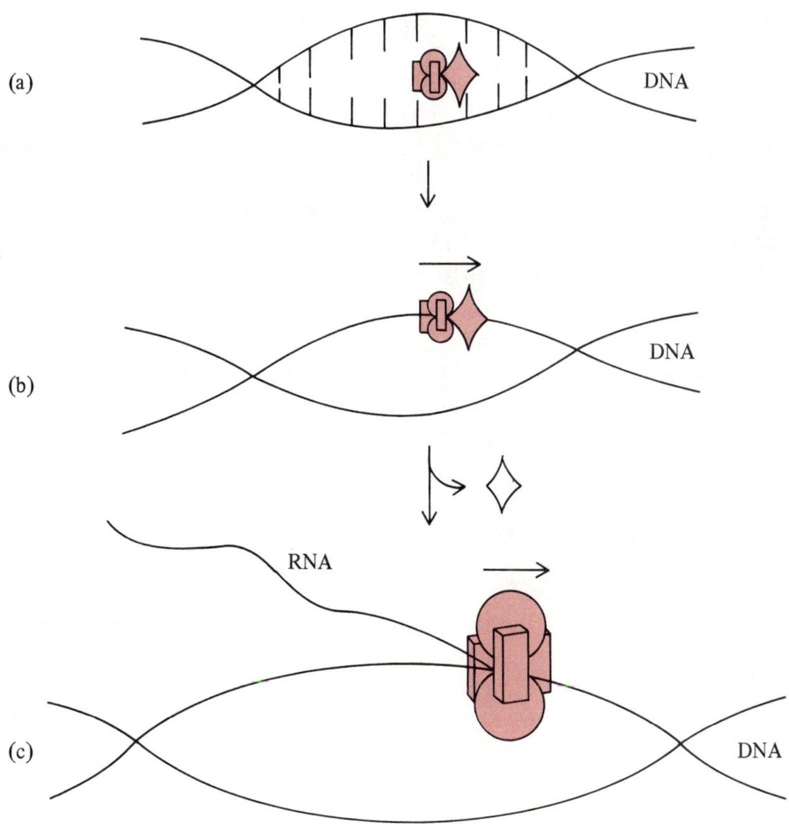

Figure 20-2
A schematic representation of the process of transcription as carried out by the *DNA directed RNA polymerase* enzyme complex. In part (a) the enzyme complex is bringing about the separation of the DNA strands sufficiently to permit attachment as shown in (b), ready to begin transcription. In part (c) the enzyme has lost one subunit and is trailing a newly synthesized RNA molecule.

the polymerase complex to dissociate from the DNA, or by recognition by the polymerase alone of a termination sequence such as –A–A–A–A–A–A–C–.

Modification of RNA Molecules

It is becoming increasingly clear that RNA molecules undergo a considerable amount of enzymatic modification or **processing** between the time of their synthesis and their action as functional units. The transfer RNAs in particular experience a great deal of change, since it is in this process that most of their unusual nitrogen heterocycles are formed. In addition, these molecules are usually formed with extra strings of nucleotides on both ends which are trimmed off to produce the functional tRNA. In some cases, such as in *E. coli*, a single gene gives rise to a section of RNA containing two tRNA sequences, joined by a short string of nucleotides that is lost during the processing into two separate tRNAs.

In eucaryotic cells there appears to be even more extensive posttranscriptional modification. A number of examples are known in which the messenger RNA as first formed consists of an RNA molecule with the real "message" at the ends of the chain, separated by a long stretch of nucleotides not useful during protein synthesis. In some fashion, this central section is removed and the two ends rejoined before the mRNA is sent to the ribosomes to take part in protein synthesis. Neither the mechanism of removal nor any rationale for the existence of these spacer regions is known at the present time. In a similar fashion, the ribosomal RNA (rRNA) molecules of the cells of higher organisms are usually made as parts of a larger precursor molecule, which is then cut, trimmed, and modified into the required units.

In addition, many (though not all) eucaryotic mRNAs are modified after transcription by the addition of a string of adenine nucleotides to the 3′- end and by addition of 7-methylguanosine to the 5′-terminal nucleotide through an unusual 5′- to 5′- triphosphate linkage. The 2′-hydroxyl group of the original terminal nucleotide is also usually methylated. The role of this "capping" process, and of the **poly-A** string on the 3′-end remains unidentified.

Reverse Transcriptase

Infection of cells by certain RNA-containing viruses leads to the production of an enzyme called **RNA directed DNA polymerase** or, more commonly, "reverse transcriptase." It catalyzes the synthesis of DNA on the basis of the nucleotide sequence of the viral RNA. The cell then uses the new DNA to produce more viral RNA. Thus information exchange can flow in either direction between DNA and RNA and both are capable of directing their own replication. There has been no evidence, however, for any backward flow of information from a protein to RNA or DNA.

20.2 Translation

An Overview

Now that we have the basic information from DNA transcribed onto a messenger RNA template, we must have a system for translating that codon sequence into an amino acid sequence. The basic features of this process can be described in a few words; the details will keep many researchers employed for years to come.

The first stage in this system involves the small RNA molecules called transfer RNAs. Each of these carries a specific **anticodon** that is complementary to one of the amino acid codons listed in Table 19-1. Each of these tRNAs can be esterified to its related amino acid by a specific enzyme that will carry out that function for one amino acid only. The specificity of these enzymes is critical in ensuring that the proper amino acid is associated with the tRNA with the proper anticodon.

The second stage takes place in association with the ribosomes. Here, each tRNA is brought in as required and associates its anticodon with the codon of the mRNA. This process brings the appropriate amino acid into position on the ribosome as each codon of the mRNA is "read" in sequence, and the amino acid joins the growing polypeptide chain in the proper sequence.

Clearly, then, we have a two-stage recognition process: first, the recognition of one amino acid and one (or more) tRNAs by the acylating enzyme; then, mutual recognition of codon and anticodon occurring at the surface of the ribosome during polypeptide synthesis.

Transfer RNAs: Structure and Function

The tRNAs are certainly the smallest of all of the nucleic acid structures. In addition, they have a few peculiar features that reflect their unique functions. Bear in mind that these molecules must be capable of the following:

a. binding a specific amino acid in a position and in a linkage from which it can readily be transferred into a growing polypeptide chain;
b. carrying in an appropriately exposed position the anticodon corresponding to its assigned amino acid;
c. providing a segment of its structure that will provide the specific acylating enzyme with a clear and unambiguous recognition signal;
d. survival for an extended period of time. Unlike the mRNAs, the transfer RNA molecule has a continuous function and thus a reason for survival in the face of hydrolytic enzymes such as the omnipresent ribonucleases.

All of these needs are addressed in the structures of the tRNAs, as illustrated in Figure 20-3. The internal hydrogen bonding leading to a fundamental cloverleaf structure provides for a rigidity of structure that permits exact location of the anticodon, regions suitable for recognition by the acylating enzyme, and further structural characteristics necessary for precise three-dimensional positioning on the ribosome during peptide synthesis.

While the role of the modified nitrogen bases in the tRNAs is not yet clear, it has been suggested that they may serve in part the role of protection from some of the specific ribonuclease enzymes that would otherwise attack tRNA.

Aminoacyl–tRNA Synthetases

Addition of the appropriate amino acid to each of the tRNA molecules is catalyzed by the aminoacyl–tRNA synthetases. Twenty of these enzymes exist in each cell, one for each of the amino acids built into proteins. The overall reaction catalyzed in each case is as follows:

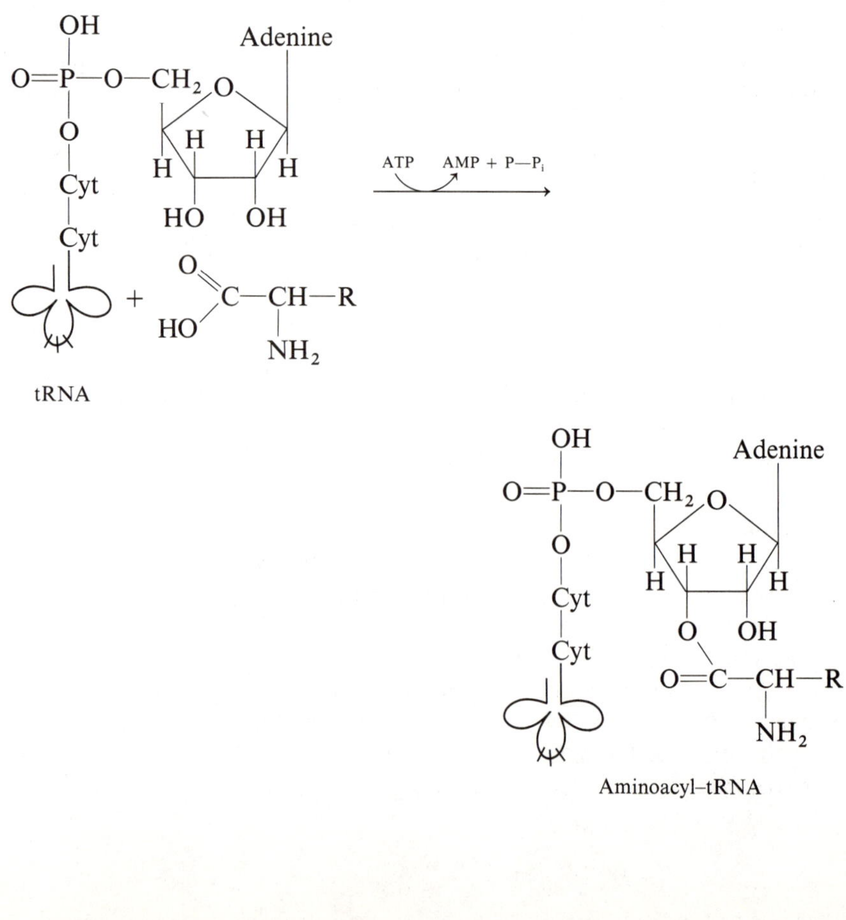

tRNA

Aminoacyl–tRNA

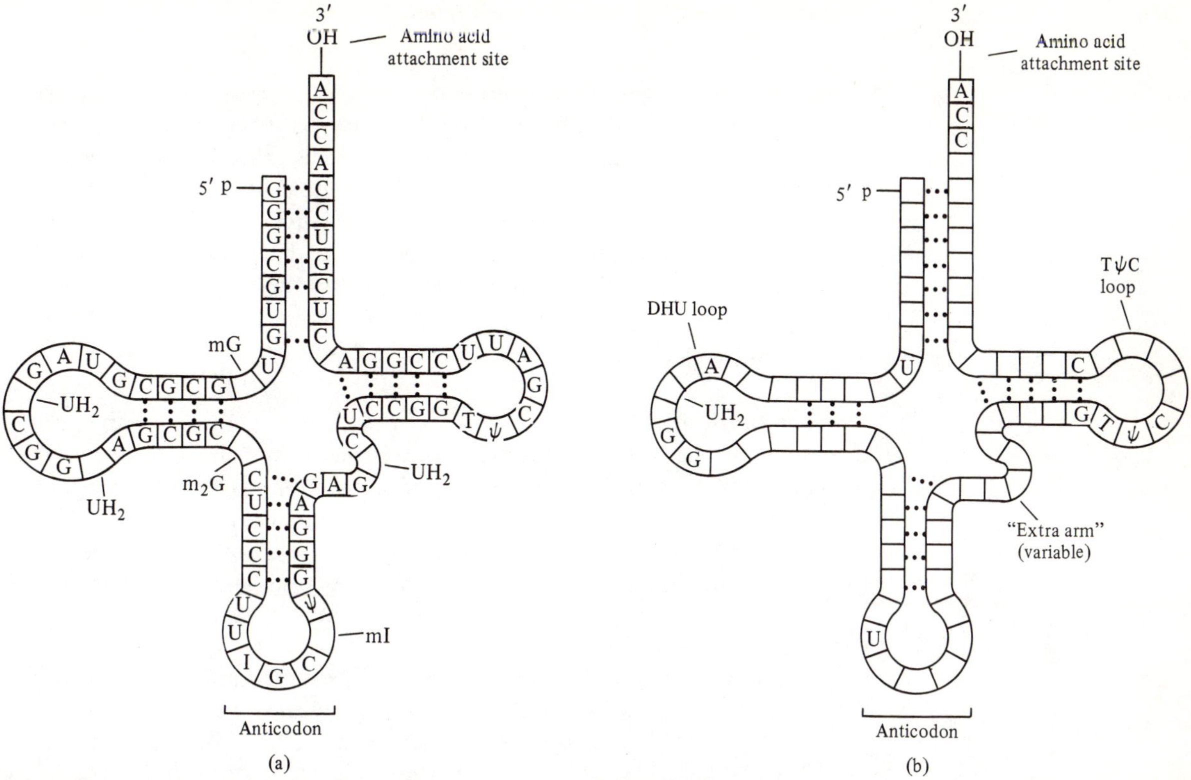

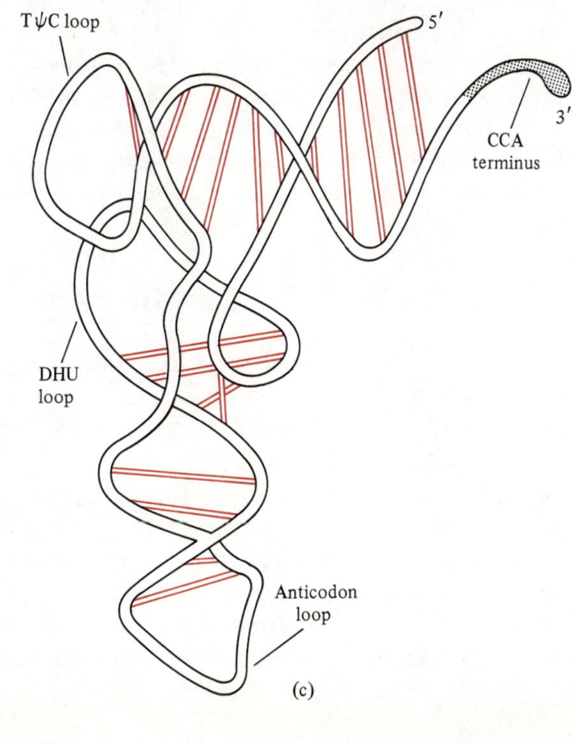

Figure 20-3
(a) The nucleotide sequence of a molecule of yeast alanine tRNA. The modified nucleosides have been represented as methylguanosine (mG), dihydrouridine (UH_2), dimethylguanosine (m_2G), inosine (I), methylinosine (mI), pseudouridine (ψ), and ribothymidine (T). (b) An illustration of the structural features and specific nucleotides that are common to tRNA molecules. (c) A drawing of the approximate three-dimensional structure of a typical tRNA.

These synthetases are, and must be, among the most specific enzymes known. Each must be capable of recognizing not just its own specific amino acid from all others, but also only certain molecules of tRNA. The recognition regions or base sequences of the tRNA molecules that permit this specificity on the part of the synthetases are not at all clear. At the present time, there is no one region of the molecule that can clearly be assigned responsibility for providing a recognition signal to the enzyme.

Ribosomes

The ribosomes are small, subcellular organelles which are distinguishable in stained cellular preparations by their relatively high concentrations of RNA. Much of the physical study of ribosomes has been carried out using the ultracentrifuge. In this system, the rate of sedimentation is often expressed in Svedberg units (S), a value that involves both the weight and the diffusion coefficient of the molecule. When two subunits unite to form a larger entity, the Svedberg values are never quite additive, although the molecular weights are.

A typical bacterial ribosome has an overall sedimentation value of 70 S and is composed of two subunits that individually show values of 30 S and 50 S. Typical eucaryotic ribosomes have an overall value of 80 S and subunits of 60 S and 40 S.

The much-studied ribosomes of *E. coli* contain about 60–70% RNA in the form of three or four molecules of different sizes (including 5 S, 16 S, and 23 S RNA). The other 30–40% of the structure is protein. Some 20 separate protein molecules have been separated from the 30 S subunit (the S series proteins) and over 30 from the larger 50 S subunit (the L series proteins). The functions of these proteins are by no means clearly established, although a number have been implicated in various phases of the protein synthesis process.

Polypeptide Synthesis at the Ribosome

For purpose of illustration, we will consider the process of polypeptide synthesis as it is believed to occur in conjunction with the ribosomes of the bacterium *E. coli*. The sequence of steps is outlined in Figure 20-4.

Initiation. The first step in protein synthesis at the level of the ribosomes involves an interaction between the smaller portion of the ribosome and the very unusual tRNA carrying the initiating amino acid (always *N*-formylmethionine in procaryotes or mitochondria, and methionine in the cytosol of eucaryotes). The formylmethionyl–tRNA binds very specifically to what is called the **peptidyl binding site**. This association requires a protein "initiation factor" that also binds GTP. The messenger RNA is then attached in such a way that its –AUG– codon lines up with the –CAU– anticodon of the tRNA. Hydrolysis of the bound GTP to GDP and P_i brings about release of the initiation factor protein and allows the larger part of the ribosome to join the smaller one. With the addition of several other proteins, this leaves a complex containing the complete ribosome, the mRNA, and the formylmethionyl–tRNA attached to the peptidyl binding site.

Elongation. Association of the two parts of the ribosome creates a new site at which an aminoacyl–tRNA can be bound. This **aminoacyl binding site** is right next to the peptidyl site, and requires the anticodon of the incoming tRNA to be aligned with the second codon of the mRNA. This binding also involves a protein which carries GTP and is released after it is hydrolyzed to GDP. Release of this protein is followed by transfer of the formylmethionine onto the free amine group of the second amino acid. The energy for this process is supplied by the relatively high bond energy of the aminoacyl linkage to tRNA. Molecules of tRNA without an attached amino acid are poorly bound to the ribosomes, and soon dissociate from the complex.

At this point, a third GTP-binding protein comes into play and brings about the **translocation** of the tRNA with dipeptide attached from the aminoacyl site to the peptidyl site. In the process, of course, GTP is hydrolyzed to GDP. Another aminoacyl–tRNA then binds at the vacant aminoacyl binding site, and the process is repeated.

Termination. Elongation of the polypeptide chain continues in this fashion until a "chain termination" codon is reached. Rather than binding a tRNA at this point, the complex binds a specific protein which recognizes the terminating codons and catalyzes hydrolysis of the bond between the polypeptide chain and the last tRNA of the sequence. Yes, activation of this protein also involves hydrolysis of GTP.

Lifetime of mRNA

The lifetime of an individual mRNA molecule is relatively short, especially in bacterial cells. It has been estimated that the average molecule of mRNA in a cell of *E. coli* involved in very active protein synthesis is responsible for the production of about eight molecules of protein before it is destroyed by the hydrolytic nucleases of the cell. The mRNAs of eucaryotes have been reported to have substantially longer lives, the estimates varying with the tissue and the organism involved.

A relatively short lifetime is essential to control of the process of protein synthesis. As we shall see in the next chapter, this control is usually expressed at the level of production of the mRNA. If the control mechanisms are to be responsive to short-term changes, the mRNA molecules must disappear within a limited period of time.

Codon/Anticodon Interactions—The Wobble Concept

As the ability to isolate and purify tRNAs has improved in recent years, it has become clear (a) that no cell contains all 61 tRNA molecules that would be needed to match the mRNA amino acid codons, and (b) that a single tRNA anticodon may pair with several mRNA codons, all coding for the same amino acid.

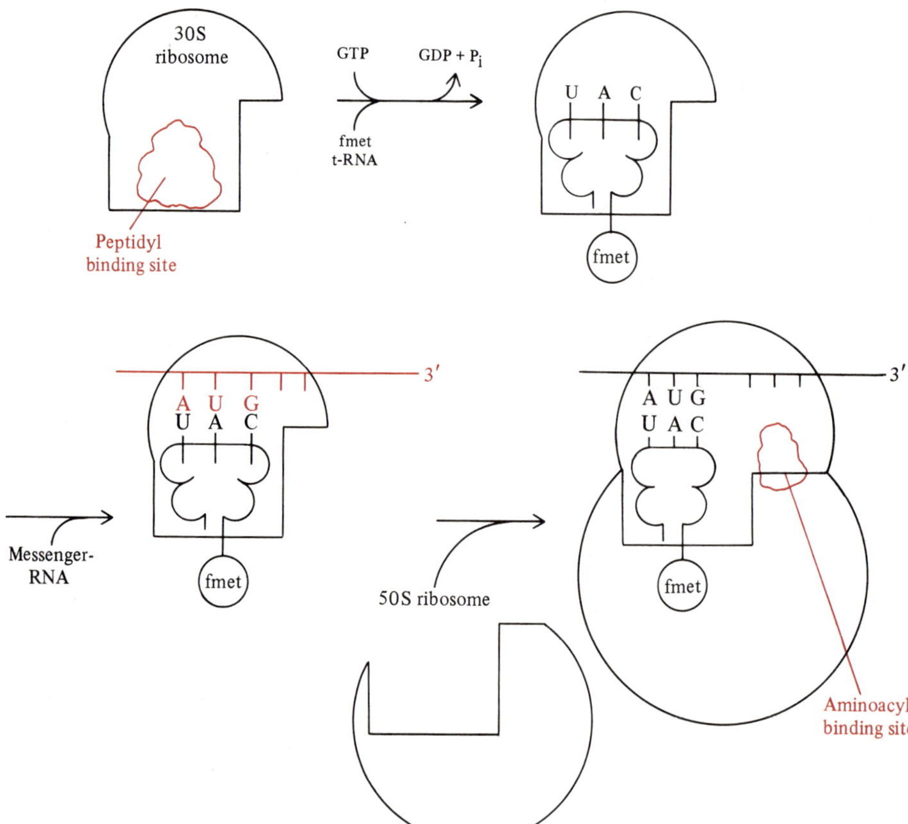

Figure 20-4
(a) Initiation of protein synthesis at the ribosome of *E. coli*. This portion of the process involves assembly of the m-RNA, the two pieces of the ribosome, and the tRNA carrying *N*-formylmethionine into a single complex. (b) The process of chain elongation and termination. The specific amino acids used to illustrate the process have no special significance.

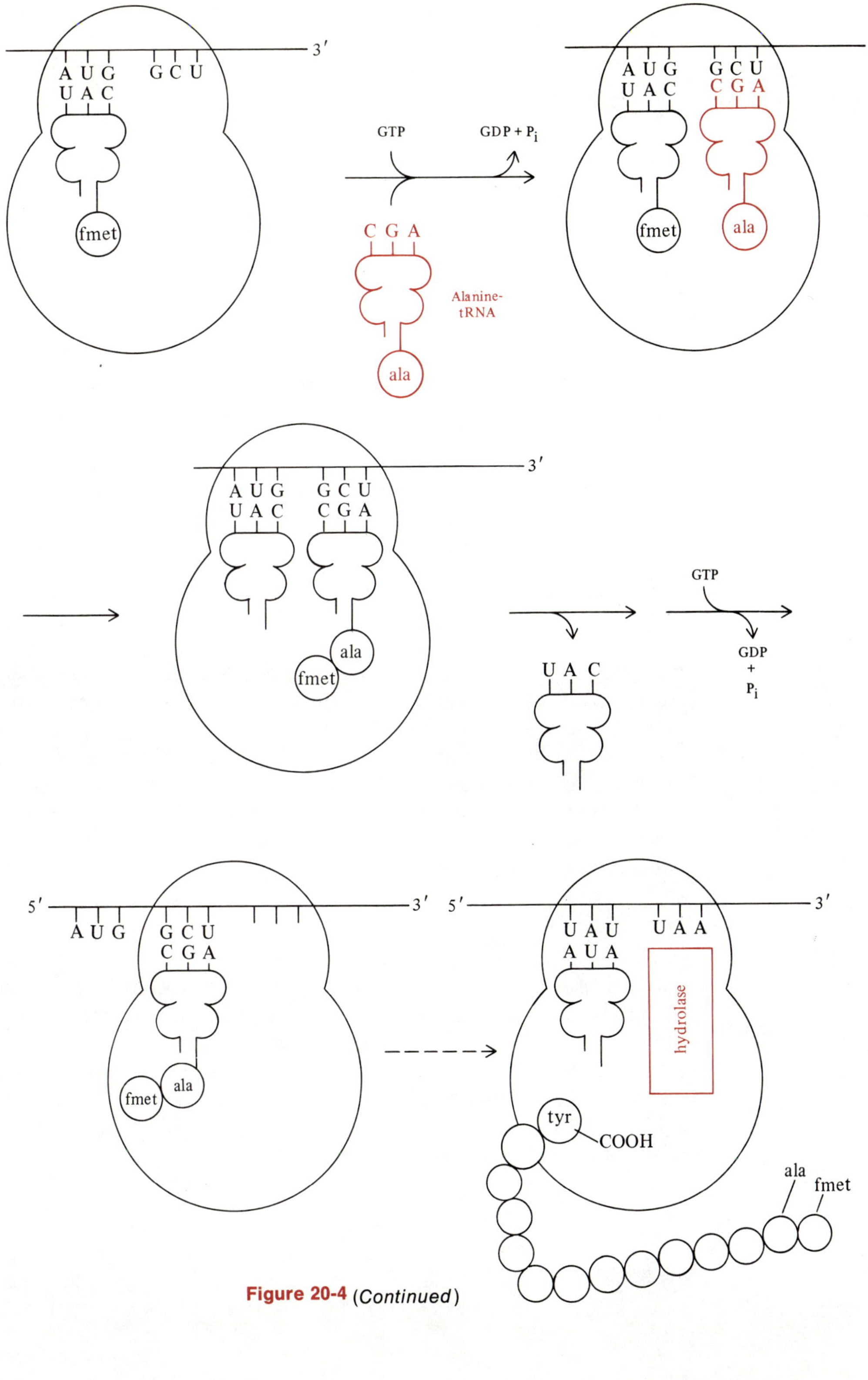

Figure 20-4 *(Continued)*

TABLE 20-1
Sterically permissible base pairs involving the third nucleotide of an mRNA codon.

Third Nucleotide of m-RNA Codon	*Corresponding Nucleotide of t-RNA Anticodon*
U	A
G	C
A or G	U
C or U	G
A, C, or U	I*

* Inosine.

Study of these phenomena led to the view that steric requirements for matching the third base of a codon are less stringent than those for the first two; that is, some "wobble" is possible. This is limited to the third base of the codon, and there are some additional steric limitations as indicated in Table 20-1. The use of the unusual purine base inosine in some tRNA anticodons appears to be a conservative device which allows for a maximum wobble in the reading of the third codon base.

Polyribosomes

Early in the studies of protein synthesis it was discovered that, in a cell undergoing rapid protein synthesis, groups of ribosomes appeared to aggregate together. These clusters were referred to as polyribosomes or polysomes, and some mechanistic importance was attached to such aggregations. It is now clear that these simply represent the fact that a messenger RNA may have more than one ribosome attached to it during the synthetic process. In fact, the number of ribosomes attached to a given molecule of mRNA will reflect the rate of protein synthesis in the cell and the length of the mRNA molecule more than any other factors.

Accessory Factors

A variety of protein factors have been isolated that are required for various phases of the protein synthesis process to proceed. Some of these are ribosomal proteins, and some are soluble proteins which may, however, normally remain fairly closely associated with the ribosome. To indicate the degree of complexity involved, the initiation process involves soluble factors IF I, II, and III and may involve any or all of the ribosomal proteins known as S 2, 12, 18, 19, 21, and 22. Elongation similarly involves factors T and G and at least ribosomal proteins L7 and L12.

Protein Synthesis in Eucaryotes

The first major difference in protein synthesis in the cytosol of eucaryotes is that methionine is used as the initiating amino acid rather than its

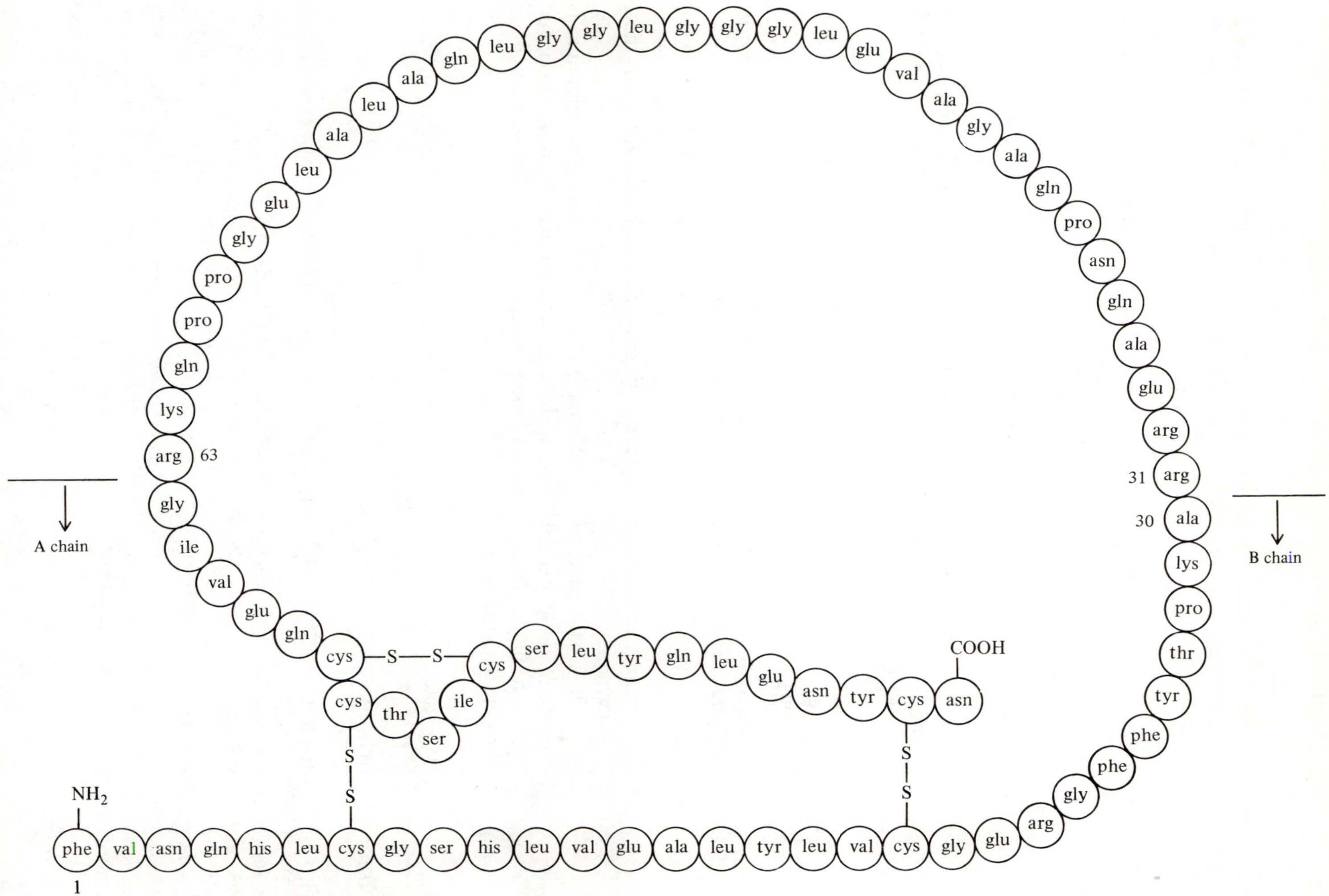

Figure 20-5
The structure of porcine proinsulin. Conversion of this molecule to active insulin requires the removal of the peptide from residue 31 through residue 63.

formyl derivative. In addition, the mRNA molecules have a longer average lifetime, and appear to be limited to carrying information for only one polypeptide chain.

There are of course differences in the number of accessory protein factors required, and in the exact nature of some of the enzymes and binding factors, but the process is otherwise remarkably similar in both types of cells.

"Grooming" the Protein Molecule

It has been clear for many years that certain proteins are synthesized in one form and then "trimmed" to their final, biologically active, form. The hormone insulin is a perfect example (Figure 20-5), as are many of the proteolytic enzymes produced by the pancreas in inactive "zymogen" forms. In recent years, it has become apparent that this is a much more general phenomenon than had been thought. Many, if not all, proteins are subject to some hydrolytic cleavage of excess amino acids or whole peptide chains. In some cases, this trimming is necessary before the protein can assume its typical "native" three-dimensional conformation.

The best illustration of this process is the simple fact that while all proteins are synthesized with methionine (or *N*-formylmethionine) as the N-terminal amino acid, very few native proteins in the cell show this same N-terminus.

It is still generally considered, however, that even this trimming is inherent in the original amino acid sequence of the genetic information, and thus the final state of the native protein requires only the specification of an amino acid sequence.

Additional Readings

Scientific American articles

O. L. Miller, Jr., "The Visualization of Genes in Action," 228(3), 34 (March 1973).

G. S. Stein, J. S. Stein and L. J. Kleinsmith, "Chromosomal Proteins and Gene Regulation," 232(2), 46 (February 1975).

H. M. Temin, "RNA-Directed DNA Synthesis," 226(1), 24 (January 1972).

A. Rich and S. H. Kim, "The Three-Dimensional Structure of Transfer RNA," 238(1), 52 (January 1978).

J. D. Watson, *Molecular Biology of the Gene*, 3rd edition (W. A. Benjamin, New York, 1976). Chapters 11, 12, and 16 are particularly related to the topics of this chapter.

H. Weisbach and S. Pestka (Eds.), *Molecular Mechanism of Protein Biosynthesis* (Academic Press, New York, 1977). A series of review articles on various aspects of genetic expression.

B. Lewin, *Gene Expression* (John Wiley, New York, 1974, 1977). This series of three volumes is a very thorough and detailed survey of the field of genetic expression and of the functions of the various nucleic acid components.

Study Questions

1. Define, or describe clearly, each of the following terms as they apply to genetic expression and related processes:

replication	mRNA
translation	rRNA
transcription	tRNA
gene	ribosome
codon	wobble
posttranscriptional modification	
reverse transcriptase	

2. Synthesis of an aminoacyl–tRNA is a two-step process involving the intermediate formation of an aminoacyl adenylate. Write equations for the two reactions involved. If you have problems, look at the reactions of fatty acid activation in Chapter 14.

3. What compound would you expect to be the source of the "formyl" group of *N*-formylmethionyl–tRNA?

4. The tRNA responsible for "initiation" of protein synthesis lacks some of the common sequences of nucleotides that are present in all other tRNA molecules as shown in Figure 20-3(b). Speculate on what characteristics the initiator tRNA must have that could be conferred on it by these differences in sequence and structure.

5. Given the reaction GDP + ATP $\rightleftharpoons$ GTP + ADP, how many molecules of ATP would have to be converted to ADP in order to supply the energy necessary for the synthesis of one molecule of porcine proinsulin (from the necessary amino acids)?

CHAPTER 21

Mechanisms of Metabolic Control

The metabolic processes of a cell are endlessly and intimately enmeshed with one another, to the point that no one reaction in a cell can take place without affecting in some way all of the others. Organization, with the attendant control of many individual reactions, is necessary if order and effective function is to be created from the chaos that would result from several thousand enzymes each carrying out its job without regard to the real needs of the cell.

Control of metabolism, of course, rapidly devolves into control of individual enzymatic reactions. If we consider the means by which a given enzymatic reaction might be affected, we find a limited list of possibilities:

1. Variation in the amount of enzyme protein that is present. This largely boils down to control of protein synthesis.
2. Variation in the effectiveness of whatever enzyme protein is present. This encompasses all of the varieties of enzyme inhibition and activation, variations in pH, temperature, cofactor concentrations, and similar effects.
3. Variations in substrate concentration. In terms of the immediate reaction, this is simply the effect of normal mass law and reaction rate. From a more cellular view, the substrate level variations may reflect selective absorption or transfer across membranes or between intracellular regions.
4. Variations in the effectiveness of the substrate molecule in its interactions with the enzyme. Many substrates exist in more than one conformation or more than one form due to ionization of various groups. Changes in pH, for example, may cause the percentage of substrate molecules that can react with the enzyme to vary greatly. This is a major contributory effect in determining the pH/activity curves of many enzymes *in vitro*, but in practical terms, within functional cells, is not a serious factor.

21.1 Control of Protein Synthesis

Control of protein synthesis is exerted almost entirely at the level of transcription. Once an mRNA molecule has been formed (and moved out of the nucleus in the case of a eucaryote), the process of synthesis is usually limited only by those factors that would affect synthesis of all proteins equally.

Much of our present knowledge of the mechanisms that control formation of the messenger RNAs has come from the study of bacterial systems. Some bacteria are capable of synthesizing certain enzymes not normally present in the organism in response to the presence in the growth medium of specific related compounds. This is usually referred to as **induction** of enzyme synthesis, and the result is an **induced** enzyme as distinguished from the **constitutive** enzyme that is always present in another organism. In the converse case, when production of an enzyme is decreased or prevented in the presence of a specific compound, the enzyme is said to be **repressed**.

As a result of extensive study of induction and repression of enzymes in *E. coli*, Jacob and Monod in 1961 proposed a model for control of protein synthesis which is shown in a modernized version in Figure 21-1.

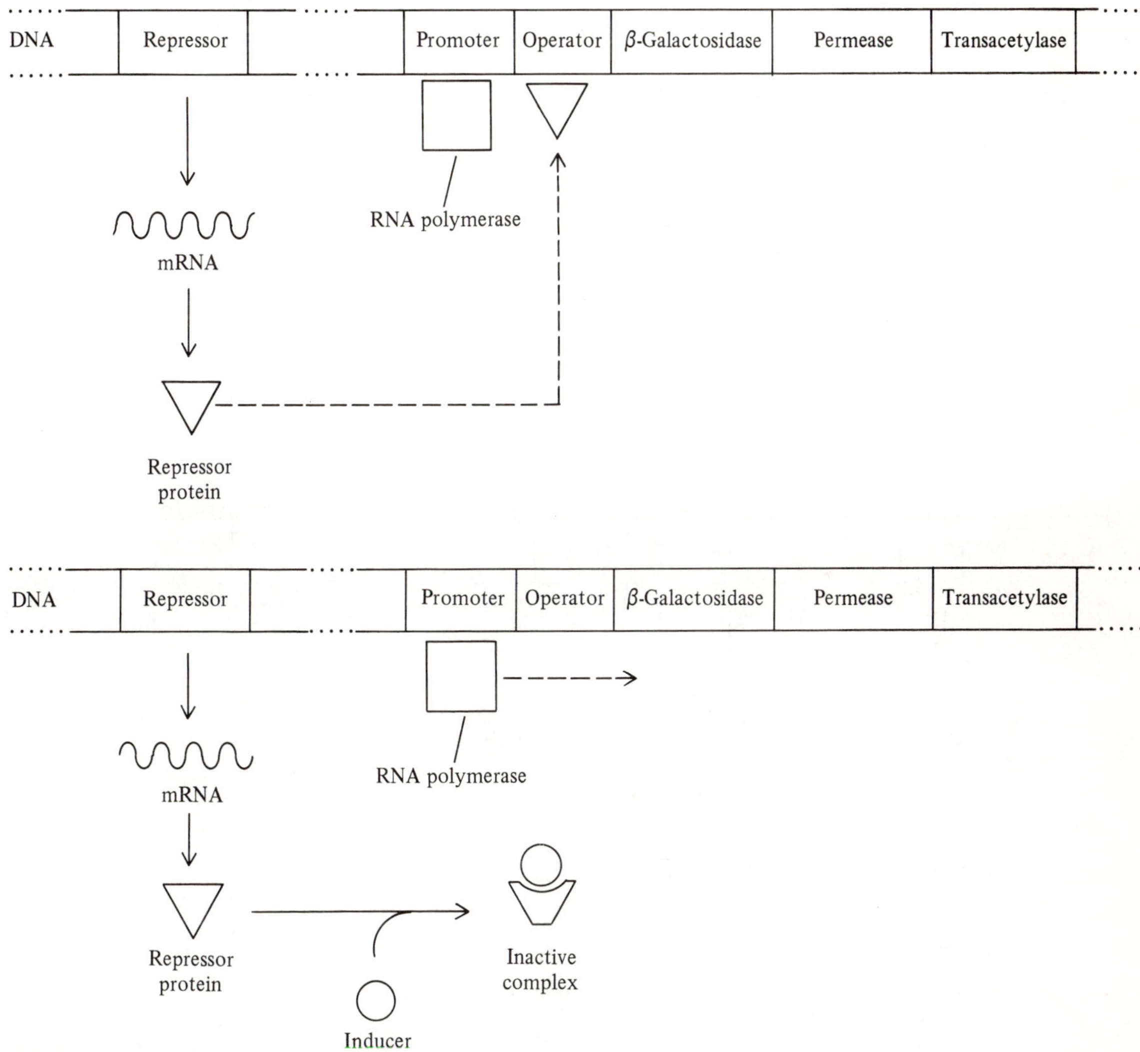

Figure 21-1
The Jacob and Monod model for control of transcription of the lac operon of *E. coli*. The top diagram represents the normal, repressed state; no transcription occurs, and no enzymes are formed. The lower drawing shows the effect of adding a compound capable of "inducing" formation of the enzymes of the lac operon.

This model deals with a section of the genome of *E. coli* that carries the information for synthesis of a group of enzymes, all related to the metabolism of lactose. It is generally referred to as the **lac operon** region of the DNA molecule, an operon being loosely defined as one or more genes coding for functional proteins, along with their associated control regions. For all practical purposes, the terms "gene" and "cistron" are used interchangeably; either refers to a segment of DNA carrying the information necessary for the production of one polypeptide chain. The lac operon can, and is, thus said to be "polycistronic."

Function of the **promoter** segment of the DNA is necessary for the attachment of the RNA polymerase. Consequent transcription of the functional genes into a molecule of mRNA requires also that the **operator** segment

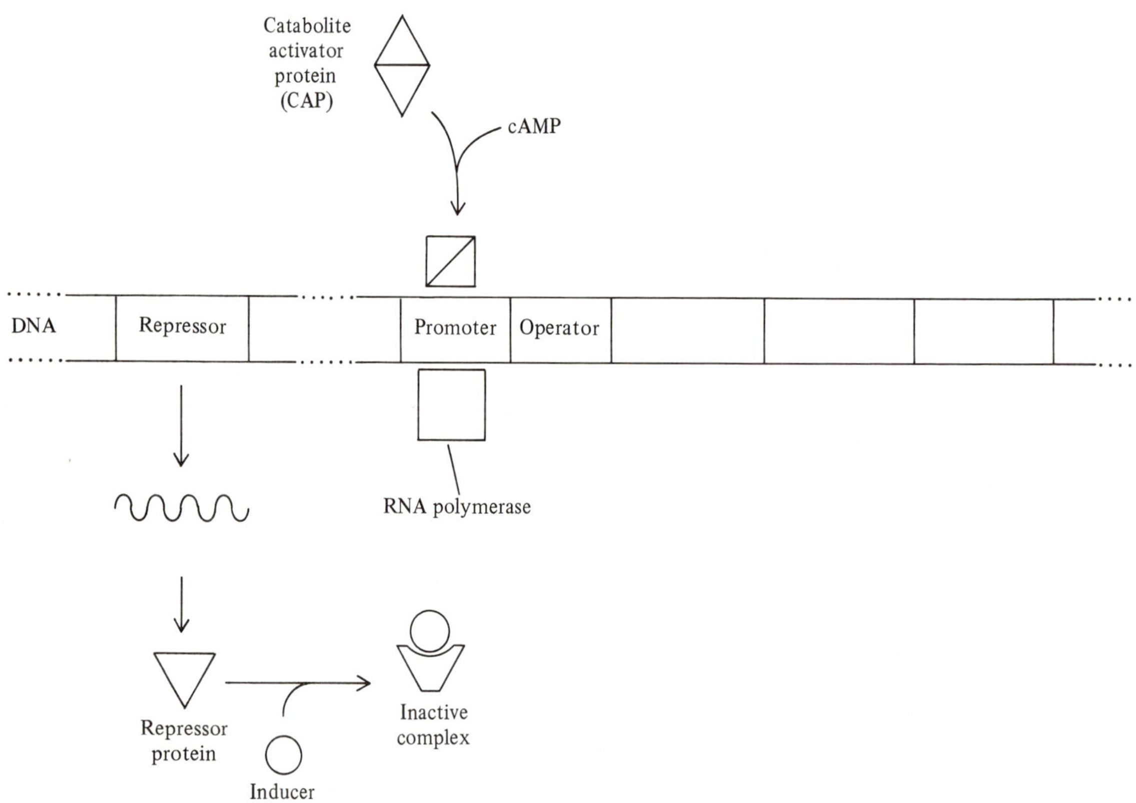

Figure 21-2
The effect of cAMP on transcription. **Cyclic**-AMP causes a conformational change in the CAP protein, which is then able to bind to the promoter region. Binding of CAP facilitates the attachment of RNA polymerase and thus encourages the transcription. This mechanism is believed to work for many operons, whether inducible or constitutive. **Cyclic**-AMP has no effect on a fully repressed operon.

is available for transcription. In *E. coli* at least, the promoter, operator, and the three enzyme genes are located in tight sequence in the DNA; the **regulatory** gene is located some distance away. The regulatory gene is responsible for production of a special molecule of mRNA that codes for the production of a **repressor** protein. This protein is generally considered to be capable of at least two allosteric conformations. In one of these conformations, it binds to the operator area and prevents the DNA polymerase from transcribing the structural genes. In the other allosteric form, it will not bind to the operator.

The function of the **inducer** molecule is to bind to the repressor and convert it to the nonbinding form, thus opening up the operon for production of the polycistronic RNA coding for the three enzymes. Similar systems have been shown to exist in which the production of a single enzyme is controlled in exactly the same fashion.

We will deal later in this chapter with some of the activities of 3′,5′-**cyclic**-AMP (cAMP) as an intermediate in the regulation of metabolism. For the moment, it is sufficient to mention that cAMP speeds up transcription of the lac operon in an otherwise derepressed situation by speeding up attachment of the RNA polymerase (Figure 21-2). It accomplishes this by reacting with a **cAMP receptor protein** (CAP), which dimerizes, attaches to the promoter region, and speeds the attachment of the RNA polymerase.

21.2 Control of Transcription in Eucaryotic Cells

Mechanisms of control of the type described above for *E. coli* have not been demonstrated for eucaryotic systems. Generally, the needs for specific enzymes do not fluctuate as rapidly in eucaryotes as in bacteria. Many of the mRNA molecules in eucaryotes have relatively long lives in recognition of this more stable need for cellular proteins. However, the formation of mRNA in all of these organisms is certainly controlled, although the exact mechanisms involved are not clear.

In the chromosomes of the eucaryotic cells, DNA is tightly associated with small, highly basic protein molecules called **histones** and folded into very compact chromosome structures. The DNA in this form of nucleoprotein complex is not available for reaction with an RNA polymerase. Before any transcription can take place, the chromosome must be unfolded, and the double-strand DNA opened up to allow binding of the polymerase and subsequent synthesis of the mRNA. While electron micrographs have for years shown swollen and enlarged areas of chromosomes caught in the process of transcription ("lamp-brush" chromosomes), the exact process by which this takes place and the factors that control it are still under intensive study.

This question of the control of the expression of the genetic complement of a cell is of course central to the question of cellular differentiation in a multicellular organism. Each cell of a human being, for example, contains exactly the same DNA. What makes a liver cell differ from a brain cell, a muscle cell, and so on, is simply the matter of which of the genes in the chromosomes are activated and used for protein synthesis and which are held permanently in an inactive form. It appears at the present time that we can look forward to having much more information about this control process available in the next few years.

21.3 Glycogen Synthesis and Utilization: A Casebook of Metabolic Control

We saw in Chapter 12 the reactions by which glycogen is synthesized and utilized in a typical animal cell. This system is affected by so many different types of control mechanisms that it provides an excellent example of the manner in which a variety of conditions within the cell (or even outside of it) are able to exert influence on the question of whether glycogen is to be synthesized, degraded, or left alone at any given time. The basic system involved is shown again in Figure 21-3.

Substrate Concentration Effects

We have commented that the most basic control mechanism is that which results from simple mass law effects due to variations in the levels

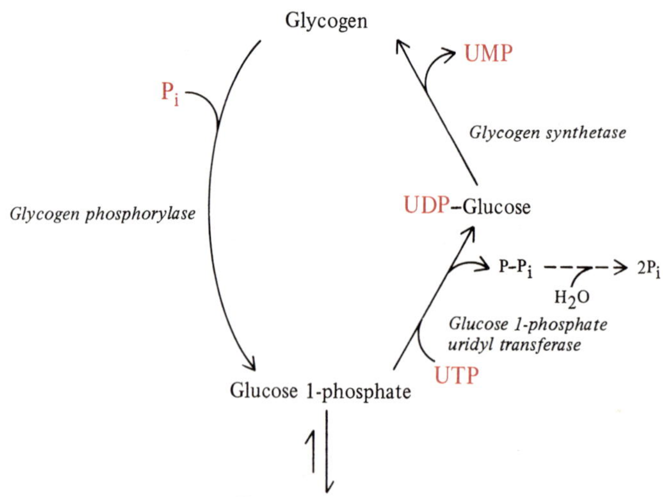

Figure 21-3
A simplified summary of glycogen synthesis and utilization.

of reactants and products. Glycogen represents in most animal cells the first-line reserve source of material to be oxidized for the production of energy. At any given time, the immediate energy state of the cell will be determined by the relative concentration of ATP as compared to ADP, AMP, and inorganic phosphate. The cell has a high content of available energy when ATP is high and the other compounds in low concentration. Conversely, use of ATP will generate ADP, AMP, and inorganic phosphate and will result in a low energy state in the cell.

Since UTP must be regenerated from UDP through the transfer of a phosphate group from ATP, synthesis of UDP–glucose will be encouraged in the presence of high concentrations of ATP and severely impaired as these concentrations fall off. On the other hand, the phosphorylase reaction requires inorganic phosphate as a reactant. When the ATP level is high, inorganic phosphate will be scarce, and the phosphorylase reaction will slow down. However, as the energy stores of the cell are used up (that is, as ATP is hydrolyzed), inorganic phosphate is produced and glucose is released from glycogen to begin its journey to CO_2, water, and useful energy.

Allosteric Control

The two key enzymes in this process are glycogen phosphorylase and glycogen synthetase. Both are subject to allosteric effects, again reflecting controls based on the energy state of the cell. The effectors known to act on these enzymes, and the nature of their actions, are shown in Figure 21-4.

When muscle activity is stimulated by nervous excitation, energy is used and one of the results is the accumulation of AMP. Addition of AMP

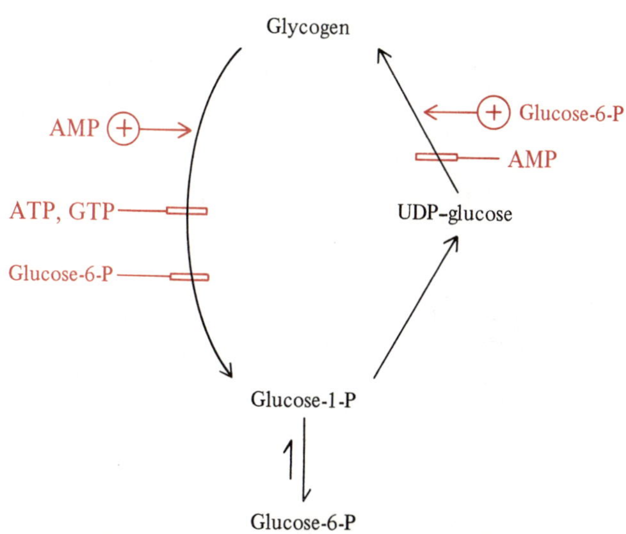

Figure 21-4
Allosteric control of glycogen synthesis and utilization. Note that only the inactive forms of phosphorylase and of glycogen synthetase are subject to activation by AMP or glucose-6-phosphate, respectively. The active phosphorylase *a*, for example, will not respond to AMP.

to inactive phosphorylase *b* converts it to an active form, thus encouraging the breakdown of glycogen.

Hormonal Control through Covalent Modification of Enzymes

Hormones are by their nature agents of communication between the units of a multicellular organism. Different hormones are capable of expressing the necessary message to cells in different ways. We will deal here with only one case—the ability of the hormone **epinephrine** (adrenaline) to promote glycogen utilization through a complex series of reactions that ends in covalent modification of the key enzymes involved. This modification can either inactivate a functional enzyme or make an active enzyme unit out of an inactive form.

Formation of Cyclic-AMP. The first step in this process occurs following the binding of the epinephrine molecule to a suitable acceptor on the surface of the target cell. This binding activates an enzyme on the inner surface of the membrane that carries out the following reaction:

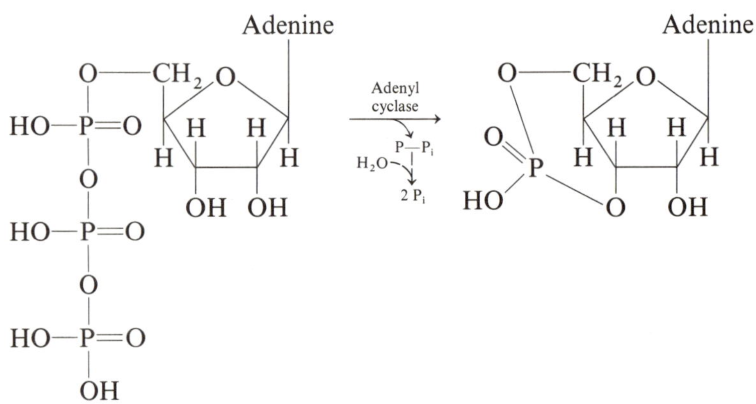

ATP

Adenosine-3′,5′-*cyclic*-phosphate (cAMP)

The cAMP thus formed appears to have many functions in different areas of metabolism in order to bring about the physiological responses that are normally produced as the result of release of epinephrine. Because this molecule carries the message inside the cell that the hormone carries outside, this is sometimes called a "second messenger" system for expression of hormonal control. We can best deal with the role of cAMP in this case with the use of the diagram in Figure 21-5.

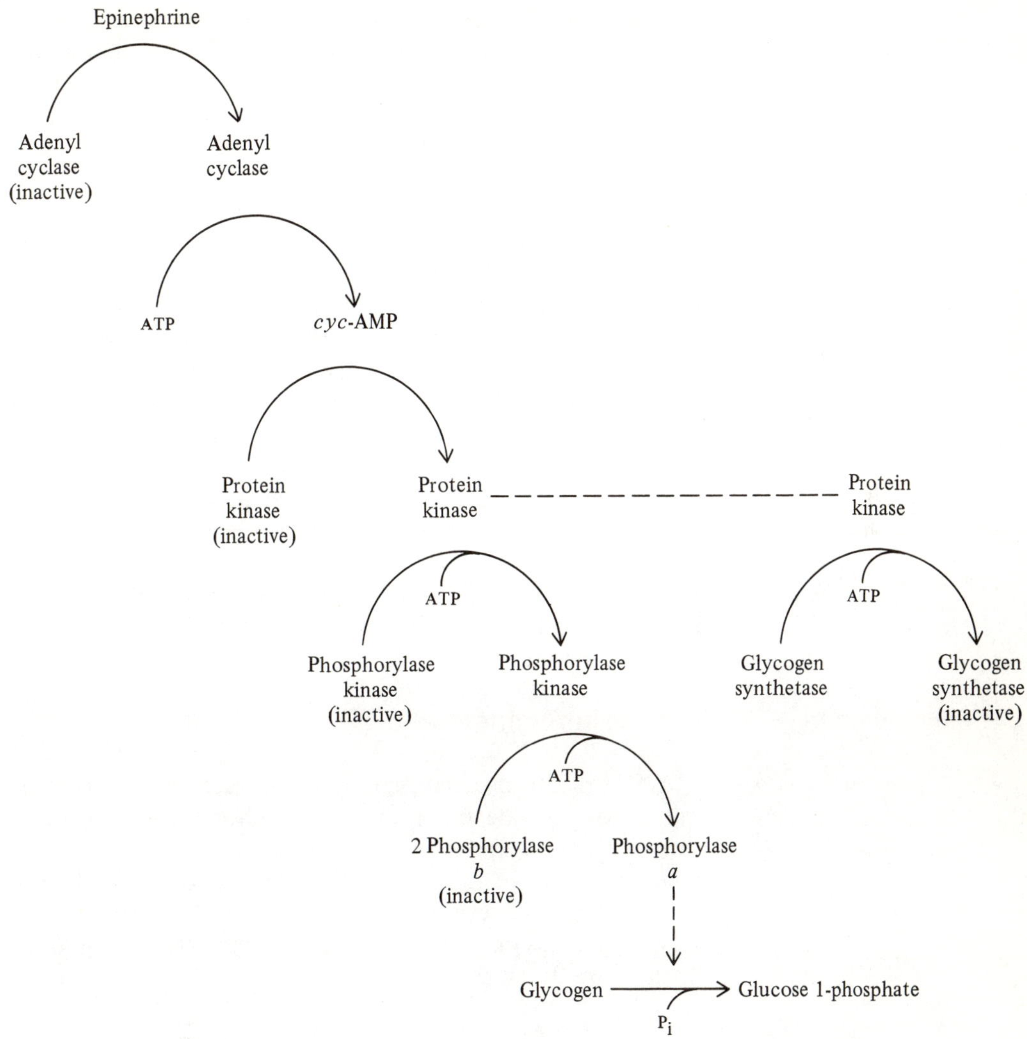

Figure 21-5

An example of hormonal control through covalent modification of enzymes. In this case, cAMP serves to carry the hormone "message" from the cell membrane to the enzymes concerned with glycogenolysis and glycogen synthesis.

Remember that in the common nomenclature of enzymes, the term "kinase" is used for any enzyme that transfers a phosphate containing group from ATP or one of the other nucleoside triphosphates. Thus, we can refer to an enzyme that transfers a phosphate group from ATP to a protein as a "protein kinase."

The Cascade Effect. At each step in the sequence of reactions shown, there is an amplification effect, since the product of one reaction serves to activate the enzyme for the next reaction. The net result of this process is that one molecule of epinephrine or similar hormone can cause a substantial number of molecules of phosphorylase *b* to be converted to the active phosphorylase *a* form. This has been called a "cascade" effect, in which a very small original action results in an increasing effect as the process continues. This is of course the type of response one gets in an avalanche, where a wrong step may result in the shifting of many tons of snow.

Note that the final protein kinase activated by this sequence of reactions has two effects. First, the phosphorylase is made more active by conversion from the inactive *b* form to phosphorylase *a*. At the same time, the glycogen synthetase is inactivated by exactly the same process that activates phosphorylase. In this case, addition of a phosphate group to the synthetase renders it inactive.

Protein Synthesis

While the mechanism remains obscure, effects of environmental temperature on the rate of production of phosphorylase protein have been known for many years. It has long been a standard procedure when isolating phosphorylase from rabbit muscle to expose the animal to temperatures close to freezing for several days. Under these circumstances, the total phosphorylase content of the muscle tissue will increase at least fourfold over the content at room temperature.

21.4 "Long-Range" Control of Metabolic Pathways

When a specific compound is synthesized by an organized sequence of reactions, there is obviously some advantage to having whatever mechanism controls the synthesis of that compound work at an early stage in the synthesis. In this way, the production of unnecessary (and perhaps harmful) intermediates can be avoided. This kind of control system has been called **feedback control** by comparison with the systems in electronic circuits that allow the output of a system to control the original input. Feedback control often

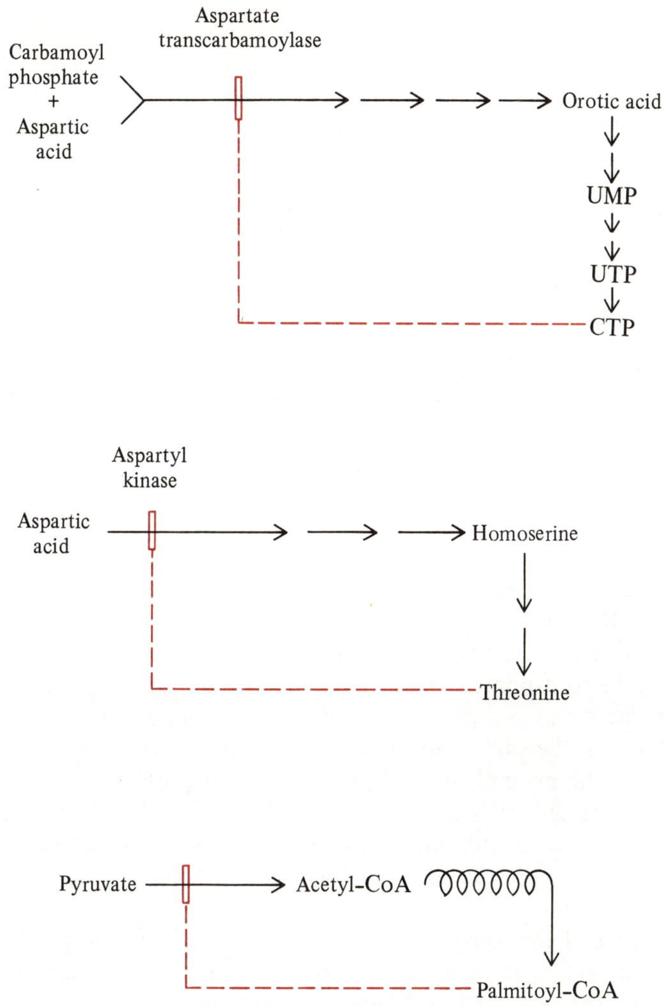

Figure 21-6
Examples of feedback inhibition of metabolic pathways.

makes use of allosteric mechanisms of inhibition or activation, since this permits an end-product to control a reaction involving substrates that bear little structural relationship to the end-products of the process. Several examples of feedback control in different areas of metabolism are shown in Figure 21-6.

When biosynthetic pathways split to produce several end-products, the control mechanisms involved can become quite complex and include both long-range control by feedback inhibition, and short-range control by inhibition of an enzyme close to the branch point. A case of this type is shown in Figure 21-7.

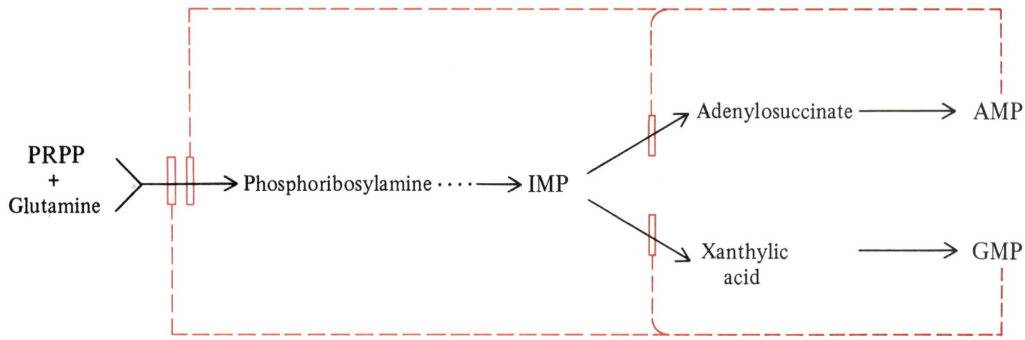

Figure 21-7
Control of a branching biosynthetic pathway. Both long-range and short-range feedback inhibition are involved in controlling biosynthesis of the various purine nucleotides.

21.5 Control of Membrane Permeability

A cell can only metabolize those compounds that can be transported across the cell membrane. Although the mechanisms are far from clear, it appears that several quite different types of hormones exert their effects by controlling the transport mechanisms by which specific compounds get into cells. The clearest case is that of insulin. This hormone is bound to specific receptor sites on the surface of muscle cells, and this binding appears to be necessary for the normal functioning of the transport systems that move glucose into the cell. It is interesting that this control is effective only for glucose, the normal circulating carbohydrate of blood. Fructose, which is not normally present in the peripheral blood, is absorbed equally well in the absence of insulin.

A number of the steroid hormones can similarly be shown to bind to membrane proteins of target cells. Just which transport phenomena are affected in these cases remains difficult to determine.

21.6 Control of Glycolysis and Related Metabolism

It is very useful to look at the multiple types and mechanisms of control that affect a total process, such as the breakdown of carbohydrate to produce energy, even though the diagrams become cluttered and rather hard to follow. Figure 21-8 attempts to show many of the relationships involved in control of glycolysis and some closely related reactions.

You will note that it is not necessary to control every reaction separately. Certain enzymes play the role of control enzymes, usually complex protein

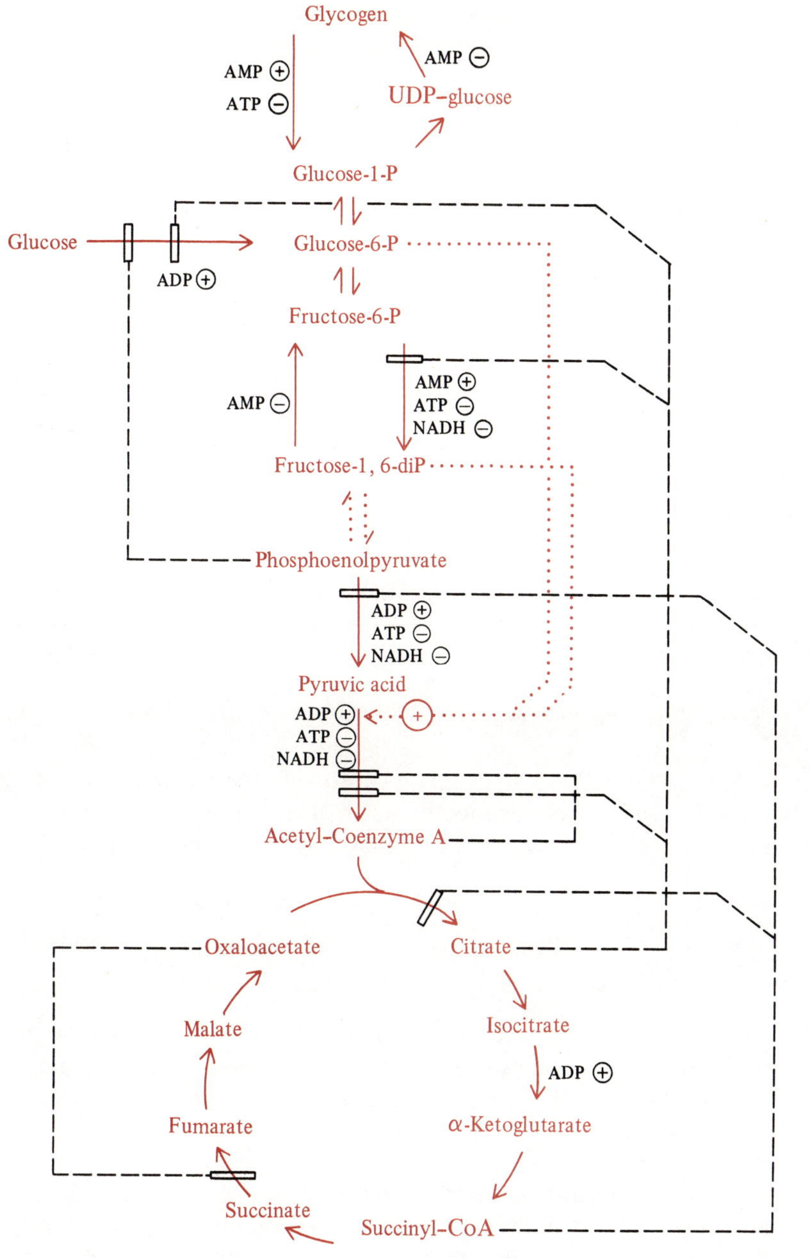

Figure 21-8
Control of glycolysis and related reactions. Shown are most of the feedback and feedforward controls involved in glycolysis and related areas of metabolism. A ⊕ symbol indicates a positive effector, and a ⊖ symbol an inhibitor.

systems showing control by some variety of the mechanisms we have discussed above. Many of the effects exerted by products of the pathway are allosteric in nature, as are most of the controls exerted by ATP and its hydrolysis products. Other compounds exert more classical types of inhibition, and of course carbohydrate metabolism is subject to multiple hormonal effects.

Additional Readings

Scientific American articles

J-P. Changeux, "The Control of Biochemical Reactions," 212(4), 36 (April 1965).

T. Maniatis and M. Ptashne, "A DNA Operator–Repressor System," 234(1), 64 (January 1976).

D. E. Koshland, Jr., "Protein Shape and Biological Control," 229(4), 52 (October 1973).

I. Pastan, "Cyclic AMP," 227(2), 97 (August 1972).

J. A. Nathanson and P. Greengard, "'Second Messengers' In the Brain," 237(2), 108 (August 1977).

D. E. Atkinson, *Cellular Energy Metabolism and Its Regulation* (Academic Press, New York, 1977). A short book exploring the mechanisms of control of energy metabolism, particularly based on the concept of the central role of the relative concentrations of adenylate nucleotides.

J. D. Sink, (Ed.), *The Control of Metabolism* (Pennsylvania State University Press, 1974). A series of chapters on various topics related to metabolic control; a considerable variation in the depth of the presentation.

E. Kun and S. Grisolia (Eds.), *Biochemical Regulatory Mechanisms in Eukaryotic Cells* (Wiley-Interscience New York, 1972). A series of very advanced articles.

Study Questions

1. In the Jacob and Monod model for control of gene expression in procaryotic cells, what is the function of (a) the repressor gene; (b) the operator gene; (c) the promoter gene; (d) an inducer molecule; (e) *cyclic*-AMP?

2. Phosphorylase *b* is an inactive form of this enzyme. What are two different ways in which this form of the enzyme can be rendered catalytically active?

3. Explain why allosteric control is a particularly suitable form of control for (a) long-range effects (feedback control) in metabolic pathways, and (b) control based on the energy state of the cell (as in glycogen synthesis and utilization).

4. We have mentioned that certain enzymes tend to act as control points and may experience several types of inhibition or activation. From the data of Figure 21-8, list those enzymes that you feel could be called "control enzymes" of carbohydrate metabolism.

5. Give one specific example of a reaction that could be described as "covalent modification of an enzyme."

GLOSSARY

A

Activation energy: The energy necessary to raise reactant molecules to the energy level of the transition state for the reaction.

Active site: That portion of an enzyme molecule that is responsible for binding substrate(s) and for promoting the catalytic process.

Active transport: Transfer of a solute through a membrane by a mechanism that requires energy expenditure; it permits concentration of a solute on one side of a membrane.

Allosteric enzymes: Oligomeric enzymes which are capable of existing in active and inactive conformations; binding of compounds at special sites remote from the active site may stabilize either of the conformations (see "effectors").

Amino acid activation: The ATP-requiring process of attaching an amino acid to the appropriate tRNA prior to use in protein synthesis.

Anabolism: Those sequences of metabolic reactions that can be considered to lead to the biosynthesis of cellular constituents rather than their breakdown.

Anaerobe: An organism that is capable of living in the absence of oxygen.

Anomers: Isomers of a single sugar that arise from the chiral nature of the glycosidic carbon atom when in a ring form (for example, α-D-glucopyranose and β-D-glucopyranose).

B

Binding sites: Those regions or specific side-chain groups within the active site of an enzyme that are responsible for binding the substrate(s).

Buffer: A solution which will tend to resist changes in pH. For maximum effectiveness, requires approximately equal concentrations of a weak acid and one of its salts (or a weak base and one of its salts).

C

Calorie: A unit of energy used primarily to measure heat transfers. One calorie is the amount of heat needed to raise 1 gram of water from 14.5 to 15.5°C. (One calorie = 4.185 joules.)

Carbohydrates: Polyhydroxyaldehydes and polyhydroxyketones and their derivatives.

Catabolism: Those reactions or pathways of metabolism that bring about the breakdown of nutrients (and in the process provide energy to the cell).

Catalyst: A substance that is capable of increasing the rate of a reaction but at the end of the reaction remains itself unchanged.

Catalytic site: The region (or the specific side-chain groups) within the active site of an enzyme that is actually involved in catalysis of the reaction.

Chemiosmotic coupling hypothesis: A theory of oxidative phosphorylation that suggests that mitochondrial electron transport results in vectorial (one-way) transport of protons across the inner membrane and the establishment of a membrane potential. These conditions provide the energy for formation of ATP and other energy-requiring membrane processes.

Chiral carbon atom: A carbon atom having four different atoms or groups of atoms attached to it.

Chlorophylls: Magnesium-containing porphyrin compounds that serve as the primary light absorbers in photosynthesis.

Chloroplasts: Highly organized special organelles of the cells of photosynthetic organisms. They are responsible for using light energy to form ATP from ADP and to reduce $NADP^+$ to NADPH.

Chromatography: A general term applied to a number of techniques which permit separation of closely related compounds on the basis of very small differences in their properties. The methods usually involve repeated partition of the compounds between some stationary phase and some moving solvent (or gas).

Codon: The sequence of three nucleotides in a nucleic acid that is required to specify one amino acid in a protein.

Coenzyme: An organic molecule required as part of the active enzyme for certain reactions. Most coenzymes are formed in part from various water-soluble (B-group) vitamins.

Cofactor: Any atom or molecule that is required in addition to the polypetide chain for activity of an enzyme.

Configuration and conformation: These terms tend to be used in a rather narrow sense. Configuration is used to refer to a particular structure—for example, to distinguish D- from L-glyceraldehyde. Conformation, on the other hand, refers to the actual three-dimensional form of a molecule at one particular time. As an example, the free aldehyde form of D-glucose has only one configuration, but many possible conformations.

β-Conformation: One possible conformation of a polypeptide chain often found in proteins. The chain is in a linear "stretched" form.

Cytochromes: Proteins carrying a heme-type molecule which serve as intermediates in the mitochondrial and photosynthetic electron transport processes.

Cytosol: The "soluble" portion of the interior of a cell, excluding the various organelles and other membranous structures.

D

Dark reactions: Those reactions which bring about the synthesis of glucose from CO_2 and H_2O using the ATP and NADPH formed by the "light reactions" of photosynthesis.

Denaturation: Any change in the three-dimensional structure of a native protein.

Digestion: Reduction of food components, by enzymatic hydrolysis, to simpler forms that can be absorbed into cells.

DPN^+ (diphosphopyridine nucleotide): An older name for NAD^+ (nicotinamide-adenine-dinucleotide).

E

Effector molecule: A molecule capable of binding to an "effector site" on an allosteric enzyme and either increasing the enzyme activity (positive effector) or decreasing it (negative effector).

Electron transport: The process by which electrons are transferred from substrate to oxygen, involving a series of intermediates (and permitting recovery of much of the energy involved in the form of ATP).

Electrophoresis: Any molecule which carries a net electrical charge will tend to move when placed (in solution) in an electrical field. Electrophoresis in gels is commonly used to separate and identify proteins.

Endergonic reaction: Any reaction for which the standard free energy change ($\Delta G°$) is positive; the equilibrium will lie in the direction of reactants.

Endonuclease: An enzyme which catalyzes hydrolysis of bonds between nucleotides somewhere within a nucleic acid chain rather than at the end of the chain.

Endoplasmic reticulum: An elaborate membranous structure within eucaryotic cells; most ribosomes are attached to this structure.

Entropy change: The amount of energy involved in a reaction that goes into changing the degree of organization of the system. This energy is not available to do work.

Epimers: When two compounds which are stereo-isomers of one another share two or more chiral centers but differ only in the configuration about one of those centers, they are said to be epimers of one another.

Essential amino acids: For any organism, those amino acids that are needed for normal operation but which cannot be synthesized by that organism must be supplied in the diet.

Eucaryotic cells: The type of cells typical of higher organisms; these have a nucleus, with surrounding membrane, which contains the DNA in the form of chromosomes, and a variety of organelles including mitochondria.

Exergonic reaction: A reaction for which the ΔG° is negative; the equilibrium constant will be greater than 1.

Exonuclease: An enzyme that removes nucleotides one at a time from the end of a polynucleotide chain.

F

Fatty acid activation: The ATP-dependent conversion of a free fatty acid to the fatty acyl-Coenzyme A thioester.

Feedback control: A mechanism of metabolic control in which the end product of a sequence of reactions is used as an effector to control an allosteric enzyme at the start of the sequence.

Fermentation: A set of metabolic reactions by means of which an anaerobe can obtain sufficient energy for survival.

Fischer projection: One means of representing the three-dimensional configurations of sugars and amino acids on a sheet of paper.

Flavoprotein enzyme: An enzyme using as a coenzyme one of the nucleotide derivatives of riboflavin (FMN or FAD).

Free energy change of a reaction: Measures the amount of the total energy change that would be available to do work.

G

Galactosemia: A hereditary malfunction of galactose metabolism capable of producing blindness and mental retardation if not identified at birth.

Gene: An amount of genetic information (DNA) sufficient to specify the amino acid sequence of one polypeptide.

Gluconeogenesis: The synthesis of glucose (and other carbohydrates) from noncarbohydrate sources such as amino acids or lipids.

Glycolysis: The metabolic pathway leading from glucose to 2 moles of pyruvic acid.

Glycoside: A full acetal or ketal resulting from the reaction of a ring form of a sugar with some alcohol.

H

Heme: A porphyrin ring with iron as the central chelated metal ion.

Hydrolysis: Splitting of one compound into two by rupture of a bond and insertion of the elements of water. Addition of water to a double bond is *not* hydrolysis.

Hydrophobic interactions: The tendency of nonpolar molecules or groups to be forced into close proximity in aqueous solutions in order to permit maximum water molecule interactions (and thus the lowest possible energy state).

I

Induction: Synthesis of an enzyme occurring only in the presence of its substrate or a closely related compound.

Inhibitor: A compound which reduces the effectiveness of an enzyme.

Isoelectric point: The pH at which a particular compound will not move in an electrical field; that is, the pH at which it carries no net charge.

Isoenzymes: Multiple forms of an enzyme occurring within a simple organism but differing in kinetic parameters and/or structure.

J

Joule: The SI unit for energy; equal to $1\,\text{kg}\cdot\text{m}\cdot\text{sec}^{-2}$; also, 4.185 joules = 1 calorie.

K

Ketone bodies: Under some circumstances, acetoacetate, β-hydroxybutyrate, and to some extent acetone, can accumulate in human tissues and blood. These compounds are often referred to as "ketone bodies."

L

Lactose intolerance: A limited capacity to digest lactose due to lack of adequate levels of digestive enzyme; very common in some parts of the world. Causes digestive upsets when milk is consumed in quantity.

Light reactions: Those reactions of photosynthesis that are dependent upon light and during which the light energy is used to form ATP from ADP and P_i and to reduce $NADP^+$ to NADPH.

Lineweaver-Burk plot: Inversion of the Michaelis-Menten equation yields a form from which a plot of $1/v$ vs. $1/[S]$ produces a straight line. This type of plot provides a means of obtaining K_m and V_{max}, as well as a means of identifying competitive and noncompetitive inhibitors.

M

Metabolism: The sum total of all of the reactions occurring within a cell.

Michaelis-Menten equation: An equation that expresses the rate of an enzyme-catalyzed reaction as a function of the substrate concentration and the enzyme-specific Michaelis constant.

Mitochondria: Highly structured organelles of eucaryotic cells which are responsible for much of the respiratory activity of the cells.

Monosaccharide: A simple sugar molecule; a carbohydrate that cannot be hydrolyzed into smaller compounds.

Mucopolysaccharides: A group of polysaccharides characterized by a high content of amino sugars and uronic acid derivatives.

Mutation: Any change in the genetic inheritance of an organism.

N

Nonheme-iron proteins: A general term applied to several kinds of protein molecules that contain iron atoms but not as part of heme rings.

Nucleoside: A compound formed from a nitrogen heterocycle and a sugar, linked together in the form of a nitrogen glycoside.

Nucleotide: A nucleoside which has a phosphate ester attached at one of the hydroxyl groups of the sugar portion.

O

Oligomeric proteins: Proteins composed of more than one polypeptide chain.

Organelles: Structures found within cells that are separated from the remainder of the cell by one or more membranes.

Oxidation: Loss of electrons (or, in a formal sense, partial loss of control of electrons).

Oxidative decarboxylation: A process in which an α-keto acid undergoes loss of its carboxyl group as CO_2 concomitant with oxidation of the remaining carbonyl group to the level of an acid.

Oxidative phosphorylation: Conversion of ADP and P_i to ATP by a process energetically coupled to the electron transport processes of mitochondria.

P

Peptide: A compound formed by the joining together of two or more amino acids through amide bonds.

Peptide bond: The rather special amide bond that results from the reaction of the carboxyl group of one amino acid with the amine group of another.

pH: The negative logarithm (base 10) of the hydrogen ion concentration in an aqueous solution.

Phosphorolysis: A reaction analogous to hydrolysis but involving a molecule of phosphoric acid rather than water.

Photophosphorylation: Conversion of ADP and P_i to ATP in a process closely linked to energy production during the electron transport processes involved in the light reactions of photosynthesis.

pK_a: The negative logarithm (base 10) of an acid-ionization constant.

Polyisoprenoid compounds: Compounds that are formed by the polymerization of 5-carbon branched chain compounds.

Polypeptide: A compound formed by the polymerization of "many" amino acids held together by peptide bonds.

Polysaccharides: Polymeric molecules consisting of "many" simple sugar units joined by glycosidic bonds.

Primary structure: The primary structure of a protein consists of the amino acid sequence of the polypeptide chain(s).

Procaryotic cells: Cells of lower (unicellular) organisms; they lack nuclear membranes and defined organelles.

Proteolytic enzymes: Those enzymes catalyzing hydrolysis of a polypeptide chain.

Q

Quaternary structure: Those elements of the structure of an oligomeric protein that arise from the association of the various polypeptide chains.

R

Reducing sugar: A sugar that in solution is present in part at least in the form of the aldehyde or α-hydroxy ketone.

Reduction: The gain of electrons.

Regulatory enzyme: An enzyme that is capable of regulating a whole pathway of metabolic reactions. These enzymes are usually allosteric enzymes, and occur early in the metabolic sequence to be controlled.

Replication: Duplication of the DNA component of a cell prior to cell division.

Respiratory chain: Another name for the mitochondrial electron transport system of oxidative cells.

Repression: Prevention of synthesis of a protein molecule for which the cell contains the necessary genetic information.

Ribosomes: Sub-cellular particles primarily responsible for protein synthesis; they are composed almost entirely of proteins and nucleic acid (mRNA) molecules.

S

Secondary structure: Most simply, this is a question of whether a given segment of a polypeptide chain is in an extended conformation (such as the β-conformation), an α-helix, or a "random" configuration.

Sequence (of a polypeptide or polynucleotide chain): The order in which individual monomeric units occur in the chain.

Steady-state systems: A system that takes in reactants at the same rate that it excretes products; all of the intermediates will remain at constant concentrations, but *not* at equilibrium concentrations.

Substrate: A compound that is a reactant in an enzyme-catalyzed reaction.

Substrate-level phosphorylations: Formation of a nucleoside triphosphate such as ATP from the diphosphate by transfer of a phosphate group from a metabolic intermediate.

T

Tertiary structure: The three-dimensional structure of a globular protein produced by folding of segments of the polypeptide chain.

Transcription: Production of an RNA molecule that is a complimentary copy of a segment of one chain of DNA.

Translation: Synthesis of a specific polypeptide chain from the coded data contained in a messenger RNA molecule.

Turnover number: The maximum number of substrate molecules reacting per molecule of enzyme in one minute of reaction.

U

Uncoupling agents: Compounds which are capable of separating the processes of electron transport and oxidative phosphorylation; electron transport continues, but no phosphorylation occurs.

V

Vitamins: Small organic molecules required in the diet of a specific organism in order for normal functions to be maintained.

W

Wobble hypothesis: A theory attempting to explain the ability of several codons to specify a single amino acid.

Z

Zwitterion form (of an amino acid): A form of an amino acid in which both the carboxyl and amino groups are charged at the same time.

Zymogen (or proenzyme): An enzymatically inactive protein which can be converted to an active enzyme (for example, trypsinogen to trypsin).

ANSWERS TO STUDY QUESTIONS

Chapter 2

1. (a) Only one form is possible

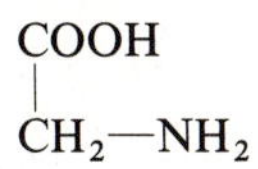

(b)

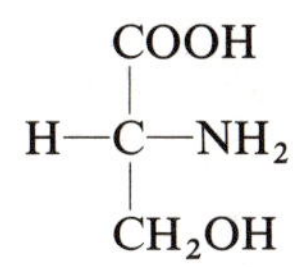

D-serine

(c)

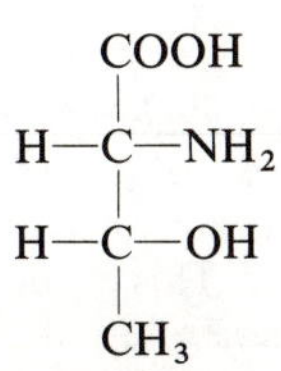

D-threonine

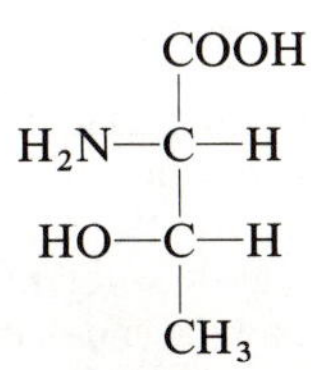

L-threonine

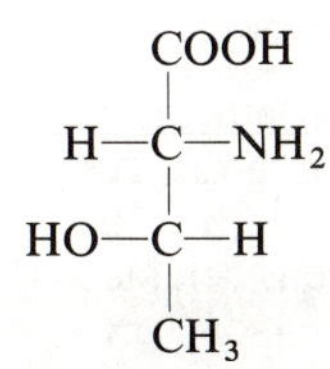

D-*allo*-threonine

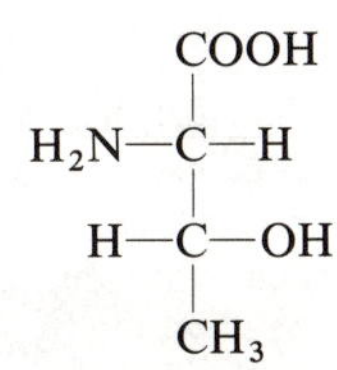

L-*allo*-threonine

2. (a)

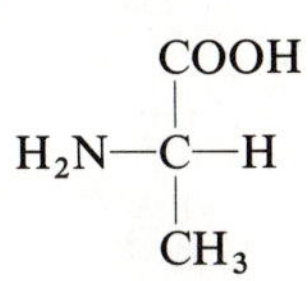

L-alanine

(b)

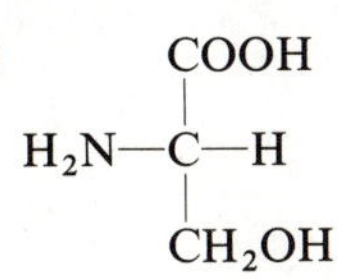

L-serine

(c)

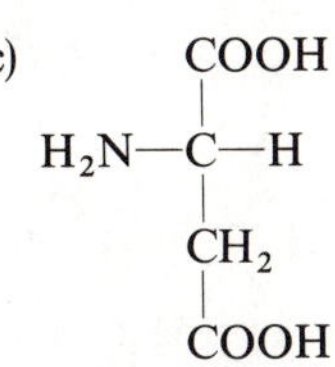

L-aspartic acid

(d)

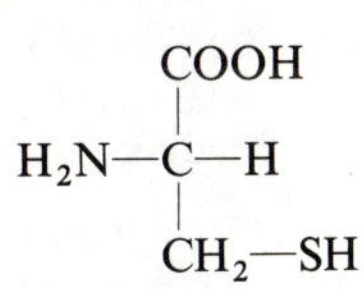

L-cysteine

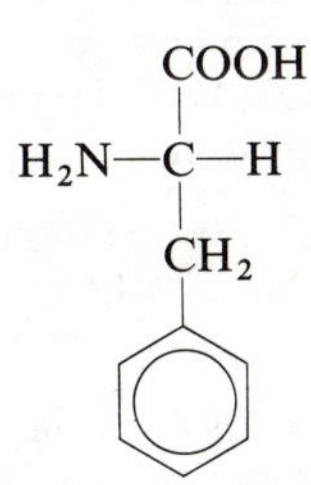

L-phenylalanine

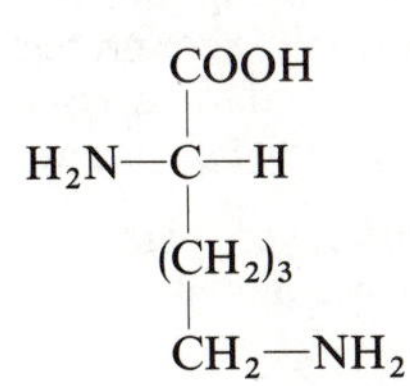

L-lysine

(e)

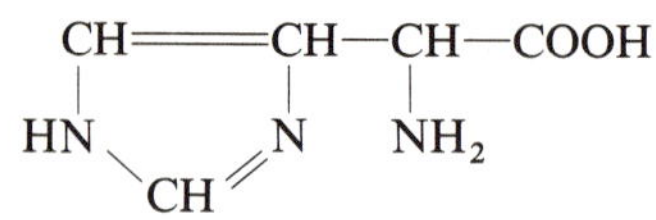

L-histidine

3.

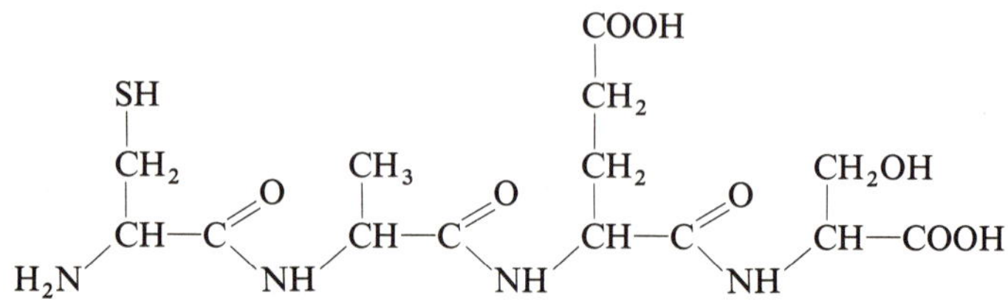

L-cysteinyl–L-alanyl–L-aspartyl–L-serine

4. The molecule will be at its isoelectric pH when all of the nine lysine residues are ionized and are matched by nine of the ten γ-carboxyl groups being ionized. You will recall that an acidic group will be 90% ionized at a point one pH unit above its pK. Given the pK_a of 4.3 for the γ-carboxyl group of glutamic acid, we obtain a value of 5.3 for the isoelectric point of our protein.

5. pro-gln-gly-arg-ser-tyr-met-gly-phe-gly-lys-thr-ala-ser-tyr-ala-arg-gly-ser-glu

6. (a) Convert the acetic acid to acetyl chloride, then react it with the methylamine.
   (b) It is possible to form the acid chloride of glycine and to react it with unchanged glycine. However, there is nothing to prevent reaction between molecules of the acid chloride, and considerable polymerization will take place. This problem can be obviated by adding some group to the amine end of one molecule and another to the carboxyl group of the second glycine to act as blocking groups during the synthetic reaction. Of course, it must be possible to remove these groups later without destroying the peptide.

Chapter 3

1. See the Glossary.

2. Peptide bonds; hydrogen bonds; hydrophobic interactions; electrostatic bonding; disulfide bonds.

3. In fibrous proteins, the polypeptide chains tend to exist as relatively long, narrow strands with little folding; this is true whether they are in a stretched conformation or in a helical form. These proteins are involved in extensive inter-chain bonding. Globular proteins tend to exist in a more spherical three-dimensional structure, with extensive folding of the chains and a great deal of intrachain bonding interaction.

4. High temperatures; very high or low pH; exposure to ultraviolet light; spreading in thin films; presence of high concentrations of salts or hydrophilic compounds such as urea.

5. If we assume a reasonable value for the energy of a single hydrogen bond in this case to be on the order of 8 kJ/mole, and given that all of the amino acids are involved fully in hydrogen bonding except a few on the ends of the chain, we can arrive at an estimate of 8 kJ per mole per bond times 1000 bonds or a total of 8,000 kJ/mole.

6.

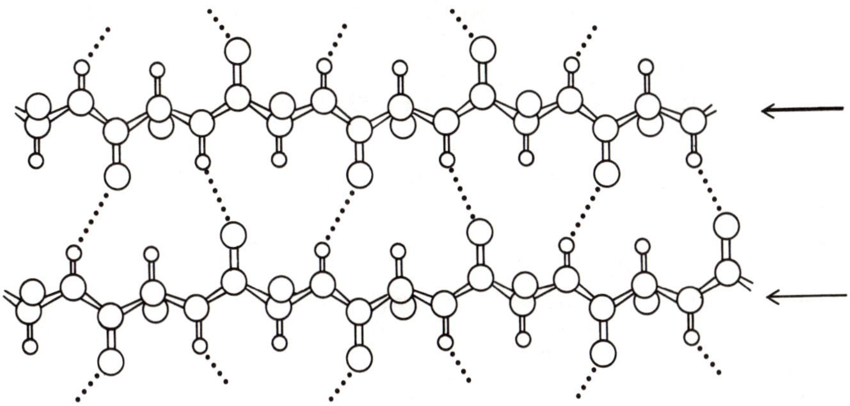

Chapter 4

1. See the Glossary: (a) not at all; (b) it lowers it; (c) not at all.

2, 3. See the Glossary.

4. Thiamin, niacin, riboflavin, pantothenic acid, pyridoxine, folic acid, vitamin B_{12}, biotin.

5. By increasing the denaturation rate, it makes it more difficult for the anaerobes to keep up the level of protein synthesis necessary for survival. The oxidative human cells have a much greater energetic capacity to meet this added demand.

6. (i) a (ii) c (iii) b (iv) b and perhaps d (v) b (vi) b.

7. The amount of enzyme would be 343 International Units or 5.72×10^{-6} katals.

8. (a) hydrolase (b) lyase (c) oxidoreductase (d) isomerase.

Chapter 5

1. (a) that the first step in the reaction is the formation of an enzyme-substrate complex; (b) that the substrate concentration is initially high compared to that of the enzyme . . . that is, that a steady-state situation will rapidly develop; (c) that the rate of the reaction is measured before any significant concentration of products can be formed.

2. If the values of k_1 and k_2 are very large compared to the value of k_3.

3. In competitive inhibition, the inhibitor acts by binding to the active site and blocking access of the substrate; a noncompetitive inhibitor does not interfere with the binding of the substrate, but attaches to the enzyme in such a way as to decrease its catalytic effectiveness.

4. (a) An apparent increase in K_m and no change in V_{max}; (b) no change in K_m but a decrease in V_{max}.

5. They are multimeric enzymes that are controlled by effector molecules that bind at specific sites remote from the active site of the enzyme and in doing so stabilize one of several possible conformations of the enzyme.

6. (a) 5×10^{-4} moles/liter; (b) $K_m = 5.0 \times 10^{-4}$ moles/liter; $V_{max} = 2.6$ μmoles/minute (uninhibited) and 1.0 μmoles/minute (inhibited); (c) the inhibition is "clearly non-competitive."

Chapter 6

1. The rate of the reaction has no effect on ΔG.

2. It is a matter of the need to supply the activation energy to get the process started. Glucose (or glucose polymers such as paper) will burn readily if started with an energy source such as a match.

3. $K_{eq} = 1.15 \times 10^5$. **4.** -21.4 kJ/mole.

5. You will have obtained a value of $+12.5$ kJ/mole for $\Delta G^{\circ\prime}$, of 6.47×10^{-3} for K_{eq}, and thus a value of 0.647 for the Cr-P/Cr ratio. (The actual value of K_{eq} in a muscle has been estimated at 1.85×10^{-2}. Use this value to get a new Cr-P/Cr ratio; then let this ratio of Cr-P/Cr decrease to 25% of its original value and look at the effect on the ATP/ADP ratio. Does this give you some idea of how creatine phosphate can supply most of its energy to the muscle before the ATP level is reduced drastically?).

Chapter 7

1. Schematically, as follows:

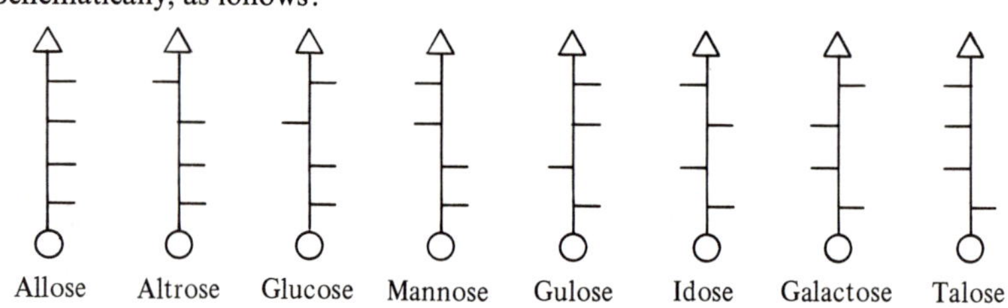

2.

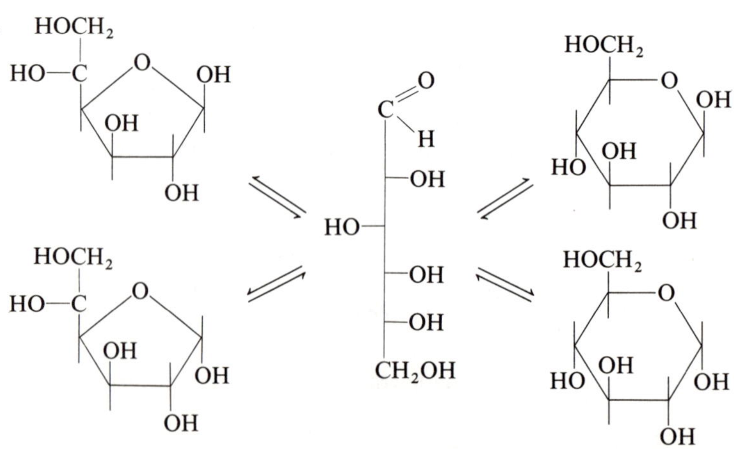

3.

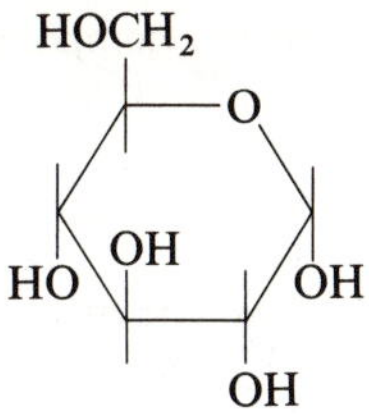

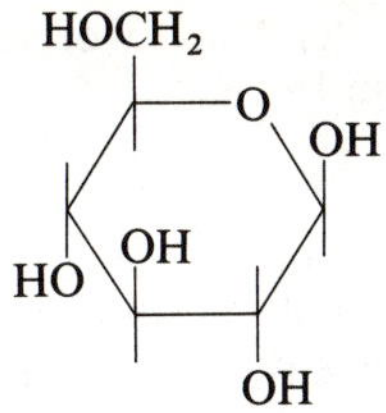

β-D-glucopyranose

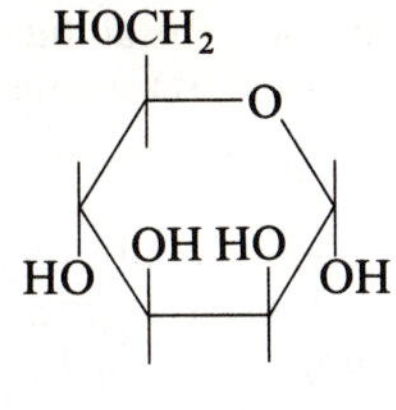

α-D-mannopyranose

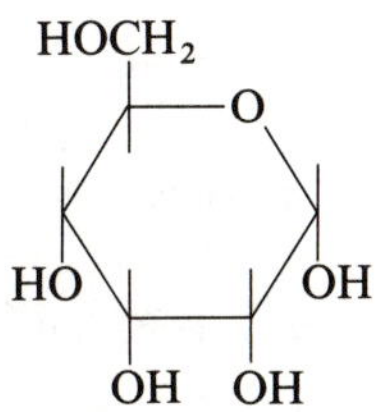

(α-D-allopyranose)

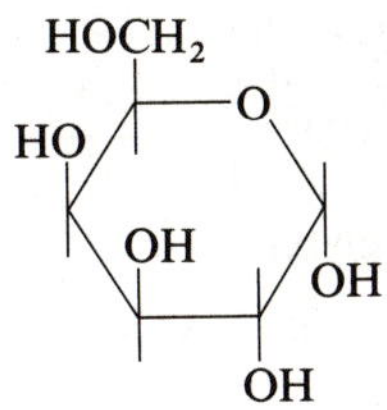

α-D-galactopyranose

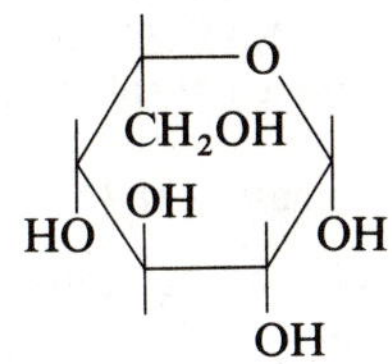

(β-L-idopyranose)

4. (a) β-D-glucopyranoside (b) α-D-xylopyranoside
(c) β-D-arabinofuranoside (d) α-D-glucopyranoside
(e) β-D-glucopyranoside

5. (a) α-D-glucopyranose (b) β-D-ribofuranose
(c) *aldehydo*-D-mannose (d) β-D-galactopyranose
(e) α-D-arabinopyranose (f) β-D-fructofuranose

6. 4(c); all of 5. **7.** Refer to drawings in the text.

8. (a)

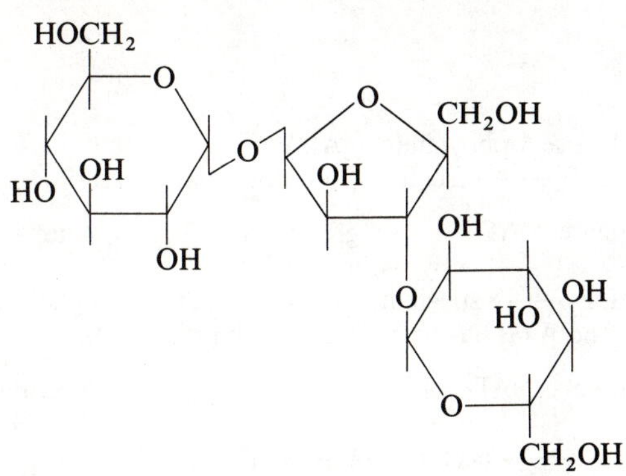

(b) From the data given, it is not possible to determine the α- or β-nature of the glycosidic bonds involved. (The structure shown in (a) is the correct one.)

Chapter 8

1. See text.

2. (a) 1.43×10^3 (b) 3.09×10^2 (c) 6.91×10^4

3. The methyl group.

4. (a) Greater than 14; (b) about 10; (c) 9 to 12 depending on the compound; (d) about 12; (e) 4.7 (f) greater than 14.

5. Look again at Figure 8-3.

6. (a) Lactic acid, ethanol (and CO_2), glycerol.
(b) In each case, NADH is generated by the glyceraldehyde 3-phosphate dehydrogenase reaction. The listed products can then be produced by the following reactions:

pyruvate + NADH + $H^+ \longrightarrow$ lactate + NAD^+
pyruvate $\longrightarrow$ acetaldehyde + CO_2 followed by
acetaldehyde + NADH + $H^+ \longrightarrow$ ethanol + NAD^+
dihydroxyacetone phosphate + NADH + $H^+ \longrightarrow$
glycerol phosphate + NAD^+
glycerol phosphate + $H_2O \longrightarrow$ glycerol + P_i

Note that production of glycerol does not allow the pathway to generate any net energy in the form of ATP.

7. See text.

8. Certainly glyceric acid 1,3-diphosphate is one of the best candidates, with phosphoenolpyruvate in the running, and glyceraldehyde 3-phosphate not far behind.

9. We have already mentioned this above in part. All of the ATP production during glycolysis comes after the formation of glyceric acid 1,3-diphosphate from glyceraldehyde 3-phosphate. If dihydroxyacetone phosphate is used to make glycerol, this energy production is lost. Just to break even, an organism would only be able to ferment half of its incoming glucose carbons to glycerol.

Chapter 9

1. (a) glucose + ATP $\longrightarrow$ glucose 6-phosphate + ADP
fructose 6-phosphate + ATP $\longrightarrow$ fructose 1,6-diphosphate + ADP

(b) glyceric acid 1,3-diphosphate + ADP $\longrightarrow$ glyceric acid 3-phosphate + ATP
phosphoenolpyruvate + ADP $\longrightarrow$ pyruvate + ATP
(also succinyl–CoA + GDP $\longrightarrow$ succinate + CoA + GTP although the actual transfer is harder to see, and ATP has to be formed from the GTP).

(c) glyceraldehyde 3-phosphate + NAD^+ + $P_i \longrightarrow$ 1,3-diphosphoglycerate
+ NADH + H^+
pyruvate + NAD^+ + CoA $\longrightarrow$ acetyl–CoA + NADH + H^+
isocitrate + $NAD^+ \longrightarrow$ α-ketoglutarate + NADH + H^+ + CO_2
α-ketoglutarate + NAD^+ + CoA $\longrightarrow$ succinyl–CoA + NADH + H^+ + CO_2
succinate + CoQ $\longrightarrow$ fumarate + $CoQH_2$
malate + $NAD^+ \longrightarrow$ oxaloacetate + NADH + H^+

(d) See the three equations given in (c).

2. (a) Any of the equations of 1(c).

(b) glucose 6-phosphate + $H_2O \longrightarrow$ glucose + P_i
sucrose + $H_2O \longrightarrow$ glucose + fructose
fructose 1,6-diphosphate + $H_2O \longrightarrow$ fructose 6-phosphate + P_i

(c) glucose 6-phosphate $\longrightarrow$ fructose 6-phosphate
glyceric acid 3-phosphate $\longrightarrow$ glyceric acid 2-phosphate
glyceraldehyde 3-phosphate $\rightleftharpoons$ dihydroxyacetone phosphate

(d) 1,3-diphosphoglycerate $\longrightarrow$ glyceraldehyde 3-phosphate + dihydroxyacetone phosphate
glyceric acid 2-phosphate $\longrightarrow$ phosphoenolpyruvate + H_2O
pyruvate $\longrightarrow$ acetaldehyde + CO_2

3. NAD: malate + NAD^+ $\longrightarrow$ oxaloacetate + NADH + H^+ niacin
FAD: succinate + CoQ $\longrightarrow$ fumarate + $CoQH_2$ riboflavin
Coenzyme A: acetyl–CoA + oxaloacetate $\longrightarrow$ citrate + CoA pantothenic acid
Thiamin pyrophosphate: pyruvate $\longrightarrow$ acetaldehyde + CO_2 thiamin
Biotinyl enzyme:
pyruvate + CO_2 + ATP $\longrightarrow$ oxaloacetate + ADP + P_i biotin

4. (a) We have mentioned that the carbons lost in the early reactions of the citric acid cycle come from the oxaloacetate; thus, the carbons from the acetyl–CoA would be retained during the one pass through the cycle. However, since succinate is a symmetrical molecule, it is not possible to distinguish between the carboxyl groups, or between the methylene groups. Thus a molecule of methyl-labeled lactate will give methyl-labeled acetyl–CoA but will then result in label in both of the methylene groups of succinate. The end product of one turn of the cycle will be 2,3-labeled oxaloacetate.
(b) Uniformly distributed within the carbons of the oxaloacetate if the incoming acetyl–CoA for this turn is not labeled; present in all carbons but more concentrated in the 2 and 3 positions if the incoming acetyl–CoA is methyl labeled.

5. Certainly succinate would predominate, with citrate a distant second and isocitrate running third.

Chapter 10

1. NAD, FMN, coenzyme Q (ubiquinone), cytochrome b, cytochrome c, cytochrome oxidase.

2. (a) Assuming the production of 36 moles of ATP per mole of glucose, glycolysis and the citric acid cycle, with associated processes, will generate 1,101.6 kJ/mole glucose. The percent efficiency is 38.4%.
(b) About 2.1%.

3. Only ions which could serve as oxidizing agents in a cell (being themselves reduced in the process) could meet this requirement. Certainly sulfate, nitrate, iron(III), chlorate, vanadate, and arsenate are possibilities. Copper(I) could theoretically serve such a function but would result in deposits of metallic copper.

4. (a) 8 (b) 3 (c) 18 (d) 12.

5. Split the box for succinic dehydrogenase into two so that the mitochondrial glycerol phosphate dehydrogenase can also feed electrons in at the level of coenzyme Q.

6. See the Glossary.

7. (a) 4 (2 as ATP and 2 as GTP); (b) it will be given off as heat.

Chapter 11

1. (i) Production of NADPH for biosynthetic purposes; (ii) synthesis of ribose 5-phosphate (needed for synthesis of nucleotides); (iii) formation of glucose units from CO_2 and water in the dark reactions of photosynthesis.

2. Almost all organisms are capable of carrying out the reactions of glycolysis. One approximately isoenergetic pathway to ribose 5-phosphate would have the following steps: (a) reaction of fructose 6-phosphate and glyceraldehyde 3-phosphate to form erythrose 4-phosphate and xylulose 5-phosphate. The xylulose 5-phosphate can be isomerized to ribulose 5-phosphate and then to ribose 5-phosphate. (The erythrose 4-phosphate can react with another fructose 6-phosphate to give sedoheptulose 7-phosphate and glyceraldehyde 3-phosphate. A transketolase reaction of these two compounds will yield xylulose 5-phosphate and ribose 5-phosphate.)

3. In a simplified form:
succinate $\longrightarrow$ furmarate $\longrightarrow$ malate $\longrightarrow$ oxaloacetate $\longrightarrow$ phosphoenolpyruvate $\longrightarrow$ 2-phosphoglycerate $\longrightarrow$ 3-phosphoglycerate $\longrightarrow$ 1,3-diphosphoglycerate $\longrightarrow$ glyceraldehyde 3-phosphate $\rightleftharpoons$ dihydroxyacetone phosphate $\longrightarrow$ fructose 1,6-diphosphate $\longrightarrow$ fructose 6-phosphate . . . then see the answer to Question 2 (above).
Using (*) to indicate carbon from the carboxyl groups of succinate and (#) for the methylene carbons you should get the pattern shown as (a) for the ribose 5-phosphate formed from fructose 6-phosphate and glyceraldehyde 3-phosphate, and the patterns (b) and (c) for those formed from sedoheptulose 7-phosphate and glyceraldehyde 3-phosphate.

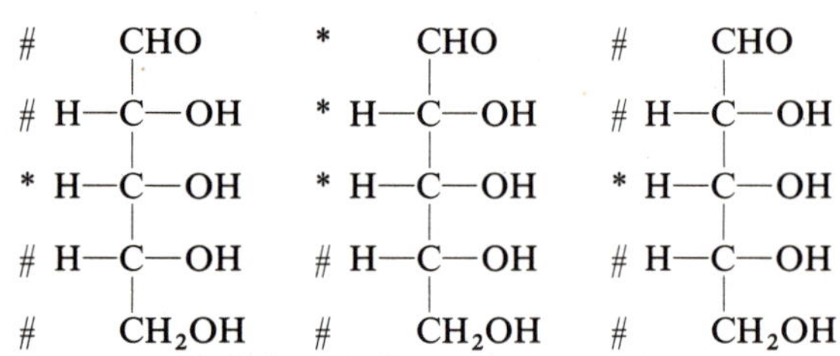

(a) (b) (c)

4. ribulose 5-phosphate + ATP $\longrightarrow$ ribulose 1,5-diphosphate + ADP
3-phosphoglycerate + ATP $\longrightarrow$ 1,3-diphosphoglycerate + ADP
1,3-diphosphoglycerate + NADPH + H^+ $\longrightarrow$ glyceraldehyde 3-phosphate + $NADP^+$ + P_i

5. (a) 36 or 38 ATPs per mole; (b) 36 ATP per mole.

6. 5 glyceraldehyde-3-P $\rightleftharpoons$ 5 dihydroxyacetone-P
5 glycerald-3-P + 5 dihydroxyacetone-P $\rightleftharpoons$ 5 fructose-1,6-diP
2 fructose-6-P + 2 glycerald-3-P $\rightleftharpoons$ 2 erythrose-4-P + 2 xylulose-5-P
2 erythrose-4-P + 2 fructose-6-P $\rightleftharpoons$ 2 sedoheptulose-7-P + 2 glycerald-3-P
2 sedoheptulose-7-P + 2 glycerald-3-P $\rightleftharpoons$ 2 ribose-5-P + 2 xylulose-5-P
4 xylulose-5-P $\rightleftharpoons$ 4 ribulose-5-P
2 ribose-5-P $\rightleftharpoons$ 2 ribulose-5-P
fructose-6-P $\rightleftharpoons$ glucose-6-P

---

12 glyceraldehyde 3-phosphate $\rightleftharpoons$ 6 ribulose 5-phosphate + glucose 6-phosphate

Chapter 12

1. The organism would have to convert some acetate to acetyl–CoA, and then use the glyoxylate shunt pathway (Figure 12-5) to get to oxaloacetate. Conversion of oxaloacetate to phosphoenolpyruvate, reversal of glycolysis, and nonoxidative formation of ribose 5-phosphate have been dealt with in Chapter 11 (see your answers to Study Questions 11.3 and 11.2).

2.

(a)

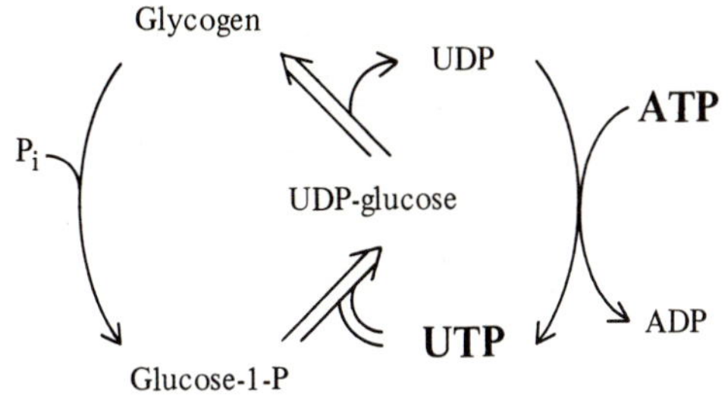

(b)

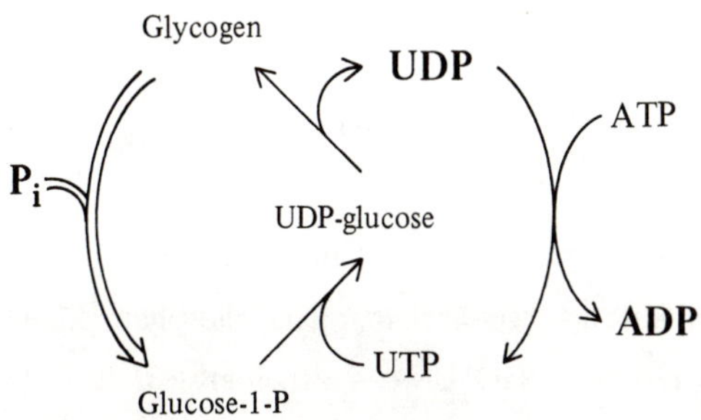

3. fructose + ATP $\longrightarrow$ fructose 1-phosphate + ADP
fructose 1-phosphate $\rightleftharpoons$ dihydroxyacetone phosphate + glyceraldehyde
glyceraldehyde + NADH + H^+ $\rightleftharpoons$ glycerol + NAD^+
glycerol + ATP $\longrightarrow$ glycerol phosphate + ADP
glycerol phosphate + NAD^+ $\rightleftharpoons$ dihydroxyacetone phosphate + NADH + H^+
dihydroxyacetone phosphate $\rightleftharpoons$ glyceraldehyde 3-phosphate
dihydroxyacetone phosphate + glyceraldehyde 3-phosphate $\rightleftharpoons$ fructose 1,6-diphosphate
fructose 1,6-diphosphate + H_2O $\longrightarrow$ fructose 6-phosphate + P_i
fructose 6-phosphate $\rightleftharpoons$ glucose 6-phosphate

4. 2 pyruvate + 2 CO_2 + 2 ATP $\longrightarrow$ 2 oxaloacetate + 2 ADP + 2 P_i
2 oxaloacetate + 2 ITP (or GTP) $\longrightarrow$ 2 phosphoenolpyruvate + 2 CO_2 + 2 IDP
2 phosphoenolpyruvate + 2 H_2O $\longrightarrow$ 2 2-phosphoglycerate + 2 3-phosphoglycerate
2 3-phosphoglycerate + 2 ATP $\longrightarrow$ 2 1,3-diphosphoglycerate + ADP
2 1,3-diphosphoglycerate + 2 NADH + 2 H^+ $\longrightarrow$ 2 glyceraldehyde 3-phosphate + 2 NAD^+ + 2 P_i
glyceraldehyde 3-phosphate $\rightleftharpoons$ dihydroxyacetone phosphate
glyceraldehyde 3-phosphate + dihydroxyacetone phosphate $\rightleftharpoons$ fructose 1,6-diphosphate
fructose 1,6-diphosphate + H_2O $\longrightarrow$ fructose 6-phosphate + P_i
fructose 6-phosphate $\rightleftharpoons$ glucose 6-phosphate

5. This again will require the use of the glyoxylate shunt pathway of Figure 12-5 to get net synthesis of oxaloacetate from acetyl–CoA. From that point, synthesis of glucose 6-phosphate would proceed as in the preceeding question.

Chapter 13

1. The possibilities are immense; any lipid compounds that might reasonably be present in animal tissues will do. These would include a wide range of triacylglycerols, specific phospholipid molecules, steroids, fat-soluble vitamins, and so on.

2. In a simple form:

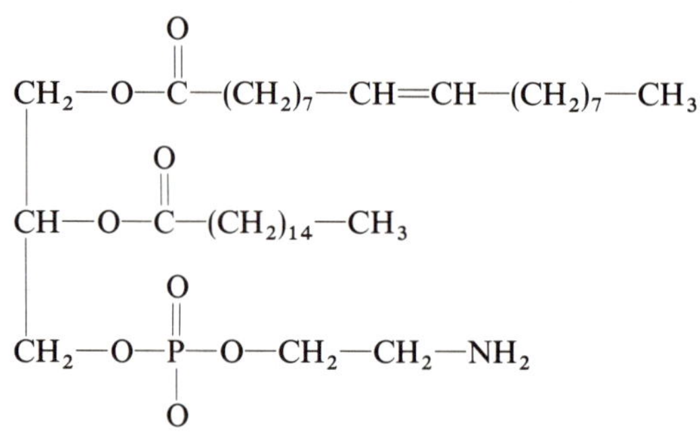

3. 1-oleioyl-2-palmitoyl phosphatidylethanolamine

4. Oleic acid, palmitic acid, glycerol, inorganic phosphate, ethanolamine.

5. Tristearoylglycerol + 3 NaOH $\longrightarrow$ 3 sodium stearate + glycerol

6. Cerebrosides.

7. Each of the vitamins listed by specific name in this question exists only (or primarily) as one compound. There are, however, several compounds having "vitamin A activity," and the same is true for D, E, and K. The letters are still commonly used to represent a summation of all of the contributions to that particular vitamin activity, rather than singling out only one of the specific compounds involved.

Chapter 14

1. -31.4 kJ/mole.

2. (a) 146 ATP per mole of stearic acid.
   (b) Primarily because a greater percentage of the long-chain fatty acid is in the fully reduced state, and thus a greater degree of oxidation is required to yield acetyl–CoA. Stearic acid requires eight trips through the β-oxidation process (or 0.44 trips per carbon atom) while a 6-carbon fatty acid requires only two (0.33 per carbon atom).

3. For maximum utilization of acetyl–CoA as an energy source, the citric acid cycle must operate a peak efficiency. This requires an optimum supply of cycle intermediates, which normally come from carbohydrate sources. In the total absence of of carbohydrates, fatty acid oxidation generally becomes incomplete.

4. Acetoacetic acid, β-hydroxybutyric acid(acetoacetic acid), and to some extent acetone.

5. (a) As carnitine esters; (b) in the form of citrate, which can move through the membrane; reversal of the condensation reaction then yields oxaloacetate and acetyl–CoA.

6. Yield for stearic acid to 9 acetyl–CoAs is 38 ATP. Cost for biosynthesis from acetyl–CoA is 8 ATP and 16 NADPH, for a total of 56 ATP equivalents.

Chapter 15

1. The answer you obtained may vary a little depending on your assumptions. Certainly, formation of the necessary 27 acetyl–CoAs from glucose will yield 162 ATP (+162 ATP) while formation of glycerol phosphate will require 3 (−3 ATP). Formation of 8 malonyl–CoAs for each fatty acid requires 3 × 8 ATP (−24 ATP). Reduction steps take 3 × 2 × 8 NADPH (−48 NADPH). Formation of the CoA esters of the fatty acids after they are synthesized will be (−6 ATP). From these figures, we get a *net* of +129 ATP and −48 NADPH; thus the overall process actually produces an excess of ATP but requires a large amount of NADPH.

2. The two oxidation steps of the phosphogluconate pathway.

3.

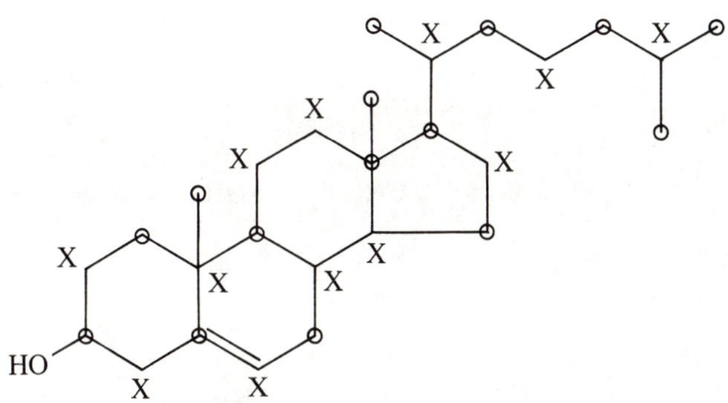

4.

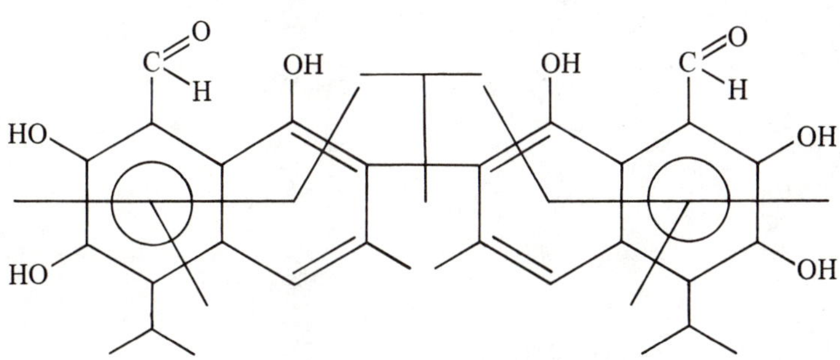

5. (a) We have commented on the tendency of highly polar or charged groups to hydrogen bond with water molecules and we have seen how the hydrogen-bonded structure of water can force nonpolar molecules or portions of molecules into a

minimum volume within the solution. In the case of bilayer membranes, the charged portion of the phospholipid faces outward to interact with the water molecules, leaving the nonpolar hydrocarbon chains to be forced into close proximity in order to minimize their disturbance of the structure of the solvent.

(b) In general the forces will be much the same, with the more polar portions of the proteins tending to lie at the surface where they can interact with the solvent, and the nonpolar portions pushed into the hydrocarbon area of the membrane. However, with as much as 70% protein, there are likely to be direct interactions between protein molecules in addition. These might include hydrogen bonding, hydrophobic interactions, electrostatic bonding, or metal ion bridging.

Chapter 16

1. Histidine, lysine, proline.

2. (a)

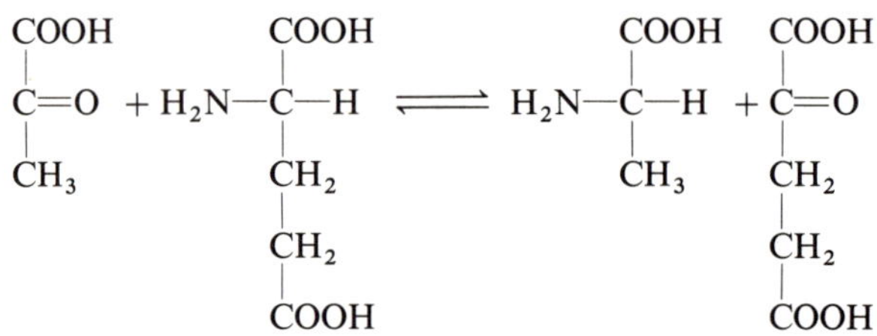

(b)

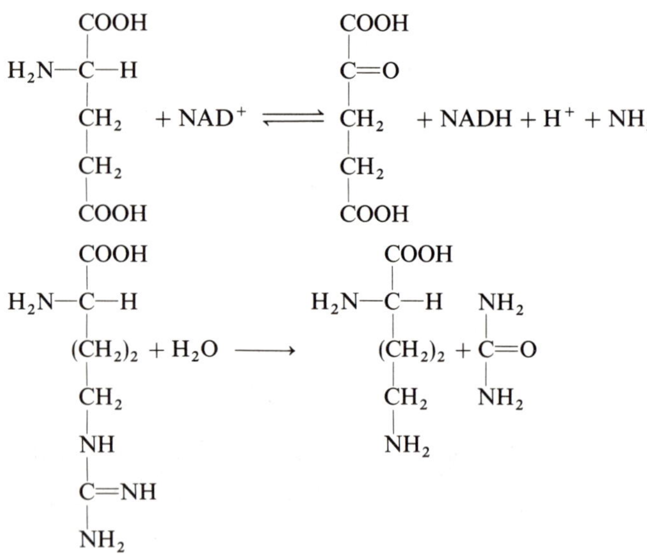

(c)

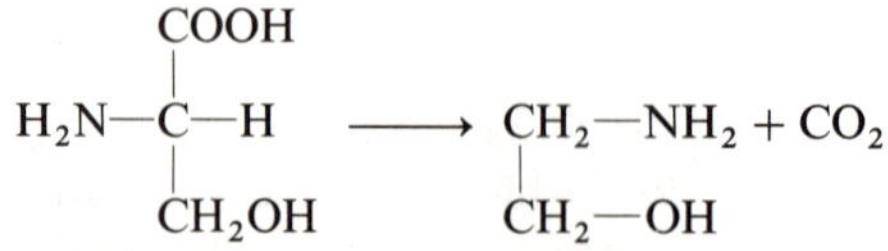

(d) See the first reaction of (b) above.

3. Ammonia, urea, uric acid.

4. alanine + α-ketoglutarate $\rightleftharpoons$ pyruvate + glutamate
glutamate + NAD^+ $\rightleftharpoons$ α-ketoglutarate + NH_3 + NADH + H^+
glutamate + oxaloacetate $\rightleftharpoons$ α-ketoglutarate + aspartate
$CO_2 + NH_3 + H_2O + 2\ ATP \longrightarrow$ carbamoyl phosphate + 2 ADP + P_i
carbamoyl phosphate + ornithine $\longrightarrow$ citrulline + P_i
citrulline + aspartate + ATP $\longrightarrow$ argininosuccinate + AMP + P–P_i
argininosuccinate $\longrightarrow$ arginine + fumaric acid
arginine + H_2O $\longrightarrow$ ornithine + urea

5.

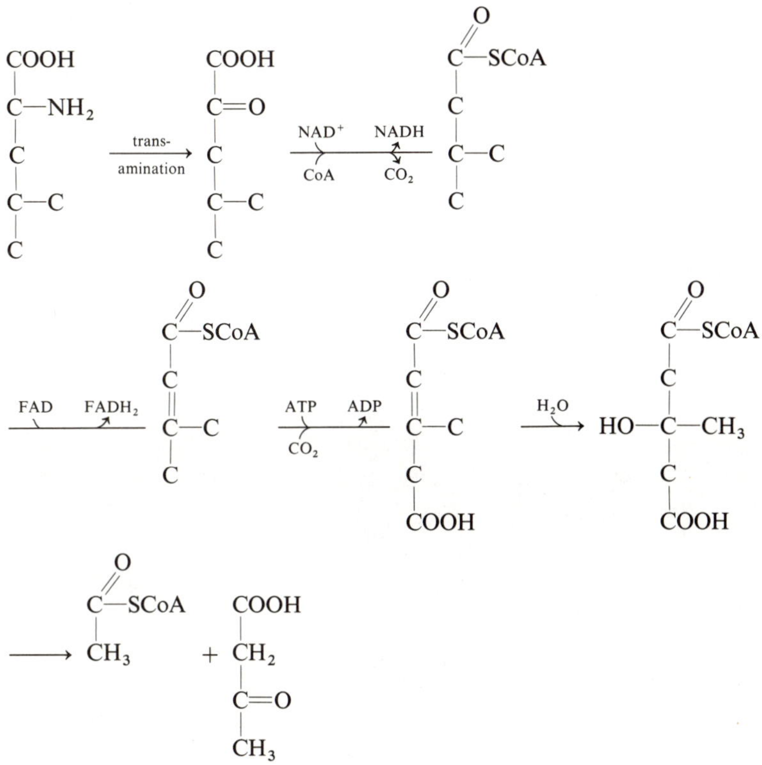

Chapter 17

1. The question is primarily one of making succinyl–CoA and glycine from the glyceric acid 3-phosphate. There are a number of possibilities, of which the following is only one: Take the 3-phosphoglycerate through the remainder of glycolysis to pyruvate; send half of it on to acetyl–CoA and with the other half make oxaloacetate (using the CO_2 from the pyruvate decarboxylation, of course). Condense these two compounds to citrate and run them through the citric acid cycle as far as succinyl–CoA. To get glycine, use the reactions of Figure 17-2 to get serine, then react the serine with tetrahydrofolic acid in the serine aldolase reaction to obtain glycine (and N^5N^{10} methylene FH_4).

2. phosphoserine + H_2O ⟶ serine + P_i
serine + α-ketoglutarate ⇌ hydroxypyruvate + glutamate
serine + FH_4 ⟶ glycine + N^5N^{10} methylene FH_4
serine ⟶ pyruvate + NH_3
serine ⟶ ethanolamine + CO_2

3. 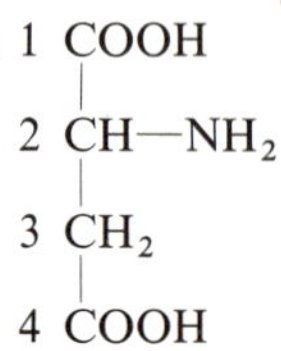

Aspartic acid

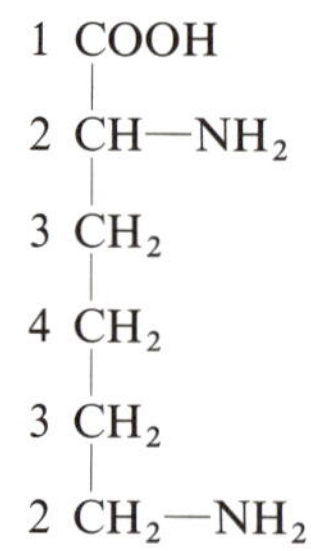

Lysine

Chapter 18

1. (a) 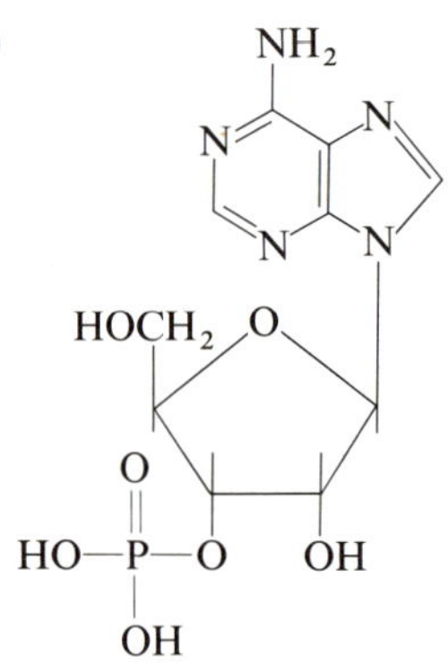

(b) 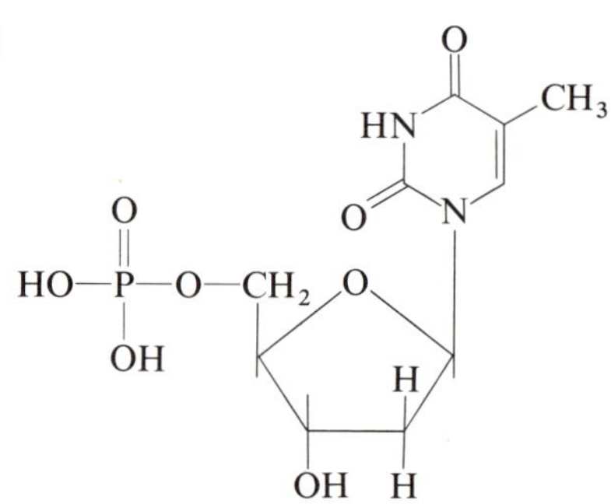

(c) 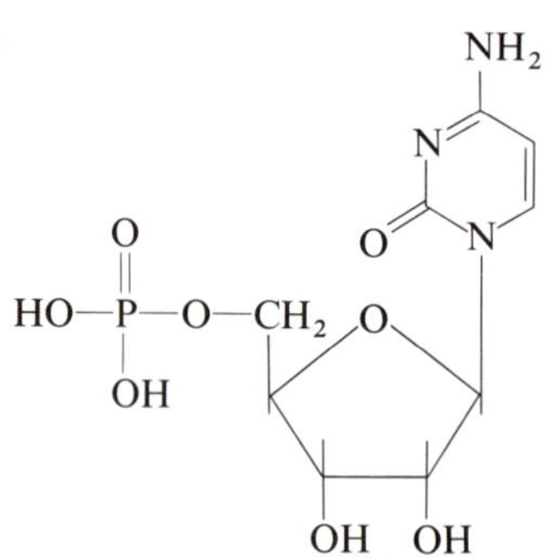

(d) 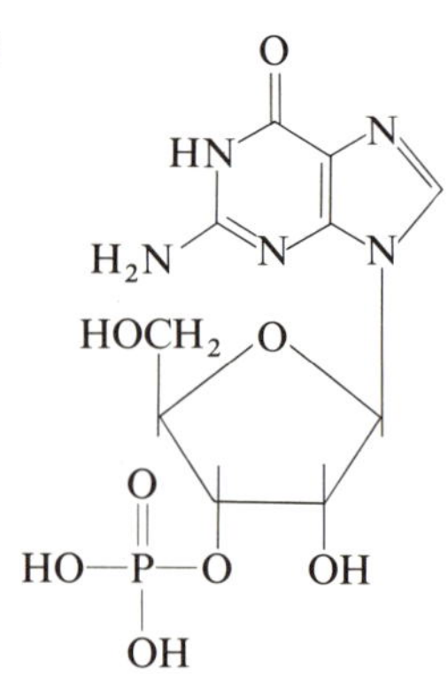

2. (a) glycine + 5-phosphoribosylamine + ATP ⟶ glycineamide nucleotide + ADP + P_i

(b) N^{10}-formyl FH_4 + 5-aminoimidazole 4-carboxamide nucleotide
$\longrightarrow$ 5-formamidoimidazole 4-carboxamide nucleotide + FH_4
(c) 5-aminoimidazole 4-carboxylate nucleotide + aspartic acid + ATP
$\longrightarrow$ 5-aminoimidazole 4-succinylcarboxamide nucleotide + ADP + P_i
(d) 5-phosphoribosyl 1-pyrophosphate + glutamine $\longrightarrow$ 5-phosphoribosylamine + glutamic acid + P–P_i
(e) See (a) or (c).
(f) inosine 5′-phosphate + aspartate + GTP $\longrightarrow$ adenylosuccinate + GDP + P_i

3. (a) The structure should be similar to that of Figure 18-7, although the base sequence will be different.
(b) C–A–G–T–T–A–C–G–T

4. 24,600.

5. (a)

(G > A)	*T–A–C–A *T–A–C–A–G–T
(A > G)	*T *T–A–C
(C + T)	*T–A *T–A–C–A–G *T–A–C–A–G–T–G
(C only)	*T–A *T–A–C–A–G–T–G

(b)

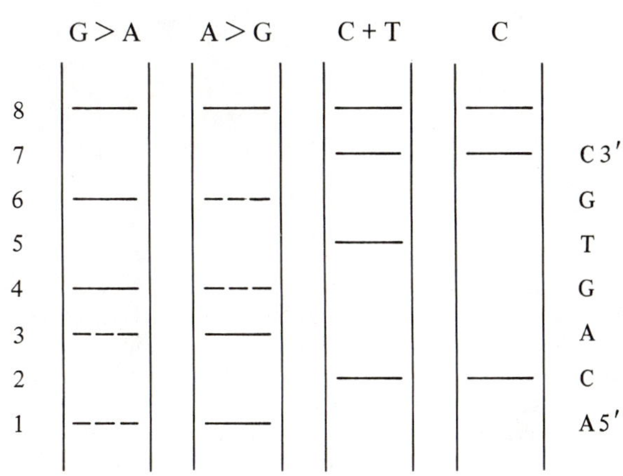

Chapter 19

1. See the Glossary.

2.

RNA	UCU	UCC	UCA	UCG	AGU	AGC
DNA	AGA	GGA	TGA	CGA	ACT	GCT

3. 5′–A–T–T–C–G–A–T–G–C–3′
3′–T–A–A–G–C–T–A–C–G–5′

4. 378.

5. (a) Half of the cells would have the original ^{15}N DNA molecules and half would have totally new DNA all made with ^{14}N.
(b) Not to my knowledge.

6. NMN + ATP $\longrightarrow$ NAD + P–P_i

Chapter 20

1. See the Glossary.

2.

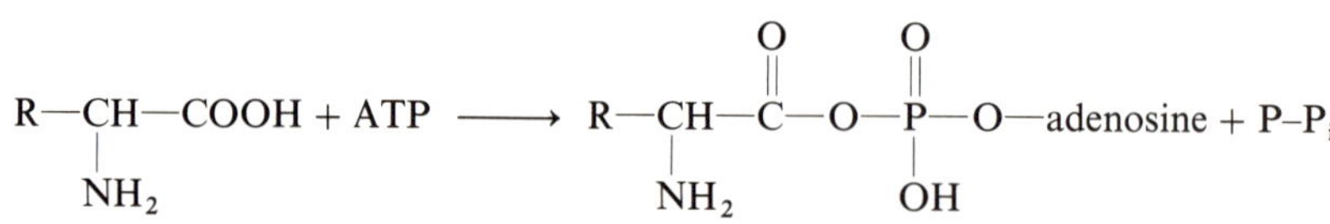

Amino acid — Aminoacyl adenylate

aminoacyladenylate + tRNA $\longrightarrow$ aminoacyl-tRNA + AMP

3. N^5-formyl FH_4, N^{10}-formyl FH_4, or N^5N^{10}-methenyl FH_4.

4. The most important feature is undoubtedly its ability to bind directly to the peptidyl binding site on the dissociated smaller portion of the ribosome. This binding starts the process of synthesis in that its prepares the stage for addition of the mRNA and for association of the two portions of ribosome. All of these characteristics are special ones not shared by the other tRNA species.

5. 167.

Chapter 21

1. (a) It codes for production of a repressor protein capable of binding to the operator gene and preventing it from functioning.
(b) It controls access of RNA polymerase complex to structural genes. The structural genes are transcribed only when the operator gene does not have a repressor protein bound to it.
(c) It binds the RNA polymerase complex in such a way that the complex can move on to the operator gene and from there to the structural genes.
(d) Cyclic-AMP binds to a special protein in such a way as to change the shape of the protein to one that can bind to the promoter and render it more effective in binding the RNA polymerase complex. The net effect is to increase the rate of transcription in a derepressed system.

2. (i) Addition of phosphate groups by transfer from ATP produces the active form, phosphorylase *a*.
(ii) Addition of the allosteric effector AMP will activate the inactive phosphorylase *b*.

3. In both cases, allosteric control allows the rate of a reaction to be controlled by changes in concentration of molecules structurally unrelated to the substrates involved. This is ideal for long-range effects (such as feedback inhibition) because it allows the complex products of a reaction sequence to slow the synthesis of the relatively small molecules at the beginning of the process. It is also possible to have one or two compounds that represent a measure of the energy state of the cell (such as ATP, ADP, AMP) control the reactions of a number of toally structurally unrelated metabolites.

4. Glycogen phosphorylase; glycogen synthetase; phosphofructokinase; pyruvate kinase; pyruvate dehydrogenase complex, hexokinase (to some limited extent).

5. phosphorylase b + 4 ATP $\longrightarrow$ phosphorylase *a* + 4 ADP

INDEX

The pages on which structures of specific compounds appear in the text are identified by **boldface** numbers.

A

B

C

D

E

F

G

H

I

Q

R

S

T

The Urea Cycle and Related Reactions

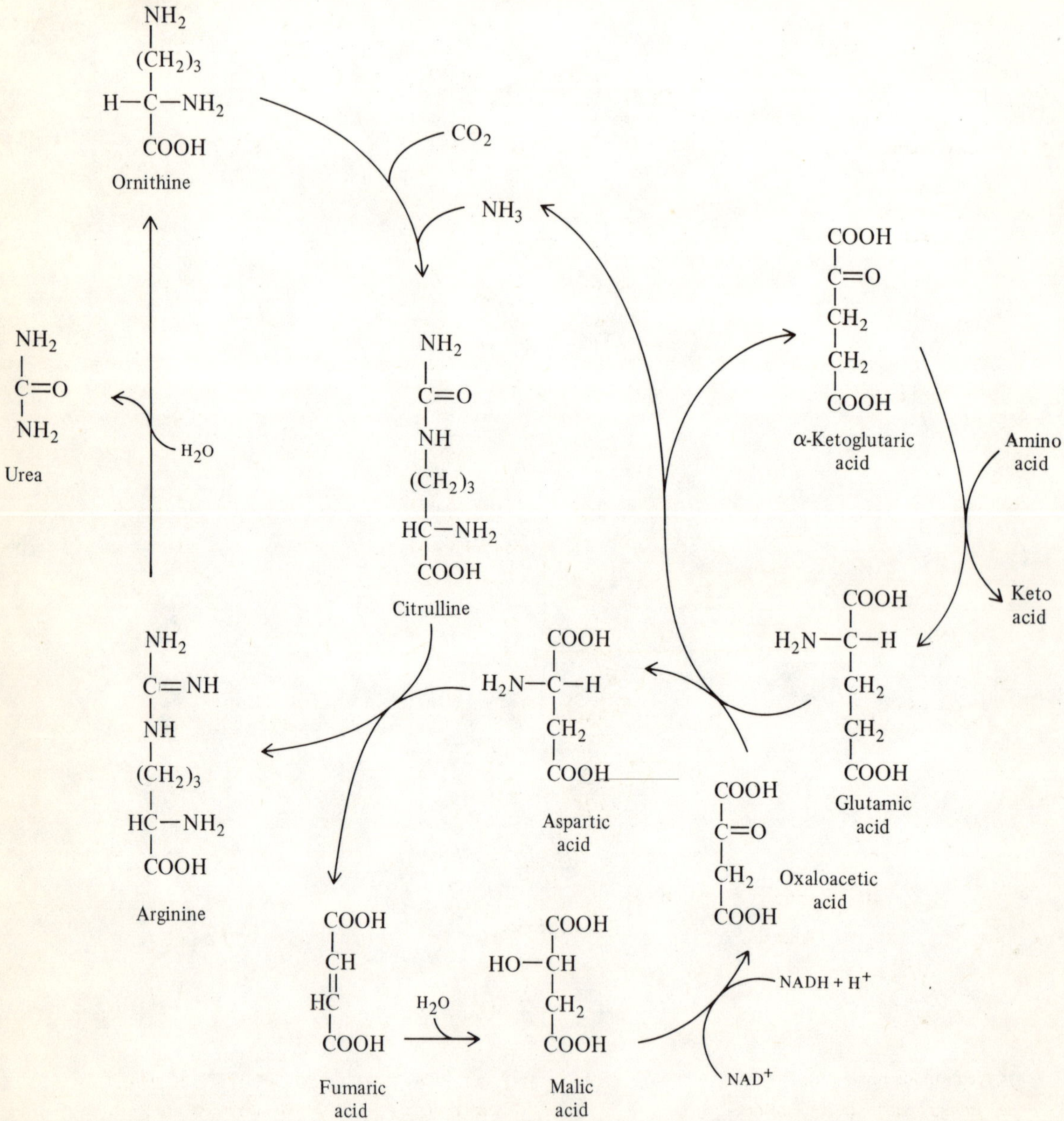